Innovative Technologies for Smart and Sustainable Infrastructure

Peerzada Danish · K. S. Nanjunda Rao ·
Manzoor Ahmad Tantary · G. Mohan Ganesh ·
A. S. Santhi
Editors

Innovative Technologies for Smart and Sustainable Infrastructure

Springer

Editors
Peerzada Danish
Department of Civil Engineering
Indian Institute of Science
Bengaluru, Karnataka, India

K. S. Nanjunda Rao
Department of Civil Engineering
Indian Institute of Science
Bengaluru, Karnataka, India

Manzoor Ahmad Tantary
Department of Civil Engineering
National Institute of Technology
Srinagar, Jammu and Kashmir, India

G. Mohan Ganesh
School of Civil Engineering
Vellore Institute of Technology
Vellore, Tamil Nadu, India

A. S. Santhi
School of Civil Engineering
Vellore Institute of Technology
Vellore, Tamil Nadu, India

ISBN 978-3-031-92800-0 ISBN 978-3-031-92801-7 (eBook)
https://doi.org/10.1007/978-3-031-92801-7

This Springer imprint is published by the registered company Springer Nature Switzerland AG
The registered company address is: Gewerbestrasse 11, 6330 Cham, Switzerland

Contents

Advanced Technologies

Key Variables Influencing the Performance of 3D Printed Concrete: A Comprehensive Analysis

Salim Barbhuiya, Bibhuti Bhusan Das, and Dibyendu Adak

Abstract This chapter examines key variables influencing 3D printed concrete performance, focusing on material, process, environmental, and geometric factors essential for achieving optimal strength and durability. It begins with an overview of 3D printed concrete, performance metrics, and the scope of the study. The chapter then delves into material composition, discussing how cement type, aggregate characteristics, additives, and water-cement ratios affect mix consistency, workability, and structural integrity. Process parameters, such as layer height, print speed, and extrusion rate, are analysed for their impact on layer adhesion and structural stability. Environmental factors—including temperature, humidity, and curing—are examined, highlighting their influence on setting time and strength. Geometric and structural considerations, like wall thickness and layer bonding, reveal the effects of design complexity on stability. The chapter concludes by synthesizing these findings, offering insights into optimizing 3D printed concrete performance through coordinated control of materials, process, and environmental conditions.

Keywords 3D printed concrete · Material composition · Process parameters · Environmental conditions · Structural performance

S. Barbhuiya (✉)
Department of Engineering and Construction, University of East London, London, UK
e-mail: s.barbhuiya@uel.ac.uk

B. B. Das
Department of Civil Engineering, NIT Karnataka, Mangaluru, Surathkal, India

D. Adak
Department of Civil Engineering, NIT Meghalaya, Shillong, India

P. Danish et al. (eds.), *Innovative Technologies for Smart and Sustainable Infrastructure*,
https://doi.org/10.1007/978-3-031-92801-7_1

1 Introduction

3D printed concrete (3DPC) is a revolutionary technology that integrates additive manufacturing into the construction industry, offering innovative solutions to traditional building methods. Originating from the concept of layer-by-layer deposition used in 3D printing, 3DPC utilizes specially designed cement-based materials extruded through a nozzle to create complex, customizable structures without the need for traditional formwork. This approach enhances efficiency, reduces material waste, and enables intricate designs that are challenging to achieve with conventional methods. The development of 3D printed concrete gained momentum in the early twenty-first century as advancements in robotics, material science, and computer-aided design (CAD) converged. Early efforts focused on overcoming technical challenges, including optimizing the mix design to balance flowability, buildability, and structural integrity. Pioneering projects demonstrated its potential in constructing housing units, bridges, and urban furniture, proving the viability of the technology for real-world applications.

Key advantages of 3DPC include its ability to reduce labour costs, construction time, and material usage while offering design flexibility and sustainability. The process minimizes waste by depositing material only where needed and allows for the incorporation of recycled or alternative materials, contributing to environmental goals. Moreover, the automated nature of the process ensures consistent quality and precision. Despite its promise, 3DPC faces challenges such as achieving uniform material performance, scaling up for large projects, and meeting regulatory standards. Additionally, the lack of standardized practices and limited understanding of long-term material behaviour pose barriers to widespread adoption. Today, research and innovation continue to refine the technology, with industry leaders and academic institutions exploring advanced mix designs, robotics, and post-processing techniques. As 3D printed concrete matures, it is poised to play a transformative role in reshaping the construction industry, driving sustainability, and enabling architectural creativity.

Key performance metrics in 3DPC are essential for evaluating the technology's effectiveness, quality, and suitability for diverse construction applications. These metrics provide a framework for optimizing processes, ensuring structural reliability, and achieving compliance with industry standards. Material properties such as compressive strength, tensile strength, and durability are critical indicators of the structural performance of 3D printed concrete. Equally important are workability and buildability, which balance the material's ability to flow smoothly during extrusion while maintaining sufficient stiffness to support successive layers. Dimensional accuracy is another critical metric, measuring how closely the printed structure aligns with the intended design in terms of geometric tolerances and surface quality. Print speed and overall productivity evaluate the time efficiency of the process, influencing its feasibility for large-scale applications. Environmental sustainability is increasingly important, with metrics such as material utilization efficiency and embodied carbon assessing the ecological impact of 3D printing. These measures focus on reducing waste and integrating eco-friendly materials. Cost-effectiveness and scalability also play significant roles, as operational costs, material expenses, and adaptability to larger projects determine the commercial viability of the technology. Finally, long-term performance metrics, including weather resistance and lifecycle analysis, ensure that 3D printed structures can endure real-world conditions reliably over time.

The purpose of this chapter is to explore the critical variables that influence the performance of 3DPC in construction applications. It examines material properties, process parameters, and environmental factors, highlighting their roles in achieving structural integrity, precision, and sustainability. The chapter aims to provide a comprehensive understanding of how these variables interact to affect outcomes like strength, durability, and printability. By analysing key factors such as mix design, extrusion rate, and curing conditions, the scope encompasses practical and theoretical insights to guide optimization efforts, enabling reliable and efficient deployment of 3DPC in diverse construction contexts.

2 Material Composition Variables

2.1 Cement Type and Composition

Cement type and composition are pivotal in determining the performance of 3DPC. While Portland cement remains the most commonly used, alternative options such as blended cements and supplementary cementitious materials (SCMs) like fly ash (FA) and ground granulated blast-furnace slag (GGBFS) are increasingly adopted to improve sustainability and performance. Achieving the optimal balance of flowability, setting time, and buildability is crucial, as the material must extrude smoothly while providing adequate support for subsequent layers. Additives such as superplasticizers and viscosity-modifying agents are frequently utilized to fine-tune the required rheological properties. Thus, the selection and proportioning of cement and other components are critical to optimizing the strength, durability, and precision of 3DPC.

Research has provided insights into the effects of different cement types and mix designs. Liu et al. [1] reported that sulphoaluminate cement significantly enhances early-age strength and stability in foamed 3DPC but requires precise mix control due to its rapid setting characteristics. Ibrahim et al. [2] investigated limestone calcined clay cement and demonstrated its potential to improve sustainability and strength, though challenges in maintaining rheological consistency remain. Xu et al. [3] emphasized the benefits of incorporating FA and GGBFS, noting their ability to refine rheological properties and enhance compressive strength, effectively balancing extrudability and load-bearing capacity. Wang et al. [4] highlighted the importance of mix proportions in particle bed printing, particularly regarding their influence on interlayer bonding and mechanical performance. Similarly, Lowke et al. [5] identified selective cement activation as a critical factor in optimizing both strength and precision in particle bed 3DPC. Together, these studies underline the importance of tailoring cement compositions to address the trade-offs between workability, structural integrity, and environmental sustainability. This delicate balance necessitates a comprehensive understanding of cement chemistry and mix design to meet specific application requirements and overcome practical challenges in the 3DPC process.

2.2 Aggregate Characteristics

Aggregate size and distribution play a crucial role in the performance of 3DPC, influencing rheology, structural integrity, and printability. Fine aggregates are often preferred because they facilitate smoother extrusion and better layer adhesion, while coarser aggregates can enhance mechanical strength but risk nozzle clogging and weak interlayer bonding. Optimal grading of aggregates improves particle packing, reducing voids and increasing material density. A well-balanced aggregate distribution is essential to achieving stability during printing and adequate buildability without compromising flowability. Careful tailoring of aggregate size and gradation is therefore critical to balancing these properties, ensuring both ease of printing and desired performance in the hardened state of 3DPC structures [6, 7].

Figure 1 illustrates the relationship between coarse aggregate (CA) and fine aggregate (FA) ratios in 3D printed concrete, highlighting a "printable zone" where mixtures achieve successful printing. Blue dots represent successfully printed mixtures, while red crosses denote failed attempts due to improper aggregate ratios. Green dots indicate mixtures printed successfully with special treatments. The diagonal boundaries define the printable zone based on specific FA and CA combinations, ensuring optimal flowability, stability, and mechanical properties. This demonstrates the importance of balanced aggregate distribution in maintaining printability and structural integrity, while excessive or imbalanced ratios lead to printing failures or require adjustment.

The impact of aggregate size and distribution on 3DPC performance has been extensively studied. Xiao et al. [7] found that combining fine and coarse aggregates enhances mechanical properties but requires precise mix control to maintain printability. Liu et al. [9] noted that excessive use of coarse aggregates can lead to pore defects, which undermine the structural integrity of the concrete. Chen et al. [10] demonstrated that optimized aggregate gradation improves extrusion consistency and minimizes defects. For mixes utilizing recycled aggregates, Ding et al. [11] highlighted the importance of balanced gradation in maintaining buildability and strength. Yu et al. [12] showed that fine grading enhances layer smoothness and bonding between layers. Furthermore, Kaliyavaradhan et al. [13] observed that sand gradation significantly affects the rheological properties of the mix, preventing segregation and ensuring uniform material flow. Collectively, these studies underscore the critical importance of aggregate size and distribution in achieving optimal fresh and hardened performance in 3DPC.

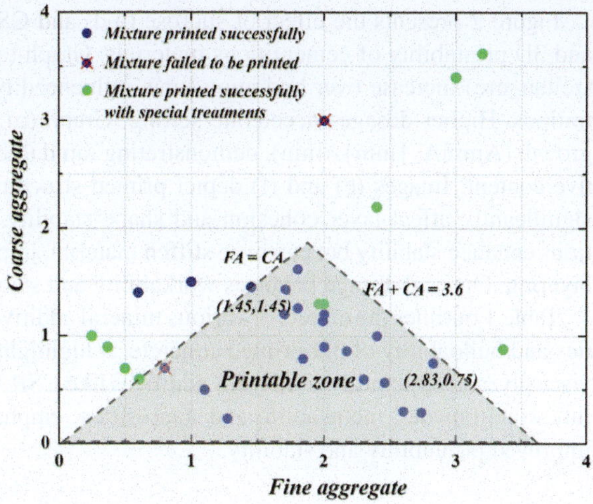

Fig. 1 Printable zone and aggregate ratio influence on 3D printed concrete performance [8]

2.3 *Additives and Admixtures*

Plasticizers, fibres, retarders, and accelerators significantly impact the performance of 3DPC. Plasticizers improve flowability, ensuring smooth extrusion without compromising structural strength [14, 15]. Fibres enhance interlayer bonding, tensile strength, and crack resistance, which are crucial for maintaining structural integrity, with their alignment during printing influencing anisotropy [16, 17, 18]. Retarders extend working time, allowing for improved handling and layer adhesion, particularly in complex designs [19]. Conversely, accelerators reduce setting time, which enhances buildability and supports faster printing speeds, especially advantageous for vertical structures [20, 21]. Achieving the precise balance of these additives is vital for optimizing rheological properties, structural performance, and efficiency, ensuring successful application in diverse 3DPC construction scenarios.

The role of plasticizers, fibres, retarders, and accelerators in enhancing 3DPC performance has been widely documented. Souza et al. [14] and Sergis and Ouellet-Plamondon [15] emphasized that plasticizers enhance workability and flowability, ensuring extrusion consistency without compromising printability. Pham et al. [18] and Ding et al. [17] highlighted the contribution of fibres, such as steel and synthetic types, in improving tensile strength and crack resistance, noting that their alignment during printing affects structural anisotropy. Retarders were shown by Seo and Lee [19] to be essential for extending workability and improving multi-layer adhesion in complex geometries. Meanwhile, accelerators were demonstrated by Rubin et al. [20, 21] to enhance buildability by reducing setting time and increasing early strength, especially for vertical applications. These findings collectively illustrate the importance of carefully balancing these additives to achieve optimal fresh and hardened performance in 3DPC.

Figure 2 presents the effect of sucrose (a-d) and CSR on rheological properties and 3D printability of cementitious materials. Graph (a) and (b) shows yield stress (τc\tau_cτc) increase over hydration time, influenced by sucrose and CSR concentrations. Higher dosages accelerate setting. Graph (c) and (d) illustrates viscosity growth (AminA_{min}Amin), demonstrating rapid thickening with increased additive content. Images (e) and (f) depict printed structures, where CSR and sucrose significantly affect layer cohesion and shape stability. Higher additive concentrations enhance stability but can over-stiffen material, impairing smooth extrusion and layering. Optimal dosage balances workability and stability for 3D-printed cement.

Table 1 outlines the effects of various mineral additives on the rheological properties and buildability of 3D-printed concrete. It highlights the impact on yield stress, viscosity, and structural stability for additives like CSR, sucrose, fly ash, silica fume, limestone powder, metakaolin, and nano-silica, emphasizing optimal dosages for improved printability and stability.

2.4 *Water-Cement Ratio*

The water-cement (w/c) ratio is a critical factor influencing the performance of 3DPC, particularly in terms of workability, strength, and durability. A lower w/c ratio enhances compressive strength and reduces shrinkage but may compromise flowability and extrusion consistency, making printing more challenging [22, 23]. Conversely, a higher w/c ratio improves flowability and pumpability but weakens structural integrity and increases the risk of drying shrinkage [24]. Achieving an optimal w/c ratio is essential to balancing printability with mechanical properties. Superplasticizers are often employed to reduce water content while maintaining workability, ensuring the material meets the specific requirements of the 3D printing process [25].

The impact of the w/c ratio on 3DPC performance has been widely studied. Wang et al. [23] demonstrated that a lower w/c ratio improves compressive strength and dimensional stability but can result in poor extrusion consistency, particularly in particle bed 3D printing. Xu et al. [24] highlighted the challenges of excessive water content, which weakens interlayer bonding and increases shrinkage risks. Shahid and Sglavo [22] emphasized that while lower w/c ratios enhance mechanical properties, they require careful management to prevent nozzle blockages. For underwater 3DPC, Wang et al. [25] noted that precise w/c ratio adjustments are crucial to maintaining cohesion and avoiding washout. Collectively, these studies underline the importance of optimizing the w/c ratio to address the dual demands of printability and mechanical performance in 3DPC applications.

Figure 3 compares the flexural (a) and compressive (b) strength of 3D-printed (P) and cast (C) concrete specimens with varying water-cement (w/c) ratios over curing times of 3, 7, and 28 days. Lower w/c ratios (0.3P and 0.3C) exhibit the highest strength, while higher ratios (0.5P and 0.5C) show reduced strength. Compressive strength consistently exceeds flexural strength across all specimens. Strength

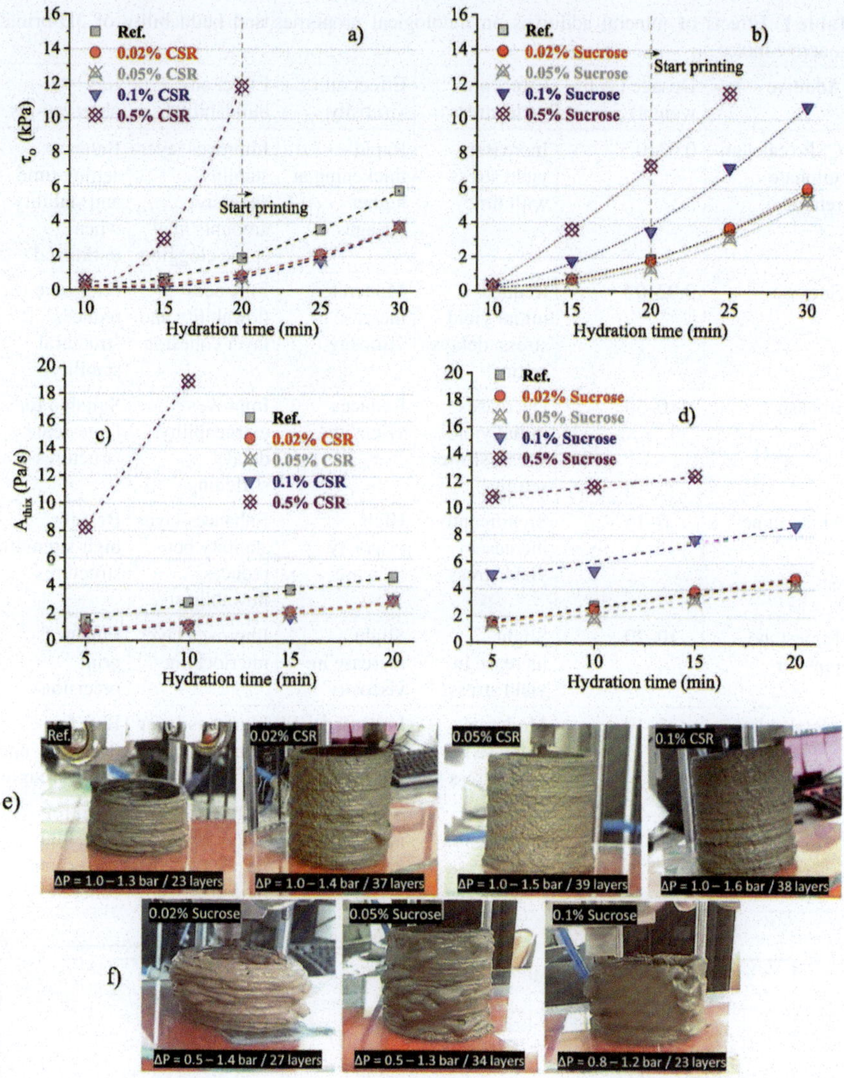

Fig. 2 Effects of sucrose and CSR on rheological properties and 3D printability of cementitious materials [14]

increases with curing time for both printed and cast specimens. Cast specimens (C) generally display higher strength than printed specimens (P) at similar w/c ratios, emphasizing the influence of curing, w/c ratio, and fabrication methods on concrete performance.

Key research trends in 3DPC emphasize optimizing the water-cement (w/c) ratio to balance printability and mechanical properties. Lower w/c ratios enhance compressive strength and dimensional stability but pose challenges in extrusion consistency

Table 1 Effects of mineral additives on rheological properties and buildability of 3D-printed concrete mixes

Additive	Dosage (% by weight)	Effect on yield stress	Effect on viscosity	Effect on buildability	Key observations
CSR (calcium sulphate retarder)	0.02–0.5	Increases yield stress with time	Rapid thickening at higher dosages	Improves layer stability; excessive amounts may cause clogging	Balances setting time and stability when optimized
Sucrose	0.02–0.5	Reduces initial yield stress, delays setting	Moderate increase in viscosity	Enhances flowability and layer cohesion	Excessive use reduces structural stability
Fly ash	10–30	Decreases initial yield stress, slower setting	Reduces viscosity	Improves extrudability, delays stiffening	Suitable for large-scale structures
Silica fume	5–15	Significantly increases yield stress	High viscosity increase	Enhances layer stability but reduces flowability	Best for high-strength structures
Limestone powder	10–20	Slight increase in yield stress	Slight increase in viscosity	Improves layer interlocking	Enhances print precision
Metakaolin	5–10	Moderate increase in yield stress	Increased viscosity	Improves layer stability	Best for durability and early strength
Nano-silica	0.5–2	Dramatic increase in yield stress	Significant increase in viscosity	Highly stable layers, risk of nozzle blockage	Used for ultra-high performance concrete

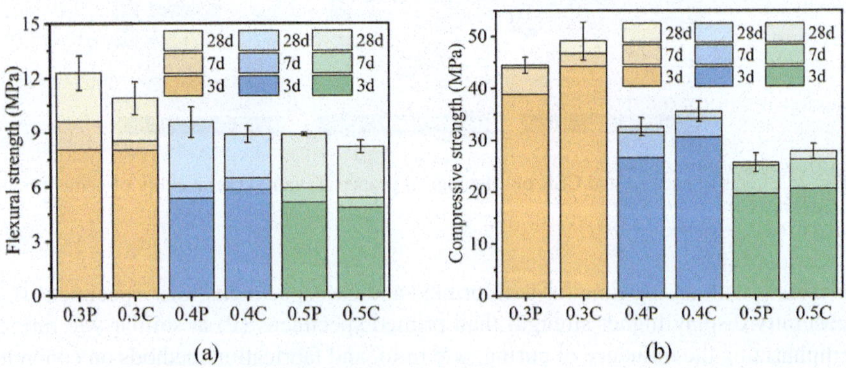

Fig. 3 Effect of w/c on mechanical properties: **a** flexural; **b** compressive strength (0.3, 0.4, 0.5 refer to the w/c; P: printed specimen; C: cast specimen) [26]

and flowability. Higher w/c ratios improve workability but weaken structural integrity and increase shrinkage risks. Advances include using superplasticizers to maintain workability while reducing water content. Challenges persist in achieving interlayer bonding and managing nozzle blockages. Research also explores w/c ratio adjustments for specialized applications, like underwater 3DPC, highlighting the critical need for tailored solutions to meet diverse performance and durability requirements.

2.5 Binder to Aggregate Ratio

The binder-to-aggregate (B/A) ratio significantly impacts the fresh and hardened properties of 3D-printed concrete, shaping its printability, buildability, and structural performance. A higher binder ratio improves flowability and extrusion, essential for maintaining layer integrity and geometric accuracy during the printing process [27, 28, 29]. However, excessive binder content can result in shrinkage and cracking in the hardened concrete, reducing long-term durability. Conversely, a lower binder ratio enhances mechanical strength and cost efficiency but may compromise workability, increasing the risk of clogging during extrusion [30, 31]. Balancing the B/A ratio is crucial to achieving the desired combination of workability, strength, and sustainability, ensuring that 3D-printed concrete aligns with modern construction performance and environmental objectives.

The importance of the binder-to-aggregate (B/A) ratio in 3DPC has been extensively studied. Wang et al. [28] demonstrated that higher binder content enhances extrudability and cohesion between layers, but noted the drawbacks of shrinkage and reduced sustainability. Şahin and Mardani-Aghabaglou [27] stressed the necessity of an optimal balance to preserve workability without sacrificing strength. Sahoo and Gupta [31] observed that excessive binder content negatively affects structural stability, particularly in alkali-activated mixtures, highlighting the trade-offs involved. Bhattacherjee et al. [30] found that higher aggregate content improves strength but increases the risk of extrusion blockages. Fernand et al. [32] confirmed that including coarse aggregates can enhance mechanical performance but complicates printability due to reduced flowability. These findings collectively underscore the need for a strategic approach to optimizing the B/A ratio, balancing the fresh and hardened properties to meet performance requirements.

The binder-to-aggregate (B/A) ratio is critical for balancing the fresh and hardened properties of 3DPC. A higher binder ratio enhances flowability, extrusion, and layer cohesion but may cause shrinkage, cracking, and reduced durability. Conversely, a lower binder ratio improves strength and cost efficiency but compromises workability, risking extrusion blockages. Optimizing this ratio is essential to balancing printability, structural performance, and sustainability. Research highlights the trade-offs between workability and strength, emphasizing the need for precise adjustments to achieve cohesion, geometric accuracy, and long-term stability in 3DPC while addressing environmental and economic construction objectives.

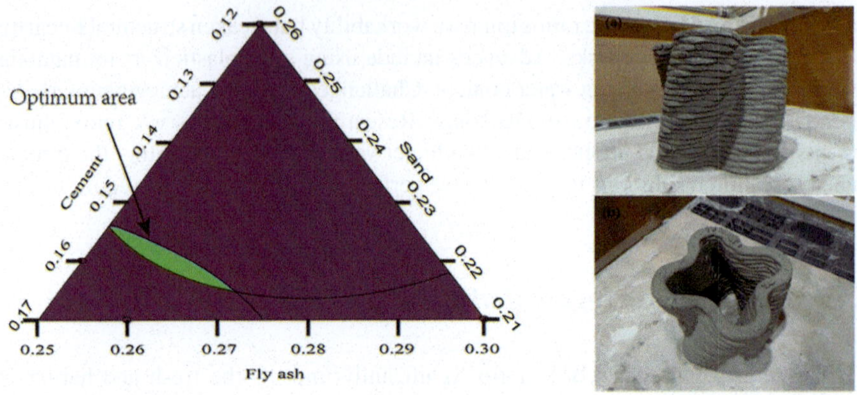

Fig. 4 Optimization of mix proportions for 3D-printed concrete: ternary analysis and structural demonstrations [33]

Figure 4 combines a ternary diagram and images of 3D-printed concrete structures. The ternary diagram (left) identifies the optimum mix proportions of cement, fly ash, and sand for 3D-printed concrete, represented by the green zone within the purple area. The balance in the green zone ensures optimal flowability, strength, and buildability. The images (right) showcase 3D-printed specimens: (a) a stacked cylindrical structure and (b) a wavy hollow structure. These demonstrate the material's ability to achieve complex geometries while maintaining stability during the printing process.

3 Process Parameters

3.1 Layer Height and Layer Adhesion

Layer height and layer adhesion are crucial parameters in 3DPC, as they directly influence structural integrity and print quality. Smaller layer heights enhance surface smoothness and interlayer bonding, improving the overall quality of the printed structure, but they also increase printing time and material usage [34, 35]. Conversely, larger layer heights expedite the printing process but may weaken layer interfaces, increasing the risk of delamination if adhesion is insufficient. Layer adhesion is affected by material rheology, extrusion speed, and the time gap between layers, requiring precise control of these factors to maintain bonding and structural continuity. Optimizing both layer height and adhesion is essential for producing durable and high-performance 3D-printed structures [36].

The role of layer height and adhesion in 3DPC performance has been widely studied. Huang et al. [34] and Zhang et al. [35] observed that smaller layer heights result in better interlayer bonding and enhanced mechanical properties. However,

Wolfs et al. [36] noted that larger layer heights can compromise adhesion, particularly when extrusion speed and layer deposition time are not properly aligned. Sanjayan et al. [37] and Nerella et al. [38] highlighted the importance of interface properties, such as moisture and alignment, with surface moisture shown to improve bonding under controlled conditions. Liu et al. [1, 9] and Wang et al. [4] emphasized that pore structure and aggregate distribution also affect adhesion quality. Skibicki et al. [39] found that stress distribution across layers critically impacts durability, especially in structural elements like beams. These studies underscore the importance of optimizing layer height and adhesion to ensure robust, sustainable, and high-performance 3D-printed concrete structures.

Layer height and adhesion are critical parameters in 3DPC that impact structural integrity and print quality. Smaller layer heights improve surface smoothness, interlayer bonding, and mechanical properties but increase printing time and material consumption. Larger layer heights expedite printing but risk weaker layer interfaces and delamination. Adhesion depends on material rheology, extrusion speed, time gaps, and interface properties such as moisture and alignment. Research emphasizes optimizing layer height and adhesion to balance printability, durability, and performance. This involves precise control of stress distribution, pore structure, and bonding quality for robust and sustainable 3D-printed concrete applications.

3.2 *Print Speed and Feed Rate*

Print speed and feed rate are critical parameters in 3DPC, significantly influencing material deposition, layer cohesion, and overall build quality. Higher print speeds increase construction efficiency but can compromise interlayer adhesion, as insufficient bonding time may lead to structural weaknesses [40, 41]. Conversely, slower speeds improve layer bonding and precision but result in longer construction times and increased costs. The feed rate, which dictates the volume of material extruded per unit time, directly impacts the uniformity of printed layers. An optimally balanced feed rate ensures consistent deposition, preventing defects such as gaps or over-extrusion [42, 43]. Coordination of print speed and feed rate is essential to achieving both structural integrity and dimensional accuracy in 3D-printed structures.

Figure 5 illustrates the relationship between print speed (mm/s) and vertical building rate (mm/min) for various nozzle diameters, ranging from 4 to 24 mm. Each coloured line represents a specific nozzle diameter, showing that as print speed increases, the vertical building rate also rises proportionally. Larger nozzle diameters produce higher vertical building rates at the same print speed compared to smaller diameters, indicating improved material deposition efficiency. The trend highlights the influence of nozzle size on construction speed, where larger diameters are advantageous for rapid building but may impact resolution and precision. Optimizing nozzle diameter and print speed is crucial for balancing speed and quality in 3D printing applications.

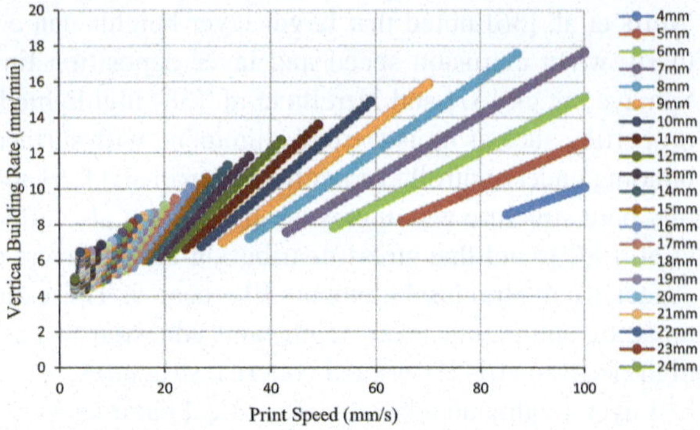

Fig. 5 Design envelope depicting the entire feasible design parameter domain and corresponding vertical building rate. Filament layer heights are specified in the legend [40]

The interplay between print speed and feed rate has been extensively studied for optimizing 3DPC performance. Zhao et al. [41] emphasized that excessive print speeds compromise cohesion and geometric precision due to inadequate bonding time, whereas slower speeds enhance interlayer adhesion but reduce productivity. Lu et al. [42] found that feed rate imbalance can result in voids or over-extrusion, affecting material consistency. Nair et al. [43] explored the relationship between print velocity and extrusion characteristics, recommending optimized rates for consistent layer deposition. Kruger et al. [40] noted the risk of plastic collapse at higher speeds, underlining the need for precise tuning of parameters. Shakor et al. [44] demonstrated that deposition velocity critically affects extrusion quality, particularly in fibre-reinforced concrete mixes. Diggs-McGee et al. [45] advocated for temporal optimization to balance print time and structural quality, ensuring integrity without excessive delays. Together, these studies underscore the importance of fine-tuning print speed and feed rate to achieve a balance between efficiency, accuracy, and material performance.

Print speed and feed rate are critical factors in 3DPC, directly affecting construction efficiency, material deposition, and structural quality. Faster print speeds enhance productivity but can compromise layer bonding due to insufficient time for proper adhesion, potentially weakening the structure. Slower speeds improve interlayer cohesion and precision but lead to longer construction durations and higher costs. The feed rate determines the material volume extruded, influencing layer uniformity. Balanced feed rates are crucial to avoid defects like gaps or over-extrusion. Achieving an optimal balance between print speed and feed rate is essential to ensure both structural integrity and efficiency.

3.3 Extrusion Rate and Nozzle Size

Extrusion rate and nozzle size are critical parameters in 3DPC, shaping material flow, layer thickness, and overall build quality. The extrusion rate dictates material deposition; excessive rates can result in over-extrusion, leading to uneven layers and poor geometric accuracy, while insufficient rates may cause gaps and weak interlayer bonding [46, 47]. Nozzle size directly influences the width and height of extruded layers, impacting surface finish and resolution. Larger nozzles enable faster construction but compromise detail precision, whereas smaller nozzles enhance accuracy and interlayer adhesion at the expense of speed [48, 49]. Optimizing these parameters is essential to ensure consistent material flow, robust bonding between layers, and structural integrity.

Table 2 compares various nozzle shapes used in 3D printing, highlighting their advantages and limitations. Circular nozzles allow flexibility but have low adhesion and surface quality. Elliptical nozzles simplify control but are prone to breakage. Rectangular nozzles offer strength and quality but struggle with complex shapes. Side blades ensure stability but risk damaging deposited layers, while groove nozzles improve interlayer adhesion.

The influence of extrusion rate and nozzle size on 3DPC has been widely studied. Ji et al. [46] emphasized that high extrusion rates facilitate continuous deposition but can reduce layer precision, while low rates risk incomplete bonding and weak structures. Nozzle size plays a crucial role in balancing speed and resolution; Zhang and Sanjayan [49] observed that larger nozzles improve construction efficiency but

Table 2 Comparison of nozzle shapes: advantages and limitations

Nozzle shape	Advantages	Limitations	References
Circular	Suitable for complex shapes, offering more freedom and easier nozzle angle adjustments	Low compactness and contact area, lack of layer adhesion leading to more voids, wavy edge causing low surface quality, and unstable flexural strength between layers	Shakor et al. [50]; Khan [51] ; McGee et al. [52]
Elliptical	Prevents filament twisting and simplifies machine control	Poor surface finish and higher susceptibility to breakage compared to rectangular nozzles	Lim et al. [53]
Rectangular	Provides high surface quality, excellent load-bearing capacity, and consistent flexural strength between layers	Lower printing quality for complex shapes, such as joints and corners	Paul et al. [54]
Side blades	Ensures high shape stability and surface finish	Can damage previously deposited filaments in crossing areas	Ji et al. [46]
Groove	Enhances interlocking between layers, significantly increasing adhesion strength by 26%	–	Zareiyan and Khoshnevis [55]

diminish detail, whereas Yang et al. [48] found that smaller nozzles enhance accuracy and surface quality. Shakor et al. [50] demonstrated that nozzle shape significantly affects reinforcement alignment and extrusion quality. Xu et al. [56] explored variable-size nozzles, showing their potential to optimize buildability in complex geometries. David et al. [57] highlighted the need for precise parameter tuning to achieve durable and high-performance printed structures. Collectively, these findings underscore the importance of optimizing extrusion rate and nozzle size to balance efficiency, accuracy, and structural performance.

Extrusion rate and nozzle size are pivotal in determining the quality, precision, and efficiency of 3DPC. High extrusion rates improve construction speed but risk over-extrusion, reducing geometric accuracy, while low rates weaken interlayer bonding. Nozzle size affects resolution and speed; larger nozzles expedite construction but compromise detail, whereas smaller nozzles enhance accuracy and adhesion at the cost of efficiency. Nozzle shape also influences material flow and structural quality, with specific designs suited for different applications. Research emphasizes balancing these parameters to achieve robust layer bonding, structural integrity, and optimized performance for complex geometries in 3DPC applications.

3.4 Printing Path and Layer Orientation

Printing path and layer orientation are critical factors that influence the structural performance, interlayer bonding, and aesthetic quality of 3DPC. The printing path determines material flow and layer alignment, with optimized paths minimizing material overlap, improving surface finish, and ensuring even stress distribution [58, 59]. Layer orientation directly affects mechanical properties, especially tensile and flexural strength. Vertically aligned layers often exhibit weaker bonding due to interlayer interfaces, whereas continuous, horizontally aligned layers promote cohesion and uniformity [60, 61]. While curved or complex paths enable innovative designs, they also pose challenges to deposition consistency. Balancing path planning and orientation is essential to achieving both structural integrity and architectural precision.

Figure 6 illustrates 3D printing processes on sloped surfaces, highlighting up-slope and down-slope printing directions. Panel (a) shows the printing apparatus depositing material along inclined paths, with "up-slope printing" moving against the slope and "down-slope printing" following the slope. Panel (b) provides a top view of the printed layers, revealing differences in material deposition patterns for up-slope and down-slope printing. Up-slope printing exhibits consistent, compact layering due to material compression, while down-slope printing shows less compactness, potentially causing weaker adhesion between layers. The images emphasize the influence of slope orientation on print quality and structural performance in 3D-printed concrete.

The effects of printing path and layer orientation on 3DPC have been extensively studied. Zahabizadeh et al. [61] found that vertically aligned layers exhibit weaker

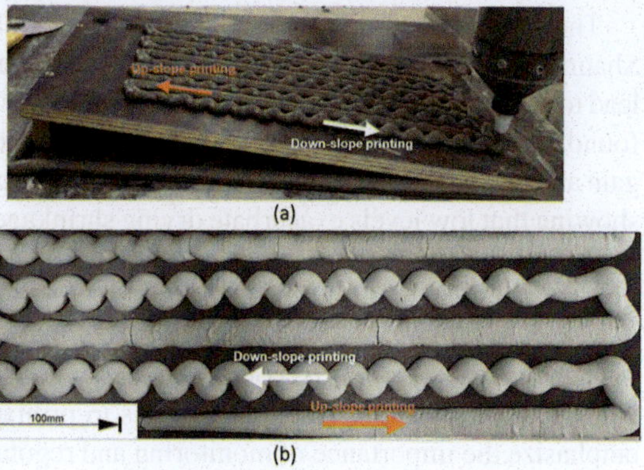

Fig. 6 Influence of inclined surfaces on the printing patterns. Individual filaments printed onto surface inclined at 7° showed straight and wavy patterns when printing in the up-slope and down-slope directions; photos taken from **a** side view and **b** top view [62, 63]

tensile properties compared to horizontal orientations, underscoring the importance of layer alignment in enhancing structural bonding. Pham et al. [59] showed that optimized printing paths improve fibre alignment, thereby enhancing strength and load distribution. Chen et al. [64] demonstrated that the inclusion of coarse aggregates significantly affects direction-dependent mechanical performance, highlighting the interplay between material composition and path design. Pan et al. [58] observed that compressive strength is influenced by path design and curing conditions, reinforcing the role of strategic path planning. Yang et al. [60] proposed a topological optimization framework, illustrating the potential of advanced path planning techniques to maximize structural efficiency and durability. These findings underscore the critical role of printing path and layer orientation in optimizing both the mechanical and aesthetic outcomes of 3D-printed concrete structures.

4 Environmental Conditions

4.1 Temperature and Humidity During Printing

Temperature and humidity are critical factors in 3DPC, significantly impacting material performance and structural integrity. High temperatures accelerate the curing process, potentially causing nozzle blockages, reduced interlayer adhesion, and increased cracking due to rapid moisture loss [65, 66]. Conversely, low temperatures slow curing, which affects buildability and prolongs printing times [67]. Humidity also plays a vital role; low humidity accelerates evaporation, increasing shrinkage and cracking, while high humidity can hinder proper setting and reduce mechanical strength [68]. Maintaining controlled environmental conditions is essential to ensure consistent extrusion, adequate interlayer bonding, and optimized mechanical properties of the printed structure.

The effects of temperature and humidity on 3DPC have been extensively studied. Shahmirzadi et al. [66] noted that high temperatures, while accelerating curing, can lead to shrinkage and cracking, negatively affecting structural integrity. Sun et al. [67] found that low temperatures slow the hydration process, reducing the rate of strength gain and extending build times. Ma et al. [68] highlighted the dual impact of humidity, showing that low levels exacerbate drying shrinkage and internal stresses, while high levels delay curing and compromise strength. Wei and Zhang [69] demonstrated that temperature and humidity also affect pore structure and mechanical performance, underlining their influence on the final material properties. Additionally, Alchaar and Al-Tamimi [65] observed that in hot climates, reduced workability and increased evaporation present challenges to printability. Collectively, these studies emphasize the importance of monitoring and regulating environmental conditions to achieve durable and high-quality 3DPC structures. Table 3 summarizes the effects of temperature and humidity on 3D printed concrete (3DPC), highlighting their impact on curing, adhesion, cracking, and mechanical strength. It emphasizes the importance of controlled environmental conditions for consistent extrusion, improved interlayer bonding, and optimized material performance, drawing insights from multiple studies and key references.

Temperature and humidity are critical factors influencing the performance and quality of 3DPC. High temperatures accelerate curing, which can lead to shrinkage, cracking, and poor interlayer adhesion, while low temperatures slow the hydration process, prolonging build times and reducing early strength gain. Humidity levels also play a significant role; low humidity accelerates evaporation, increasing shrinkage and internal stresses, whereas high humidity delays curing, compromising mechanical strength. These environmental factors affect extrusion consistency, interlayer bonding, and pore structure, underscoring the need for controlled conditions. Optimized temperature and humidity management is essential for achieving durable, high-quality 3DPC structures.

4.2 Curing Conditions

Curing conditions play a crucial role in determining the properties of 3DPC, influencing its strength, durability, and dimensional stability. Proper curing ensures adequate hydration of cementitious materials, optimizing mechanical performance [70, 67]. Factors such as temperature, humidity, and curing duration are pivotal in controlling hydration quality and rate. Elevated temperatures accelerate curing but risk cracking due to rapid water evaporation, whereas low temperatures slow hydration, compromising early-age strength [67, 28]. Controlled humidity minimizes shrinkage and prevents cracking, maintaining structural integrity. In 3D printing, the layer-by-layer approach magnifies these effects, as inadequate curing between layers can weaken interlayer adhesion. Tailored curing protocols are essential for achieving consistent quality in 3DPC structures.

Table 3 Impact of temperature and humidity on 3D printed concrete (3DPC) performance

Environmental factor	Condition	Effects on 3DPC	Key references
Temperature	High	– Accelerates curing process – Increases nozzle blockages – Reduces interlayer adhesion – Causes rapid moisture loss, shrinkage, and cracking	Shahmirzadi et al. [66]; Alchaar and Al-Tamimi [65]
	Low	– Slows curing process – Reduces hydration rate – Prolongs printing time – Affects buildability	Sun et al. [67]
Humidity	Low	– Increases evaporation – Exacerbates drying shrinkage – Promotes internal stresses, increasing cracking	Ma et al. [68]
	High	– Slows curing process – Reduces mechanical strength – Hinders proper setting	Ma et al. [68]; Wei and Zhang [69]
Combined impacts	High temperature and humidity	– Alters pore structure – Reduces workability – Challenges printability in hot climates	Alchaar and Al-Tamimi [65]; Wei and Zhang [69]
	Regulated conditions	– Ensures consistent extrusion – Enhances interlayer bonding – Optimizes mechanical properties	

The impact of curing conditions on 3DPC has been extensively studied. Federowicz et al. [70] highlighted the risk of shrinkage and dimensional instability caused by improper curing. Sun et al. [67] demonstrated the synergistic effects of temperature and humidity, using modified maturity models to show how optimized curing accelerates strength development. Steam curing, investigated by Wang et al. [29] and Yao et al. [71], was found to enhance mechanical properties and interlayer bonding, though anisotropy remains a challenge. Li et al. [62, 63] confirmed the direct impact of curing conditions on interlayer adhesion, emphasizing the need for precise control. Geopolymer-based alternatives, studied by Bong et al. [72] and Nematollahi et al. [73], revealed that ambient curing provides a sustainable approach for

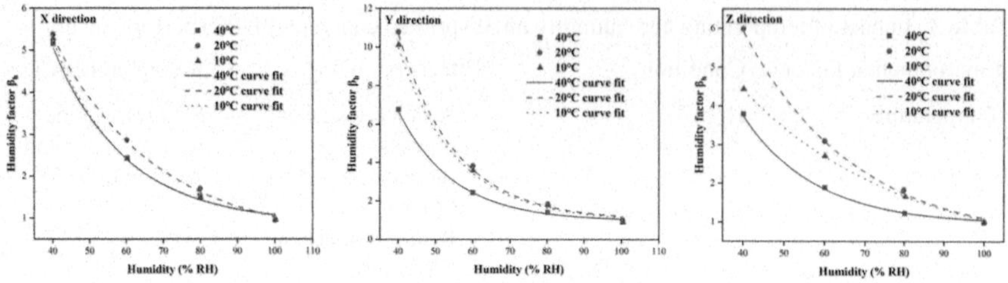

(a) X-direction (b) Y-direction (c) Z-direction

Fig. 7 Effect of humidity and temperature on humidity factor (F_h) across X, Y, and Z directions [67]

3D-printed concrete, reducing energy consumption while maintaining performance. Collectively, these findings underscore the importance of optimizing curing conditions to achieve durable and high-quality 3DPC structures. Figure 7 illustrates the relationship between humidity (% RH) and humidity factor (F_h) in three spatial directions (X, Y, Z) at different temperatures (40 °C, 20 °C, 10 °C). Curve fits highlight varying humidity effects across directions and temperatures.

Curing conditions significantly influence the strength, durability, and dimensional stability of 3DPC. Proper curing ensures sufficient hydration of cementitious materials, optimizing mechanical properties and preventing shrinkage or cracking. High temperatures accelerate curing but risk water loss and cracking, while low temperatures slow hydration and compromise early strength. Controlled humidity is critical to minimizing shrinkage and maintaining structural integrity. The layer-by-layer nature of 3DPC amplifies the importance of curing, as inadequate interlayer adhesion weakens the structure. Advances include tailored curing protocols, steam curing to enhance bonding, and ambient curing for geopolymers, balancing sustainability and performance in 3DPC applications.

4.3 *Moisture Exposure Post-Printing*

Moisture exposure post-printing plays a pivotal role in the durability and mechanical properties of 3DPC. The curing process relies on controlled moisture to ensure proper hydration of cement particles, forming a dense and durable microstructure. Excessive moisture can lead to swelling, reduced strength, and surface defects caused by uneven curing, while insufficient moisture results in incomplete hydration, causing cracks and brittleness [65, 74]. Prolonged exposure to moisture may also accelerate chemical attacks, such as sulphate or chloride ingress, further compromising structural integrity [75]. Therefore, effective moisture management post-printing is essential for optimizing curing and preventing degradation due to environmental factors.

The influence of moisture exposure post-printing on 3DPC has been widely studied. Zhang et al. [74] observed that temperature fluctuations coupled with moisture variations negatively affect interlayer adhesion, weakening structural integrity. Alchaar and Al-Tamimi [65] found that excessive heat-induced moisture loss reduces strength, while Moelich et al. [76] demonstrated that surface moisture is critical for enhancing interlayer bond strength. Ler et al. [75] highlighted the role of controlled curing in reducing porosity and improving durability. Additionally, Ramakrishnan et al. [77] emphasized that in-line activation and precise hydration control significantly enhance mechanical properties. These studies collectively underscore the importance of managing moisture exposure to achieve durable, high-performance 3D-printed concrete structures.

Moisture exposure post-printing is critical to the durability and mechanical properties 3DPC. Controlled moisture ensures proper hydration, forming a dense microstructure, while excessive moisture leads to swelling, reduced strength, and surface defects. Insufficient moisture causes incomplete hydration, resulting in cracks and brittleness. Prolonged exposure to moisture also increases vulnerability to chemical attacks, such as sulphate and chloride ingress, compromising structural integrity. Research emphasizes the importance of precise hydration control, in-line activation, and surface moisture for enhancing interlayer adhesion and reducing porosity. Effective post-printing moisture management is essential for producing durable and high-performance 3DPC structures.

Table 4 summarizes the effects of moisture exposure on 3D printed concrete (3DPC) after printing, emphasizing the importance of post-printing conditions for material performance. Prolonged high moisture exposure promotes hydration, enhancing strength and reducing shrinkage and cracking. Moderate moisture balances hydration and drying, creating optimal curing conditions. In contrast, low moisture accelerates drying, causing shrinkage, cracking, and compromised structural integrity. Cyclic wet-dry conditions induce microcracking and fatigue, reducing durability over time. Submerged curing slows carbonation and enhances compressive strength, making it ideal for specific applications like marine environments. The table highlights the need for controlled moisture conditions to optimize 3DPC properties.

4.4 *Effect of Environmental Conditions on Setting Time and Strength*

Environmental conditions critically influence the setting time and strength of 3DPC. Elevated temperatures speed up cement hydration, reducing setting time but increasing the risk of nozzle blockage and layer delamination, as noted by Alchaar and Al-Tamimi [65]. Cold conditions slow hydration, delaying strength gain and potentially weakening bond layers, which Wolfs et al. [36] emphasize are affected by process parameters. Low humidity accelerates evaporation, leading to insufficient curing and weaker interlayer adhesion, while high humidity extends setting

Table 4 Influence of post-printing moisture exposure on 3D printed concrete properties

Moisture condition	Effects on 3DPC	Key outcomes	References
Prolonged high moisture	Promotes continued hydration enhances long-term strength reduces shrinkage and cracking	Improved durability and mechanical performance	Moelich et al. [78]; Zhu et al. [79]
Moderate moisture exposure	Balances hydration and drying supports optimal curing reduces surface defects	Best condition for strength and surface finish	Sanjayan et al. [37] ; Li et al. [80, 81]
Low moisture (drying)	Accelerates drying causes surface shrinkage induces cracking and internal stresses	Compromised structural integrity and durability	Moelich et al. [82]; Shahmirzadi et al. [66]
Cyclic wet-dry conditions	Causes microcracking promotes fatigue and reduced durability	Increased wear over time	Van Tittelboom et al. [83]; Ler et al. [75]
Submerged curing	Slows carbonation maintains pore saturation reduces internal stresses	Optimized compressive strength and durability	Bong et al. [72]; Li and Zhang [80]

time, affecting buildability, as shown by Sanjayan et al. [37]. Wind further exacerbates moisture loss, causing shrinkage and warping. Proper control of environmental conditions, such as using accelerators, retarders, or controlled curing environments, is crucial to ensure uniform setting, optimal strength, and consistent quality in 3DPC structures, a strategy supported by Huang et al. [34].

Environmental conditions, such as temperature and humidity, critically influence the setting time and strength of 3DPC. High temperatures accelerate the setting process, potentially reducing interlayer adhesion and increasing brittleness, as observed by Alchaar and Al-Tamimi [65]. Surface moisture plays a crucial role in interlayer strength, with insufficient moisture leading to poor bonding and cracks, highlighted by Sanjayan et al. [37]. Wolfs et al. [36] noted that process parameters significantly affect interlayer adhesion under varying conditions. Furthermore, Huang et al. [34] demonstrated that optimized printing parameters can mitigate environmental impacts on mechanical properties. Effective environmental management enhances performance and durability.

Environmental conditions significantly affect the setting time and strength of 3DPC. High temperatures accelerate hydration, reducing setting time but risking interlayer delamination and brittleness, while cold conditions delay strength development and weaken layer bonds. Low humidity causes rapid evaporation, leading to poor curing and adhesion, whereas high humidity extends setting time, impacting buildability. Wind exacerbates moisture loss, increasing shrinkage and warping. Proper environmental control, including the use of accelerators, retarders, and controlled curing environments, ensures uniform setting, optimal strength, and consistent quality. Optimized printing parameters further mitigate these effects, enhancing the performance and durability of 3D-printed concrete structures.

5 Geometric and Structural Considerations

5.1 Wall Thickness and Print Geometry

Wall thickness and print geometry significantly influence the performance and durability of 3DPC. According to Nguyen-Van et al. [84], thicker walls provide greater structural stability and load-bearing capacity, but they also increase material usage and printing time. Conversely, thin walls, while more resource-efficient, can compromise stability and risk cracking or deformation. Print geometry, including curvature, angles, and layer alignment, affects stress distribution and interlayer bonding, as noted by Xu et al. [85]. Complex geometries require precise control of nozzle path and extrusion rate to avoid defects like voids or weak joints, as demonstrated by Comminal et al. [86]. Optimizing wall thickness and geometry is essential for balancing material efficiency, structural integrity, and aesthetic appeal in 3DPC construction.

Wall thickness and print geometry are pivotal in the performance of 3D-printed concrete (3DPC) structures. Nguyen-Van et al. [84] highlighted that thin-walled structures demand precise modelling to balance material efficiency and mechanical stability. Marais et al. [87] demonstrated how wall cavities improve thermal performance, showcasing geometry's impact on functionality. Furthermore, Comminal et al. [86] linked processing parameters with layer geometry, emphasizing the role of extrusion in achieving consistent dimensions. Liu et al. [88] provided strength calculation methods for varying wall designs under diverse loads. Geometry precision, as Xu et al. [85] noted, is critical to minimizing defects, thereby enhancing overall structural integrity.

Wall thickness and print geometry are crucial for optimizing the performance and durability of 3DPC structures. Thicker walls enhance structural stability and load-bearing capacity but increase material usage and printing time, while thinner walls offer resource efficiency but risk instability and cracking. Print geometry, including curvature and angles, significantly influences stress distribution and interlayer bonding. Precise control of nozzle paths and extrusion rates is critical for maintaining consistent dimensions and avoiding defects like voids or weak joints. Innovations in wall cavity design and advanced strength modelling methods demonstrate progress in balancing material efficiency, structural integrity, and functionality in 3DPC.

5.2 Layer Bond Strength and Interlayer Adhesion

Layer bond strength and interlayer adhesion are critical for the structural integrity of 3DPC. These parameters determine the cohesion between successive layers during the printing process. Factors such as material composition, printing speed, layer deposition time, and environmental conditions significantly influence adhesion [76, 78, 36]. Poor bonding can result in weak points, reducing load-bearing capacity and durability. Strategies like optimizing the mix design, enhancing surface roughness, and applying bonding agents can improve interlayer strength [89, 90]. Additionally, time-dependent properties such as setting time and rheology play pivotal roles, requiring precise control to ensure seamless integration and structural stability [91].

Layer bond strength and interlayer adhesion are pivotal to the structural integrity of 3DPC. Adhesion is significantly influenced by process parameters, including layer deposition time and surface moisture [76, 78, 36]. Fresh material properties such as rheology and setting time are also critical to maintaining layer cohesion [91]. Enhancing bond strength can be achieved by modifying material compositions with polymers and applying surface treatments [89, 90]. Studies on vertical stacking and environmental factors underscore their impact on interlayer properties [35]. Strengthening interlayer bonds is essential for improved durability and structural performance [92, 93]. Figure 8 illustrates the relationship between yield stress and time in 3D printed concrete (3DPC) to optimize interlayer bonding strength. Initially, at t_{min}, low yield stress (τ_{min}) can lead to layer collapse. Over time, yield stress increases, reaching an optimal range between t_{min} and t_{max}, ensuring well-bonded layers with sufficient structural integrity. Beyond t_{max}, excessive yield stress (τ_{max}) results in poor interlayer bonding strength due to insufficient adhesion. The figure emphasizes the critical timing of material placement to balance structural stability and interlayer adhesion in 3DPC.

Layer bond strength and interlayer adhesion are crucial for ensuring the structural integrity of 3DPC. These properties depend on factors like material composition, printing speed, deposition time, and environmental conditions. Poor bonding can create weak points, reducing load-bearing capacity and durability. Key advancements include optimizing mix designs, incorporating polymers, and using surface treatments to enhance adhesion. Fresh material properties, such as rheology and

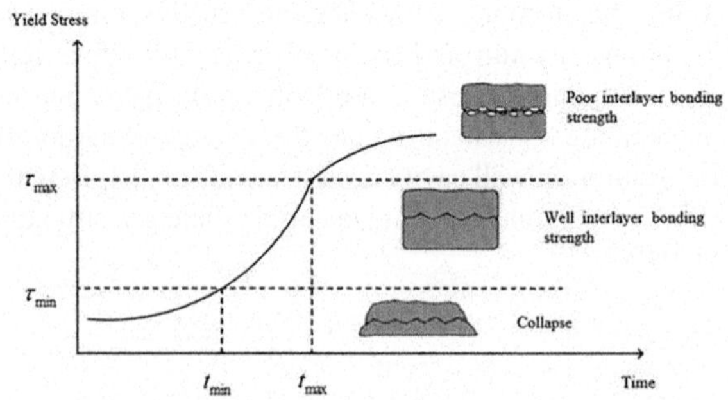

Fig. 8 Yield stress evolution and its impact on interlayer bonding strength in 3D printed concrete [94]

setting time, also play critical roles in maintaining cohesion during the printing process. Research on vertical stacking and environmental influences further highlights their impact on interlayer strength, emphasizing the need for precise control to ensure durable, high-performance structures.

5.3 Structural Stability During Printing

Structural stability during 3D printing of concrete is a critical factor for ensuring the integrity of printed structures. To achieve this, the material must effectively support its own weight and maintain shape during layer-by-layer deposition. This requires balancing early-age strength and workability, as the concrete must flow smoothly through the nozzle while stiffening quickly enough to prevent collapse [95, 96]. Factors influencing stability include mix design, such as the inclusion of admixtures like superplasticizers and accelerators, which directly affect the material's performance [97]. Furthermore, optimizing printing parameters such as speed and layer intervals plays a vital role in achieving proper interlayer bonding and preventing excessive deformation [98]. Continuous monitoring and adjustments during the printing process are crucial to ensure a structurally sound and geometrically accurate final product.

The structural stability of 3D-printed concrete is shaped by various factors, including material properties, layer interfaces, and process parameters. Liu and Sun [95] noted that weak layer interfaces could compromise stability, emphasizing the importance of bond strength control. Gomaa et al. [98] highlighted the significance of advanced printing systems and ultra-high-performance concrete in maintaining structural integrity under complex geometries. Cho et al. [99] identified foam stability as a crucial element for printable foamed concrete. Vallurupalli et al. [96] reviewed how flow behaviour and process-induced variations impact stability, while Liu et al. [97] demonstrated the stabilizing effects of additives such as hydroxypropyl methylcellulose and silica fume on foam concrete rheology. These studies collectively underscore the importance of material and process optimization for stable 3D-printed concrete structures.

Structural stability in 3D-printed concrete (3DPC) is crucial to maintaining integrity during layer-by-layer construction. Achieving this requires balancing early-age strength and workability; the material must flow smoothly while stiffening quickly to avoid collapse. Optimizing mix designs with admixtures like superplasticizers and accelerators enhances stability by improving rheology and early strength. Printing parameters such as speed and layer intervals are vital for proper interlayer bonding and minimizing deformation. Advances in ultra-high-performance concrete, foam stability techniques, and additives like silica fume and cellulose have further improved stability. Continuous monitoring and process adjustments are essential for producing durable, geometrically precise structures.

6 Concluding Remarks

The performance of 3D printed concrete (3DPC) depends on material composition, process variables, environmental conditions, and structural design factors. Optimizing cement formulations, such as incorporating supplementary materials like fly ash and ground granulated blast-furnace slag (GGBFS), improves sustainability and rheological properties while maintaining strength. Aggregate size and gradation also play a pivotal role, with fine aggregates ensuring smoother extrusion and coarse aggregates enhancing mechanical integrity. Additives like plasticizers, fibres, and accelerators improve flowability, tensile strength, and crack resistance, vital for structural reliability.

Process parameters, including print speed, layer height, and extrusion rate, require precise control to ensure consistent material deposition, strong interlayer adhesion, and dimensional accuracy. Strategic planning of print paths and layer orientation minimizes stress concentrations and optimizes load distribution in complex geometries. Environmental conditions, particularly temperature and humidity, are critical to controlling curing rates, avoiding shrinkage, and ensuring structural stability. Tailored curing protocols, such as steam curing or controlled hydration, prevent defects like delamination and cracking while enhancing mechanical performance. Advanced technologies like real-time monitoring and topological optimization can address challenges such as weak interlayer bonding and geometric inaccuracies. By integrating these approaches, 3DPC can achieve superior printability, durability, and sustainability, advancing its role as a transformative technology in the construction industry.

References

1. Liu, H., Liu, C., Bai, G., Wu, Y., He, C., Zhang, R., & Wang, Y. (2022a). Influence of pore defects on the hardened properties of 3D printed concrete with coarse aggregate. *Additive Manufacturing, 55*, Article 102843.
2. Ibrahim, K. A., van Zijl, G. P., & Babafemi, A. J. (2023). Influence of limestone calcined clay cement on properties of 3D printed concrete for sustainable construction. *Journal of Building Engineering, 69*, Article 106186.
3. Xu, Z., Zhang, D., Li, H., Sun, X., Zhao, K., & Wang, Y. (2022). Effect of FA and GGBFS on compressive strength, rheology, and printing properties of cement-based 3D printing material. *Construction and Building Materials, 339*, Article 127685.
4. Wang, X., Naito, C., Fox, J. T., & Bocchini, P. (2024b). Impact of mix proportions on particle bed 3D printed concrete properties. *Construction and Building Materials, 419*, Article 135441.
5. Lowke, D., Talke, D., Dressler, I., Weger, D., Gehlen, C., Ostertag, C., & Rael, R. (2020). Particle bed 3D printing by selective cement activation–Applications, material and process technology. *Cement and Concrete Research, 134*, Article 106077.
6. Liu, C., Xiong, Y., Chen, Y., Jia, L., Ma, L., Deng, Z., Wang, Z., Chen, C., Banthia, N., & Zhang, Y. (2022). Effect of sulphoaluminate cement on fresh and hardened properties of 3D printing foamed concrete. *Composites Part B: Engineering, 232*, Article 109619.

7. Xiao, J., Lv, Z., Duan, Z., & Hou, S. (2022). Study on preparation and mechanical properties of 3D printed concrete with different aggregate combinations. *Journal of Building Engineering, 51*, Article 104282.
8. An, D., Zhang, Y. X., & Yang, R. (2024). Incorporating coarse aggregates into 3D concrete printing from mixture design and process control to structural behaviours and practical applications: A review. *Virtual and Physical Prototyping, 19*(1), Article e2351154.
9. Liu, H., Liu, C., Wu, Y., Bai, G., He, C., Yao, Y., Zhang, R., & Wang, Y. (2022b). 3D printing concrete with recycled coarse aggregates: The influence of pore structure on interlayer adhesion. *Cement and Concrete Composites, 134*, Article 104742.
10. Chen, Y., Zhang, Y., Zhang, Y., Pang, B., Zhang, W., Liu, C., Liu, Z., Wang, D., & Sun, G. (2023). Influence of gradation on extrusion-based 3D printing concrete with coarse aggregate. *Construction and Building Materials, 403*, Article 133135.
11. Ding, Y., Tong, J., Zhang, M., Guo, S., Zhang, Y., & Zhao, Y. (2024) Effect of aggregate gradation on the properties of 3D printed recycled coarse aggregate concrete. *Journal of Wuhan University of Technology-Mater. Sci. Ed., 39*(6), 1541–1554.
12. Yu, S., Sanjayan, J., & Du, H. (2020). Impact of particle size and grading on aggregate-bed 3D concrete printing. In: *Second RILEM international conference on concrete and digital fabrication: Digital concrete 2020* (Vol. 2, pp. 557–563). Springer International Publishing.
13. Kaliyavaradhan, S. K., Ambily, P. S., Shekar, D., & Sebastian, S. (2024). Effect of sand gradations on the fresh properties of 3D printable concrete. *Magazine of Concrete Research*, 1–10.
14. Souza, M. T., Ferreira, I. M., de Moraes, E. G., Senff, L., Arcaro, S., Pessôa, J. R. C., Ribeiro, M. J., & de Oliveira, A. P. N. (2022). Role of chemical admixtures on 3D printed Portland cement: Assessing rheology and buildability. *Construction and Building Materials, 314*, Article 125666.
15. Sergis, V., & Ouellet-Plamondon, C. M. (2022). Fractional factorial design to study admixtures used for 3D concrete printing applications. *Materials Letters, 324*, Article 132697.
16. Chen, Y., Zhang, Y., Pang, B., Wang, D., Liu, Z., & Liu, G. (2022). Steel fiber orientational distribution and effects on 3D printed concrete with coarse aggregate. *Materials and Structures, 55*(3), 100.
17. Ding, T., Xiao, J., Zou, S., & Zhou, X. (2020). Anisotropic behavior in bending of 3D printed concrete reinforced with fibers. *Composite Structures, 254*, Article 112808.
18. Pham, L., Tran, P., & Sanjayan, J. (2020). Steel fibres reinforced 3D printed concrete: Influence of fibre sizes on mechanical performance. *Construction and Building Materials, 250*, Article 118785.
19. Seo, E. A., & Lee, H. J. (2024). Influence of chemical admixtures on buildability and deformation of concrete for additive manufacturing. *Case Studies in Construction Materials, 21*, Article e03871.
20. Rubin, A. P., Hasse, J. A., & Repette, W. L. (2021). The evaluation of rheological parameters of 3D printable concretes and the effect of accelerating admixture. *Construction and Building Materials, 276*, Article 122221.
21. Rubin, A. P., Quintanilha, L. C., & Repette, W. L. (2023). Influence of structuration rate, with hydration accelerating admixture, on the physical and mechanical properties of concrete for 3D printing. *Construction and Building Materials, 363*, Article 129826.
22. Shahid, M., & Sglavo, V. M. (2024). Effect of water-to-quick setting cement ratio and aggregate size on mechanical properties and dimensional accuracy of binder jetting 3D-printed bodies. *Open Ceramics*, 100704.
23. Wang, Y., Qiu, L. C., Chen, S. G., & Liu, Y. (2024a). 3D concrete printing in air and under water: A comparative study on the buildability and interlayer adhesion. *Construction and Building Materials, 411*, Article 134403.
24. Xu, Z., Zhang, D., Li, H., Jin, H., Cai, X., & Tian, Z. (2023). Mix design and 3D printing correlation performance of PCC-based 3DPM. *Composite Structures, 320*, Article 117229.
25. Wang, Y., Qiu, L. C., Hu, Y. Y., Chen, S. G., & Liu, Y. (2023). Influential factors on mechanical properties and microscopic characteristics of underwater 3D printing concrete. *Journal of Building Engineering, 77*, Article 107571.

26. Ma, L., Jia, Z., Chen, Y., Jiang, Y., Huet, B., Delaplace, A., Zhang, Y., & Zhang, Q. (2024). Water loss and shrinkage prediction in 3D printed concrete with varying w/c and specimen sizes. *Cement and Concrete Composites, 149*, Article 105523.
27. Şahin, H. G., & Mardani-Aghabaglou, A. (2022). Assessment of materials, design parameters and some properties of 3D printing concrete mixtures; a state-of-the-art review. *Construction and Building Materials, 316*, Article 125865.
28. Wang, X., Jia, L., Jia, Z., Zhang, C., Chen, Y., Ma, L., Wang, Z., Deng, Z., Banthia, N., & Zhang, Y. (2022a). Optimization of 3D printing concrete with coarse aggregate via proper mix design and printing process. *Journal of Building Engineering, 56*, Article 104745.
29. Wang, B., Yao, X., Yang, M., Zhang, R., Huang, J., Wang, X., Dong, Z., & Zhao, H. (2022b). Mechanical performance of 3D printed concrete in steam curing conditions. *Materials, 15*(8), 2864.
30. Bhattacherjee, S., Jain, S., & Santhanam, M. (2023). Developing 3D printable and buildable limestone calcined clay-based cement composites with higher aggregate content. *Construction and Building Materials, 376*, Article 131058.
31. Sahoo, P., & Gupta, S. (2024). 3D printable earth-based alkali activated "ink": Effect of alkali concentration and binder-to-aggregate ratio. *Journal of Building Engineering, 98*, Article 111208.
32. Fernand, M., Li, Y., Qian, Q., Chi, Y., & Yang, Z. (2024). Effects of coarse aggregates on 3D printability and mechanical properties of ultra high performance fiber reinforced concrete. *Journal of Building Engineering, 96*, Article 110516.
33. Liu, Z., Li, M., Weng, Y., Wong, T. N., & Tan, M. J. (2019). Mixture Design Approach to optimize the rheological properties of the material used in 3D cementitious material printing. *Construction and Building Materials, 198*, 245–255.
34. Huang, X., Yang, W., Song, F., & Zou, J. (2022). Study on the mechanical properties of 3D printing concrete layers and the mechanism of influence of printing parameters. *Construction and Building Materials, 335*, Article 127496.
35. Zhang, Y., Yang, L., Qian, R., Liu, G., Zhang, Y., & Du, H. (2023). Interlayer adhesion of 3D printed concrete: Influence of layer stacked vertically. *Construction and Building Materials, 399*, Article 132424.
36. Wolfs, R. J. M., Bos, F. P., & Salet, T. A. M. (2019). Hardened properties of 3D printed concrete: The influence of process parameters on interlayer adhesion. *Cement and Concrete Research, 119*, 132–140.
37. Sanjayan, J. G., Nematollahi, B., Xia, M., & Marchment, T. (2018). Effect of surface moisture on inter-layer strength of 3D printed concrete. *Construction and Building Materials, 172*, 468–475.
38. Nerella, V. N., Hempel, S., & Mechtcherine, V. (2019). Effects of layer-interface properties on mechanical performance of concrete elements produced by extrusion-based 3D-printing. *Construction and Building Materials, 205*, 586–601.
39. Skibicki, S., Szewczyk, P., Majewska, J., Sibera, D., Ekiert, E., Chung, S. Y., & Sikora, P. (2024). The effect of interlayer adhesion on stress distribution in 3D printed beam elements. *Journal of Building Engineering, 87*, Article 109093.
40. Kruger, J., Cho, S., Zeranka, S., Viljoen, C., & van Zijl, G. (2020). 3D concrete printer parameter optimisation for high rate digital construction avoiding plastic collapse. *Composites Part B: Engineering, 183*, Article 107660.
41. Zhao, Y., Yang, G., Zhu, L., Ding, Y., Guan, X., Wu, X., & Yang, Z. (2022). Effects of rheological properties and printing speed on molding accuracy of 3D printing basalt fiber cementitious materials. *Journal of Materials Research and Technology, 21*, 3462–3475.
42. Lu, B., Li, M., Wong, T. N., & Qian, S. (2021). Effect of printing parameters on material distribution in spray-based 3D concrete printing (S-3DCP). *Automation in Construction, 124*, Article 103570.
43. Nair, S., Panda, S., Tripathi, A., & Neithalath, N. (2021). Relating print velocity and extrusion characteristics of 3D-printable cementitious binders: Implications towards testing methods. *Additive Manufacturing, 46*, Article 102127.

44. Shakor, P., Nejadi, S., Sutjipto, S., Paul, G., & Gowripalan, N. (2020). Effects of deposition velocity in the presence/absence of E6-glass fibre on extrusion-based 3D printed mortar. *Additive Manufacturing, 32*, Article 101069.
45. Diggs-McGee, B. N., Kreiger, E. L., Kreiger, M. A., & Case, M. P. (2019). Print time vs. elapsed time: A temporal analysis of a continuous printing operation for additive constructed concrete. *Additive Manufacturing, 28*, 205–214.
46. Ji, G., Xiao, J., Zhi, P., Wu, Y. C., & Han, N. (2022). Effects of extrusion parameters on properties of 3D printing concrete with coarse aggregates. *Construction and Building Materials, 325*, Article 126740.
47. Nair, S. A., Panda, S., Santhanam, M., Sant, G., & Neithalath, N. (2020). A critical examination of the influence of material characteristics and extruder geometry on 3D printing of cementitious binders. *Cement and Concrete Composites, 112*, Article 103671.
48. Yang, L., Sepasgozar, S. M., Shirowzhan, S., Kashani, A., & Edwards, D. (2023a). Nozzle criteria for enhancing extrudability, buildability and interlayer bonding in 3D printing concrete. *Automation in Construction, 146*, Article 104671.
49. Zhang, N., & Sanjayan, J. (2023). Extrusion nozzle design and print parameter selections for 3D concrete printing. *Cement and Concrete Composites, 137*, Article 104939.
50. Shakor, P., Nejadi, S., & Paul, G. (2019). A study into the effect of different nozzles shapes and fibre-reinforcement in 3D printed mortar. *Materials, 12*(10), 1708.
51. Khan, M. A. (2020). Mix suitable for concrete 3D printing: A review. *Materials Today: Proceedings, 32*, 831–837.
52. McGee, W., Ng, T. Y., Yu, K., & Li, V. C. (2020). Extrusion nozzle shaping for improved 3DP of engineered cementitious composites (ECC/SHCC). In: *Second RILEM international conference on concrete and digital fabrication: Digital concrete 2020* (Vol. 2, pp. 916–925). Springer International Publishing.
53. Lim, S., Buswell, R. A., Le, T. T., Austin, S. A., Gibb, A. G., & Thorpe, T. (2012). Developments in construction-scale additive manufacturing processes. *Automation in Construction, 21*, 262–268.
54. Paul, S. C., Tay, Y. W. D., Panda, B., & Tan, M. J. (2018). Fresh and hardened properties of 3D printable cementitious materials for building and construction. *Archives of Civil and Mechanical Engineering, 18*, 311–319.
55. Zareiyan, B., & Khoshnevis, B. (2017). Effects of interlocking on interlayer adhesion and strength of structures in 3D printing of concrete. *Automation in Construction, 83*, 212–221.
56. Xu, J., Ding, L., Cai, L., Zhang, L., Luo, H., & Qin, W. (2019). Volume-forming 3D concrete printing using a variable-size square nozzle. *Automation in Construction, 104*, 95–106.
57. David, M., Freund, N., Dröder, K., & Lowke, D. (2023). The effects of nozzle diameter and length on the resulting strand properties for shotcrete 3D printing. *Materials and Structures, 56*(8), 157.
58. Pan, Z., Si, D., Tao, J., & Xiao, J. (2023). Compressive behavior of 3D printed concrete with different printing paths and concrete ages. *Case Studies in Construction Materials, 18*, Article e01949.
59. Pham, L., Lu, G., & Tran, P. (2022). Influences of printing pattern on mechanical performance of three-dimensional-printed fiber-reinforced concrete. *3D Printing and Additive Manufacturing, 9*(1), 46–63.
60. Yang, W., Wang, L., Ma, G., & Feng, P. (2023b). An integrated method of topological optimization and path design for 3D concrete printing. *Engineering Structures, 291*, Article 116435.
61. Zahabizadeh, B., Pereira, J., Gonçalves, C., Pereira, E. N., & Cunha, V. M. (2021). Influence of the printing direction and age on the mechanical properties of 3D printed concrete. *Materials and Structures, 54*, 1–22.
62. Li, Q., Gao, X., Su, A., & Lu, X. (2023a). Interlayer adhesion strength of 3D-printed cement-based materials exposed to varying curing conditions. *Journal of Building Engineering, 74*, Article 106825.

63. Li, S., Nguyen-Xuan, H., & Tran, P. (2023b). Digital design and parametric study of 3D concrete printing on non-planar surfaces. *Automation in Construction, 145*, Article 104624.
64. Chen, Y., Zhang, Y., Pang, B., Liu, Z., & Liu, G. (2021). Extrusion-based 3D printing concrete with coarse aggregate: Printability and direction-dependent mechanical performance. *Construction and Building Materials, 296*, Article 123624.
65. Alchaar, A. S., & Al-Tamimi, A. K. (2021). Mechanical properties of 3D printed concrete in hot temperatures. *Construction and Building Materials, 266*, Article 120991.
66. Shahmirzadi, M. R., Gholampour, A., Kashani, A., & Ngo, T. D. (2021). Shrinkage behavior of cementitious 3D printing materials: Effect of temperature and relative humidity. *Cement and Concrete Composites, 124*, Article 104238.
67. Sun, B., Dominicus, R., Dong, E., Li, P., Ye, Z., & Wang, W. (2024). Predicting the strength development of 3D printed concrete considering the synergistic effect of curing temperature and humidity: From perspective of modified maturity model. *Construction and Building Materials, 427*, Article 136291.
68. Ma, L., Zhang, Q., Jia, Z., Liu, C., Deng, Z., & Zhang, Y. (2022). Effect of drying environment on mechanical properties, internal RH and pore structure of 3D printed concrete. *Construction and Building Materials, 315*, Article 125731.
69. Wei, Y., & Zhang, H. (2024). Influence of temperature and humidity on mechanical properties of calcined oyster-shell powder-modified 3D-printed concrete. *ACS Omega, 9*(37), 39180–39187.
70. Federowicz, K., Kaszyńska, M., Zieliński, A., & Hoffmann, M. (2020). Effect of curing methods on shrinkage development in 3D-printed concrete. *Materials, 13*(11), 2590.
71. Yao, X., Lyu, X., Sun, J., Wang, B., Wang, Y., Yang, M., Wei, Y., Elchalakani, M., Li, D., & Wang, X. (2023). AI-based performance prediction for 3D-printed concrete considering anisotropy and steam curing condition. *Construction and Building Materials, 375*, Article 130898.
72. Bong, S. H., Xia, M., Nematollahi, B., & Shi, C. (2021). Ambient temperature cured 'just-add-water' geopolymer for 3D concrete printing applications. *Cement and Concrete Composites, 121*, Article 104060.
73. Nematollahi, B., Bong, S. H., Xia, M., & Sanjayan, J. (2020). Digital fabrication of 'just-add-water' geopolymers: Effects of curing condition and print-time interval. In: *Second RILEM international conference on concrete and digital fabrication: Digital concrete 2020* (Vol. 2, pp. 93–102). Springer International Publishing.
74. Zhang, Y., Tao, Y., Godinho, J. R., Ren, Q., Jiang, Z., Van Tittelboom, K., & De Schutter, G. (2025). Layer interface characteristics and adhesion of 3D printed cement-based materials exposed to post-printing temperature disturbance. *Cement and Concrete Composites, 155*, Article 105837.
75. Ler, K. H., Ma, C. K., Chin, C. L., Ibrahim, I. S., Padil, K. H., Ab Ghafar, M. A. I., & Lenya, A. A. (2024). Porosity and durability tests on 3D printing concrete: A review. *Construction and Building Materials, 446*, Article 137973.
76. Moelich, G. M., Kruger, J., & Combrinck, R. (2021). Modelling the interlayer bond strength of 3D printed concrete with surface moisture. *Cement and Concrete Research, 150*, Article 106559.
77. Ramakrishnan, S., Kanagasuntharam, S., & Sanjayan, J. (2022). In-line activation of cementitious materials for 3D concrete printing. *Cement and Concrete Composites, 131*, Article 104598.
78. Moelich, G. M., Kruger, P. J., & Combrinck, R. (2021). The effect of restrained early age shrinkage on the interlayer bond and durability of 3D printed concrete. *Journal of Building Engineering, 43*, Article 102857.
79. Zhu, L., Zhang, M., Zhang, Y., Yao, J., Yang, G., Guan, X., & Zhao, Y. (2023). Research progress on shrinkage properties of extruded 3D printed cement-based materials. *Journal of Building Engineering*, 107394.
80. Li, L. G., & Zhang, G. H. (2024). Feasibility of underwater 3D printing: Effects of anti-washout admixtures on printability and strength of mortar. *Journal of Building Engineering, 96*, Article 110434.

81. Li, J. N., Singh, A., Zhao, Y., Sun, J., Tam, V. W., & Xiao, J. (2024). Enhancing thermo-mechanical and moisture properties of 3D-Printed concrete through recycled ultra-fine waste glass powder. *Journal of Cleaner Production, 480*, Article 144121.
82. Moelich, G. M., Kruger, J., & Combrinck, R. (2020). Plastic shrinkage cracking in 3D printed concrete. *Composites Part B: Engineering, 200*, Article 108313.
83. Van Tittelboom, K., Mohan, M. K., Šavija, B., Keita, E., Ma, G., Du, H., Kruger, J., Caneda-Martinez, L., Wang, L., Bekaert, M., & Wangler, T. (2024). On the micro-and meso-structure and durability of 3D printed concrete elements. *Cement and Concrete Research, 185*, Article 107649.
84. Nguyen-Van, V., Nguyen-Xuan, H., Panda, B., & Tran, P. (2022). 3D concrete printing modelling of thin-walled structures. *Structures, 39*, 496–511.
85. Xu, J., Buswell, R. A., Kinnell, P., Biro, I., Hodgson, J., Konstantinidis, N., & Ding, L. (2020). Inspecting manufacturing precision of 3D printed concrete parts based on geometric dimensioning and tolerancing. *Automation in Construction, 117*, Article 103233.
86. Comminal, R., da Silva, W. R. L., Andersen, T. J., Stang, H., & Spangenberg, J. (2020). Influence of processing parameters on the layer geometry in 3D concrete printing: experiments and modelling. In: *RILEM international conference on concrete and digital fabrication* (pp. 852–862). Springer International Publishing.
87. Marais, H., Christen, H., Cho, S., De Villiers, W., & Van Zijl, G. (2021). Computational assessment of thermal performance of 3D printed concrete wall structures with cavities. *Journal of Building Engineering, 41*, Article 102431.
88. Liu, H., Wang, Y., Zhu, C., Wu, Y., Liu, C., He, C., Yao, Y., Wang, Y., & Bai, G. (2025). Design of 3D printed concrete masonry for wall structures: Mechanical behavior and strength calculation methods under various loads. *Engineering Structures, 325*, Article 119374.
89. Cao, J., Shengzhao, E., Yang, Y., Shi, Y., Chai, J., & Xu, Z. (2024). A strategy for the improvement of the bonding performance of 3D-printed concrete interlayer interfaces. *Journal of Building Engineering, 97*, Article 110675.
90. Wang, L., Tian, Z., Ma, G., & Zhang, M. (2020). Interlayer bonding improvement of 3D printed concrete with polymer modified mortar: Experiments and molecular dynamics studies. *Cement and Concrete Composites, 110*, Article 103571.
91. Panda, B., Noor Mohamed, N. A., Paul, S. C., Bhagath Singh, G. V. P., Tan, M. J., & Šavija, B. (2019). The effect of material fresh properties and process parameters on buildability and interlayer adhesion of 3D printed concrete. *Materials, 12*(13), 2149.
92. Babafemi, A. J., Kolawole, J. T., Miah, M. J., Paul, S. C., & Panda, B. (2021). A concise review on interlayer bond strength in 3D concrete printing. *Sustainability, 13*(13), 7137.
93. Hosseini, E., Zakertabrizi, M., Korayem, A. H., & Xu, G. (2019). A novel method to enhance the interlayer bonding of 3D printing concrete: An experimental and computational investigation. *Cement and Concrete Composites, 99*, 112–119.
94. Şahin, H. G., & Mardani, A. (2024). How does rheological behaviour affect the interlayer-bonding strength of 3DPC mixtures? *Journal of Adhesion Science and Technology, 38*(9), 1353–1377.
95. Liu, X., & Sun, B. (2021). The influence of interface on the structural stability in 3D concrete printing processes. *Additive Manufacturing, 48*, Article 102456.
96. Vallurupalli, K., Farzadnia, N., & Khayat, K. H. (2021). Effect of flow behavior and process-induced variations on shape stability of 3D printed elements–a review. *Cement and Concrete Composites, 118*, Article 103952.
97. Liu, C., Wang, X., Chen, Y., Zhang, C., Ma, L., Deng, Z., Chen, C., Zhang, Y., Pan, J., & Banthia, N. (2021). Influence of hydroxypropyl methylcellulose and silica fume on stability, rheological properties, and printability of 3D printing foam concrete. *Cement and Concrete Composites, 122*, Article 104158.

98. Gomaa, S., Irizarry, E. M., Ahmed, A., Rosa, R. M., Ahmed, H., Burroughs, J., Kreiger, E., Liu, J., Troemner, M., & Cusatis, G. (2024). 3D printing of ultra-high-performance concrete: Shape stability for various printing systems. *Construction and Building Materials, 456*, Article 139039.
99. Cho, S., Van Rooyen, A., Kearsley, E., & Van Zijl, G. (2022). Foam stability of 3D printable foamed concrete. *Journal of Building Engineering, 47*, Article 103884.

Numerical Modeling of Geopolymer Concrete Short Columns Reinforced with GFRP Bars Under Different Loading Conditions

T. N. Guruprasad, M. H. Prashanth, L. Manjunath, and M. C. Narasimhan

Abstract Infrastructure developments are on the increase all around the world. Building these modern infrastructural facilities requires large amounts of steel and concrete. India is one of the world's largest producers of steel and cement. However, to produce one ton each of steel and cement, 1.85 tons and one ton, respectively, of CO2 is released into the environment, contributing immensely to global warming. Again, embedded steel bar corrosion is the main threat to the durability of the reinforced concrete elements. So, Glass fiber-reinforced polymer bars and geopolymer concrete can be employed in RC elements, like columns, for enhanced sustainability. A sustainable and green alternative to OPC concrete is geopolymer concrete (GPC), which uses waste residues such as GGBS, flyash, and aggregates to prevent extraction of raw materials and reduce demand for disposal of waste. The current investigation aims to find the effectiveness of using short geopolymer concrete columns reinforced with GFRP bars. The present study performs a 3D finite element analysis over geopolymer concrete-based short columns with GFRP bars. A representative with a total 1200 mm height and 160 mm × 260 mm cross-section is considered for the GFRP bars reinforced columns under different conditions of loading. Results of FEA analysis carried out using the commercial ANSYS software show that reinforcing the RC short columns with GFRP bars is effective in that they exhibit 10% more buckling strength and 40% more in compression under concentric loading as compared

T. N. Guruprasad (✉) · M. H. Prashanth · M. C. Narasimhan
National Institute of Technology Karnataka, Surathkal, Mangalore, India
e-mail: guruprasadtn.217cv503@nitk.edu.in

M. H. Prashanth
e-mail: mh_prashanth@nitk.edu.in

M. C. Narasimhan
e-mail: mattur@nitk.edu.in

T. N. Guruprasad
Department of Civil Engineering, Sri Siddhartha Institute of Technology, SSAHE, Tumkur, Karnataka, India

L. Manjunath
Nitte Meenakshi Institute of Technology, Bangalore, India
e-mail: manjunath.l@nmit.ac.in

P. Danish et al. (eds.), *Innovative Technologies for Smart and Sustainable Infrastructure*,
https://doi.org/10.1007/978-3-031-92801-7_2

to eccentric loading and also the deformation in the eccentrically loaded column is 50% more when compared to concentrically loaded GFRP RC short column.

Keywords Short columns · Finite element modelling · Geopolymer concrete · GFRP bars

1 Introduction

The world is growing rapidly due to the increase in urbanization, which requires a lot of infrastructure development, buildings, and industries. For this, a large amount of steel and cement is required. To produce one ton of steel and cement, 1.85 tons and one ton of CO2 are emitted into the atmosphere, respectively [8]. Around 7% of the world's yearly carbon emissions are attributed to the cement industry [10]

The columns of a structure are more important than the other structural members since their failure may result in the collapse of adjacent structural elements. The steel bar corrosion is the major threat to the Reinforced Cement Concrete columns durability. Therefore, improving the durability of the structures has become a critical issue in civil engineering [13, 16]. New building materials have emerged over the last 30 years, including engineered cementitious composites, high-performance concrete, self-compacting concretes, and fiber-reinforced polymers [1]. The diffusion of chloride ions in concrete can cause corrosion, which occurs when chloride-containing additives are used in the concrete mix, when chloride-contaminated aggregate is used, or when chloride from the outside, especially in ocean engineering structures. A thin corrosion-preventing oxide film on the steel surface could be destroyed by the chloride ions [6]. The tensile stress is generated in concrete when the volume of rust generated is 2–4 times greater than the volume of concrete. Because of this, the concrete cover could crack and spall, and the concrete will fail to adhere to the steel bars, which will result in a drastically reduced load-bearing capacity [17].

GFRP is a composite material of polyester resins, longitudinal glass fibers, and matrix material. Advantages of GFRP bars in RCC elements include their high tensile strength, higher corrosion resistance, lightweight, non-conductive, electromagnetic transparency, and higher fatigue resistance. They are maintenance-free, easy to transport and install, and have a longer life span [13]. GFRP bars have applications in all types of construction, including those in Aerospace, Defense, Automotive Industry, Power Generation, Marine engineering, Mining and tunneling, Highways, Transport Infrastructure, High-rise building construction, and earth and earth Retaining structures [3]. Compared to steel reinforcement, GFRP reinforcement bars have a lesser elastic modulus, and it requires greater lateral restraint to prevent buckling [5]. FRPs behave linearly almost up to failure. They generally undergo brittle failure in that they fail suddenly without sufficient warning. In addition to their brittle nature, FRPs have poor transverse or shear strength, making the concept of stress redistribution and plasticity often overlooked [15].

As a building material, Portland cement is the most important material that is used almost in every construction [8] [9]. Mineral admixtures like silica fume, flyash, ground granulated blastfurnace slag (GGBF), metakaolin, and rice husk ash, have been used for several years now. Indian thermal power plants produce a large amount of fly ash, billion tons annually, which is disposed of in landfills, ponds, or rejected into rivers. This is posing serious environmental problems. In addition to bottom ash and blast furnace slag, other wastes are also being used [2]. Initially, cement was partially replaced with mineral admixtures, with useful results being obtained, but later, in 1978, Professor Davidovits introduced a new form of concrete called geopolymer concrete produced by the blending of silica and alumina materials with a solution of alkaline silicate [7, 11, 13, 17]. The geopolymer concrete gains strength quickly when compared to OPC concrete without curing. Some of the advantages of geopolymer concrete are cutting the world's carbon, low permeability, fireproof, eco-friendly, lower costs, better compressive strength, good thermal stability, low creep and shrinkage, best bond with reinforcing steel, excellent resistance of acid and sulfates [1]. Despite these advantages of geopolymer concrete, there are some challenges in applying them on a large scale on the site. They include special handling requirements of the activating solutions, skilled labor, high chemical costs, and a lack of uniformity. [3, 7, 11, 12, 17]. Because of the formation of zeolitic, which is an amorphous material.

during geo-polymerization, geopolymers remain stable in different types of aggressive fluids for up to 270 days. Construction projects in coastal zones benefit from the enhanced protection of embedding reinforcing bars against aggressive fluids [4]. A geopolymer material is very effective at adsorbing heavy metals, dyes, and radioactive pollutants, which is very beneficial for society [17].

FEM is a numerical technique for splitting up an object's structure into parts or elements and connecting them at nodes. As a result, engineers, developers, and other designers can conduct finite element analysis using a set of algebraic equations generated by FEM [14]

In the present investigation, FE analysis is carried out to analyze RC short columns with eccentric loading conditions using ANSYS software.

2 Materials

A column of size 160 mm × 260 mm with a height of 1200 mm is considered for simulation. This column is made of GFRP bars reinforced geopolymer concrete with bars along a longitude of 14 mm in diameter and 8 mm in diameter stirrups. The physical parameters of the reinforced bars assumed in FEM simulations are given in Table 1.

The geopolymer concrete binder is a combination of GGBS and flyash. Then the mix is assumed to consist of 6.5% alkaline solution, 15% binder, 24.9% fine aggregates, 47.3% coarse aggregates, 0.1% superplasticizer, and 6.1% water. The alkaline solution was 12 M with a sodium silicate to sodium hydroxide ratio of 3.2.

Table 1 GFRP bars physical properties [3]

Type of bar	Outside bar diameter (mm)	Inside diameter of the bar (mm)	Area of bar (mm^2)	Modulus of elasticity (GPa)	Tension strength (MPa)	Strain under tension
Longitudinal bar	14.0	13.4	136.80	59.0	931	0.016
Stirrup	8.4	7.2	40.30	55.0	651	0.011

3 Finite Element Modeling

The geometry of the three-dimensional modeling of the GFRP-reinforced short column is drawn using AutoCAD. The geometrical properties of the reinforced column conform to IS 456:2000. The column has a total length of 1200 mm and cross section of 160 mm × 260 mm. For columns with eccentric loading, the corbel is provided at the bottom and top with a length of 200 mm and the column length in corbels is 800 mm. SOLID65 is a 3-D modeling program that is used to model solids with or without reinforcing bars. With eight nodes, this element possesses 3 DOF per translation node in the three nodal directions: x, y, and z. This element can be deformed plastically, cracked orthogonally, and crushed. The finite element analysis is performed through ANSYS software by importing geometry files. The models which are considered for the present investigation are shown in Fig. 1.

The convergence mesh study was conducted to choose the finest mesh size of 10 mm [4], which was adopted with mechanical-type meshing for the GFRP-reinforced short column. After meshing the model, the details of the generated nodes and elements are given in Table 2. Meshed models considered for the present study are shown in Fig. 2. GFRP-reinforced short-column models were subjected to eccentric loading and concentric loading. Pinned–pinned boundary conditions were used to find the deformation and the reinforced concrete short column buckling behavior. The representation of considered boundary conditions for the analysis is shown in Fig. 3.

4 Results and Discussion

4.1 Load versus Displacement Curve

To find the compressive strength of glass fiber polymer-reinforced short columns subjected to eccentric and concentric loading, Static Non-linear analysis was performed. Two models were considered for the FEM analysis.

The load–displacement plot obtained for the 2 models from finite element analysis (one for concentric loading and another for eccentric loading) is shown in Figs. 4 and

Fig. 1 Geometry of models considered for **a** Concentric loading and **b** Eccentric loading

Table 2 Model nodes and elements

Type of loading	Nodes	Elements
Concentric	42,752	24,029
Eccentric	187,506	42,614

5. It has been observed that the ultimate load is 1357 kN and the maximum displacement is 7.37 mm for GFRP-reinforced short column under concentric loading and under eccentric loading, the ultimate load is 800 kN and the maximum displacement is 14.57 mm. The deformed shapes of the models are displayed in Figs. 6 and 7.

The variation of stress along the length of the GFRP-reinforced short column for concentrically loaded and eccentrically loaded models is shown in Figs. 8 and 9, respectively.

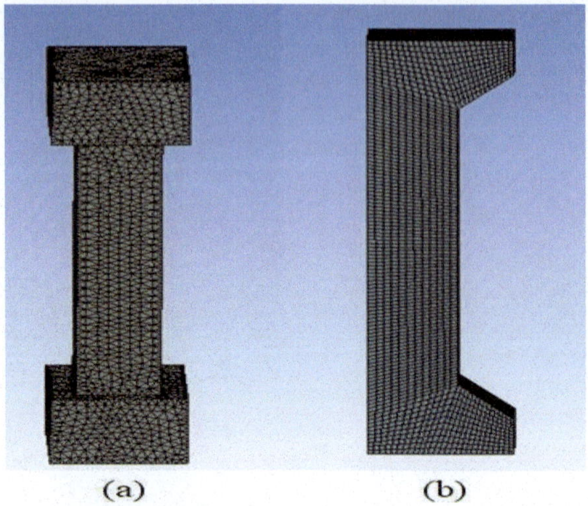

Fig. 2 Meshing of models considered for **a** Concentric loading and **b** Eccentric loading

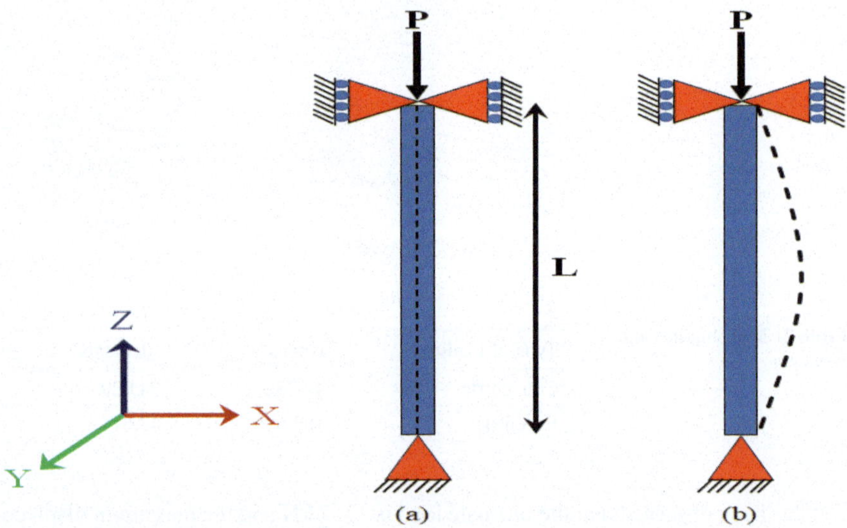

Fig. 3 Schematic representation of boundary conditions considered for GFRP-reinforced short column **a** Before loading **b** After loading

4.2 Buckling Analysis of Eigenvalue

The analysis was performed to predict the effect of buckling over the RCshort column due to compressive load for pinned–pinned boundary conditions. The critical buckling load obtained for model 1 and model 2 are given in Table 3. The buckling shape of the RC short columns is demonstrated in Figs. 10. and 11.

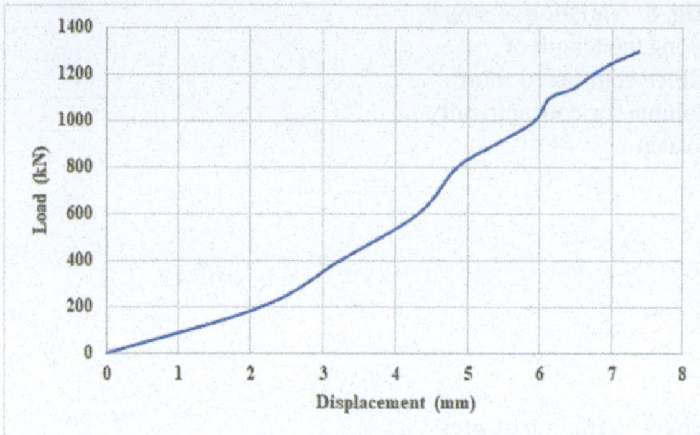

Fig. 4 Load versus displacement graph of concentrically loaded model

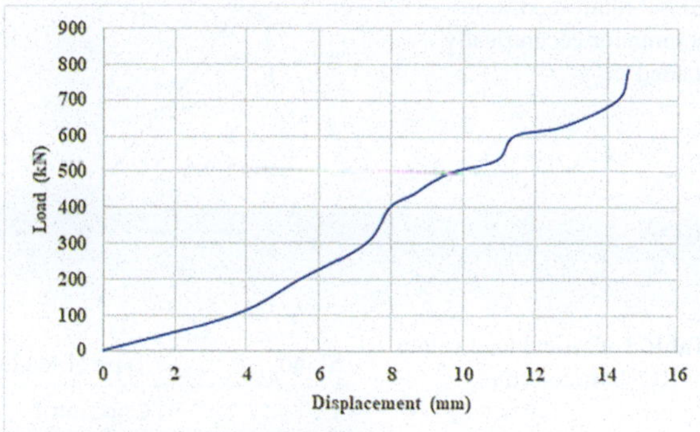

Fig. 5 Load versus displacement graph of eccentrically loaded model

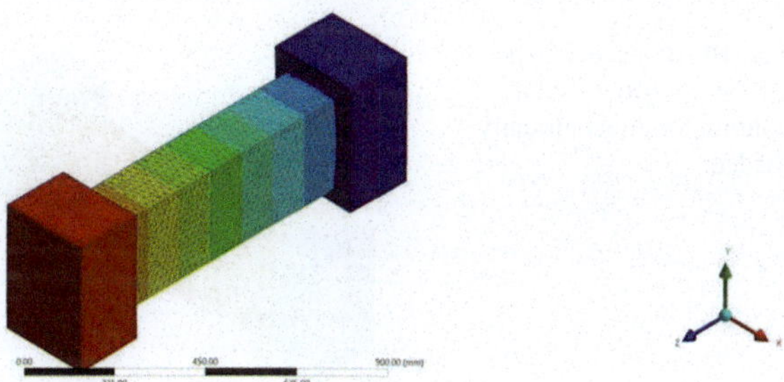

Fig. 6 Deformed shape of the concentrically loaded model

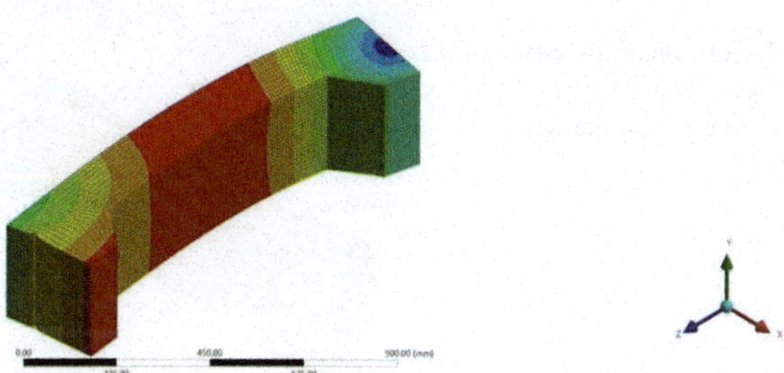

Fig. 7 Deformed shape of the eccentrically loaded model

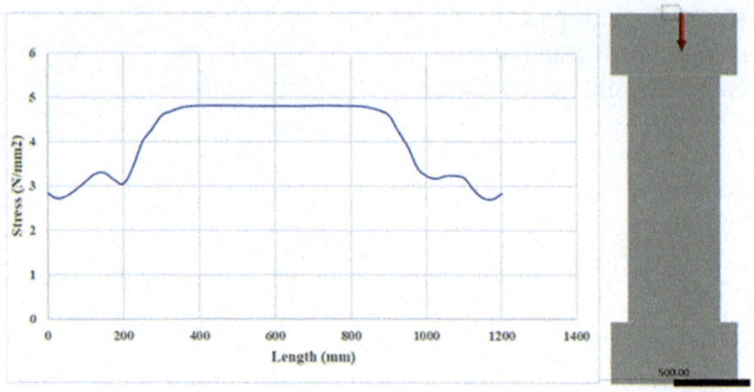

Fig. 8 Variation of stress along the length of GFRP-reinforced short column for concentrically loaded

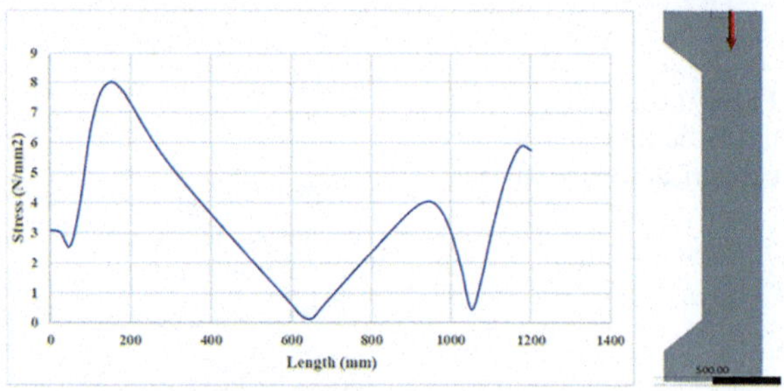

Fig. 9 Variation of stress along the length of GFRP-reinforced short column for eccentrically loaded

Table 3 Critical load values of RC short column

Sl. no	Type of loading	Pcr (kN)
1	Concentric	1423
2	Eccentric	1285

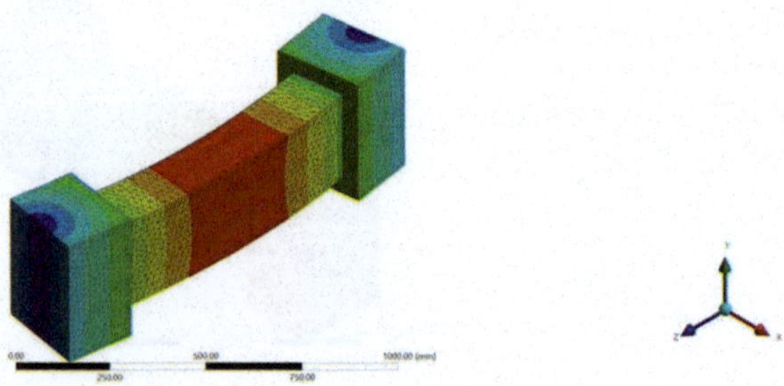

Fig. 10 Buckling shape of GFRP-reinforced short column for concentrically loaded

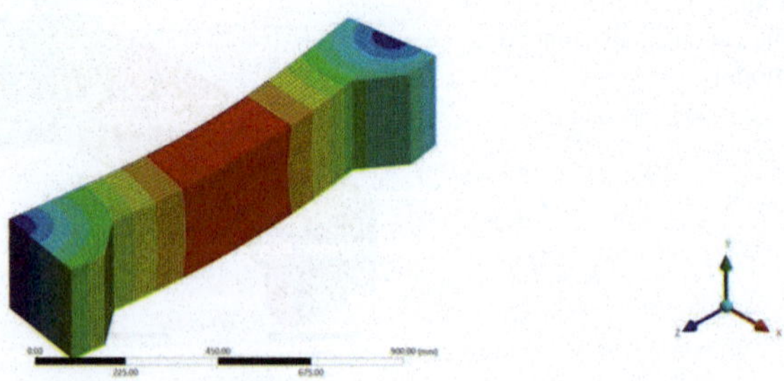

Fig. 11 Buckling shape of GFRP-reinforced short column for eccentrically loaded

5 Conclusions

The present work investigates the short column of geopolymer concrete reinforced with GFRP bars under various loading conditions to analyze the deformation and critical buckling load by performing finite element analysis using ANSYS software. From the results obtained, the following conclusions are made.

- GFRP-reinforced short column subjected to concentric loading has 40% more ultimate load-bearing capacity when compared to the RC short column subjected to eccentric loading.
- Deformation of the GFRP-reinforced short column for concentrically loaded is less when compared to the RC short column for eccentrically loaded.
- The GFRP-reinforced short column under concentric loading has a 10% more critical buckling load-carrying capacity than the RC shot column under eccentric loading.

References

1. Ahmad, H., Sheikh, M. N., & Hadi, M. N. S. (2022). Experimental investigation on the behavior of hollow-core glass fiber-reinforced concrete columns with GFRP bars. *Journal of Composites for Construction, 26*. https://doi.org/10.1061/(ASCE)cc.1943-5614.0001183
2. Danda, U. K., Himath Kumar, Y., & Sarath Chandra Kumar, B. (2020). Experimental study on reinforced geopolymer concrete columns using GGBS. *Materials Today Proceedings, 33,* 632–636. https://doi.org/10.1016/j.matpr.2020.05.607
3. Dong, M., Elchalakani, M., Karrech, A., et al. (2019). Glass fiber-reinforced polymer circular alkali- activated fly ash/slag concrete members under combined loading. *Engineering Structures, 199,* 109598. https://doi.org/10.1016/j.engstruct.2019.109598
4. Elchalakani, M., Karrech, A., Dong, M., et al. (2018). Experiments and finite element analysis of GFRP reinforced geopolymer concrete rectangular columns subjected to concentric and eccentric axial loading. *Structures, 14,* 273–289. https://doi.org/10.1016/j.istruc.2018.04.001
5. Elchalakani, M., & Ma, G. (2017). Tests of glass fiber reinforced polymer rectangular concrete columns subjected to concentric and eccentric axial loading. *Engineering Structures, 151,* 93–104. https://doi.org/10.1016/j.engstruct.2017.08.023
6. Firdous, R., Stephan, D., & Djobo, J. N. Y. (2018). Natural pozzolan-based geopolymers: A review on mechanical, microstructural and durability characteristics. *Construction and Building Materials, 190,* 1251–1263. https://doi.org/10.1016/j.conbuildmat.2018.09.191
7. Hadi, M. N. S., Ali, S., & Neaz Sheikh, M. (2021). Experimental study of GFRP-reinforced geopolymer concrete columns under different loading conditions. *Journal of Composites for Construction, 25.* https://doi.org/10.1061/(ASCE)cc.1943-5614.0001164
8. Huzaifa, M., Wankhede, N., Ahire, D., et al. A review on development of flyash based high strength geopolymer concrete. ISSN 2250-3021.
9. Islam, A., Alengaram, U. J., Jumaat, M. Z., et al. (2015). Engineering properties, and carbon footprint of ground granulated blast-furnace slag-palm oil fuel ash-based structural geopolymer concrete. *Construction and Building Materials, 101,* 503–521. https://doi.org/10.1016/j.conbuildmat.2015.10.026
10. Maranan, G. B., Manalo, A. C., Benmokrane, B., et al. (2016). Behavior of concentrically loaded geopolymer-concrete circular columns reinforced longitudinally and transversely with

GFRP bars. *Engineering Structures, 117,* 422–436. https://doi.org/10.1016/j.engstruct.2016.03.036
11. Nath, S. K., & Kumar, S. (2019). Role of alkali concentration on reaction kinetics of fly ash geopolymerization. *Journal of Non-Crystalline Solids, 505,* 241–251. https://doi.org/10.1016/j.jnoncrysol.2018.11.007
12. Nath, S. K., & Kumar, S. (2020). Role of particle fineness on engineering properties and microstructure of fly ash derived geopolymer. *Construction and Building Materials, 233,* 117294. https://doi.org/10.1016/j.conbuildmat.2019.117294
13. Nhabih, H. T., Hussein, A. M., & Salman, M. M. (2020). Study a structural behavior of eccentrically loaded GFRP reinforced columns made of geopolymer concrete. *Civil Engineering Journal (Iran), 6,* 563–575. https://doi.org/10.28991/cej-2020-03091492
14. Nuruddin, M. F., Malkawi, A. B., Fauzi, A., et al. (2016) Geopolymer concrete for structural use: Recent findings and limitations. *IOP Conference Series: Earth and Environmental Science, 133.* https://doi.org/10.1088/1757-899X/133/1/012021
15. Raza, A., Alomayri, T., & Berradia, M. (2022). Rapid repair of partially damaged GFRP-reinforced recycled aggregate concrete columns using FRP composites. *Mechanics of Advanced Materials and Structures, 29,* 6070–6086. https://doi.org/10.1080/15376494.2021.1972368
16. Sreenath, S., Balaji, S., & Raja Mohan, K. S. (2017). Behavior of axially and eccentrically loaded short columns reinforced with GFRP bars. *IOP Conference Series: Earth and Environmental Science.* Institute of Physics Publishing.
17. Zhang, Z., Provis, J. L., Reid, A., & Wang, H. (2014). Geopolymer foam concrete: An emerging material for sustainable construction. *Construction Build Materials, 56,* 113–127. https://doi.org/10.1016/j.conbuildmat.2014.01.081

Helical Stiffener—An Innovative Way of Strengthening Open-Top Vertical Cylindrical Steel Tanks Against Wind-Induced Buckling

Soumya Mukherjee and Dilip Kumar Singha Roy

Abstract Wind-induced buckling occurs mostly in case of empty or partially filled thin-walled storages due to moderate to severe wind events. Vertical buckling-ripples of the short shells are happened by high membrane stress, whereas tall shells face horizontal buckling-ripples due to high axial stress under bending by transverse wind load. In case of intermediate shells shear stress is dominant under the influence of aerodynamic load, which reflects on the combined pattern of horizontal and inclined buckling ripples. Ring and vertical stiffeners resist the wind-induced buckling. This study explores the applicability of helical-stiffener for the strengthening of empty, open-top, short, cylindrical steel-tanks against wind-induced buckling. The stiffener is designed according to the American Petroleum Institute. The performance of the helical stiffener has been measured by wind-induced buckling load-multiplier (λ) for the variation of basic wind-speed (V_b), height to diameter (H/D) ratio of shell, radius to thickness (r/t) ratio of shell-wall, gauge length (L_g), and number (N_{HS}) of the helical-stiffener. It is evident that decreasing gauge length of the helical-stiffener can provide greater stiffness when Vb, H/D and r/t ratio of shell-wall increase. It is very much economic to apply low gauge length to the helical-stiffener rather than increasing wall-thickness when V_b and H/D ratio increase, as it ensures economy through less material consumption.

Keywords Cylindrical steel tank · Wind induced buckling · Wind pressure · Helical stiffener · Buckling load multiplier

S. Mukherjee · D. K. S. Roy (✉)
Department of Civil Engineering, National Institute of Technology Durgapur, Durgapur, West Bengal, India
e-mail: dsingharoy.ce@nitdgp.ac.in

S. Mukherjee
e-mail: mukherjee.s@iitpkd.ac.in; sm.19ce1104@phd.nitdgp.ac.in

S. Mukherjee
Department of Civil Engineering, Indian Institue of Technology Palakkad, Kanjikode West, Palakkad, Kerala, India

P. Danish et al. (eds.), *Innovative Technologies for Smart and Sustainable Infrastructure*,
https://doi.org/10.1007/978-3-031-92801-7_3

1 Introduction

Among various external factors aerodynamic oscillation, stagnation pressure of wind and transverse wind-load are accordingly responsible for ovalling, buckling, bending, and overturning of empty thin-walled steel shells required for storage [1–5]. To maintain the stability of shell by avoiding the overturning the base of the shell is anchored with anchor bolts, which can be further strengthened with anchor straps [6, 7]. Vertical stiffener is applied to counter the bending under transverse wind load and axial stress induced buckling of shell [6]. The performance of the vertical stiffener can be further enhanced by providing corrugated shell-course as the shell wall [8]. Whereas, ring stiffener is used to resist the ovalling under the hoop stress induced circumferential compression and membrane stress induced buckling of storage shells [6, 9]. The unstiffened shell height and equivalent wall thickness based on the variation of different shell course thicknesses are the main criteria of understanding the requirement of ring-stiffeners [10, 11]. Simplified modeling of the equivalent wall thickness through materialistic nonlinear analysis to strengthen the tank against uniform, thermal, and wind load induced buckling should be 24, 27, and 40 times of bottom shell course, respectively but this simplification can amplify the risk of destabilized buckling of the tank by 14%, 7% and 12% accordingly [12]. Top stiffeners are used to maintain the circular shape of a cylindrical shell at the upper portion and intermediate stiffeners secure the shape-integrity of a cylindrical shell's intermediate portion [9]. Greater size of the top ring stiffener needs to be provided than that of an intermediate stiffener [13, 14]. Further exploration showed that the reduced size of top stiffener similar to the size of intermediate stiffener is suffice to stiffen shell [15, 16]. Empirical design formulations are developed for assessing the unstiffened shell height, shear force along radius, moment of bending, second moment of area, stresses at ring stiffener and tributary height or the spacing between consecutive intermediate stiffeners [18–21]. However, it is documented that simplified modeling of ring-stiffener affects it performance in stabilizing a tank-shell against wind-induced buckling [22].

Shao-Wen [23] introduced application of spiral stiffener against uniform axial, lateral, and torsional load induced buckling of composite cylindrical shell. Shao-Wen [23] also optimized the structural weight of the stiffener based on the spiral angle. Progressive studies on the effect of elastic foundation on the buckling of spirally stiffened cylindrical shells show that Winkler foundation produces better response with 30° spiral angle and higher buckling strength against axial loading is provided by ceramic shell in comparison to metal shell [24]. In case of thermal buckling external spiral stiffener performs better by increasing the critical buckling temperature than that for internal spiral stiffener, but the spiral angle needs to be < 60° [25]. Linear distribution of thermal load exhibits higher resistance against large deflection of functionally graded shell, compared to uniformly distributed thermal load, which gets more increased due to the application of damping at the elastic foundation [26, 27]. Whereas, under torsional load in a thermal environment, buckling capacity of spirally stiffened shell is higher than that of orthogonally stiffened shell [28]. Further analysis shows that Donnel's shell theory combined with Von Karman's analogy of geometrical nonlinearity better describes torsional buckling problem of spirally stiffened shell under thermal condition, which can be solved by Galerkin method, when the spiral stiffener's specification is considered based on Lekhnitskii's technique of smeared stiffeners [29, 30]. These literatures highlight the extensive exploration of ring-stiffeners and a serious lacuna in investigating the performance of helical stiffener for avoiding and minimizing wind-induced ovalling and buckling of thin-walled cylindrical storage shells, respectively.

By carefully exploring these literatures, two key aspects are evident. First, extensive usage of ring-stiffener is there for protecting vertical cylindrical storage shell against wind-induced buckling. Secondly, mechanism of spiral-stiffener is a developing field of research, which is limited to axial, thermal and torsional buckling resistance of horizontal functionally graded shell on elastic foundation. These two aspects have drawn the authors' attention for studying whether spiral or helical stiffener as intermediate stiffener, is applicable against wind-induced buckling of steel-made, vertical, thin-walled tank-shell. The focus is on buckling load multiplier as the stability parameter and buckling-induced radial deformation for assessing damaged wall-portion. This study does not introduce design specification for the helical stiffener and uses the API 650 [31] guidelines for calculating the second moment of area of the helical stiffener. The authors study the wind-induced buckling response of helically stiffened, open-top tank for the variation of tank's height (H): diameter (D), radius (r): wall-thickness (t), helical stiffener's gauge length (L_g): tank-height (H) ratios and basic wind speed (V_b). This study is carried out through Multiphysics coupling of computational fluid dynamics (CFD) and structural mechanics (SM). To simulate the aerodynamic environment, CFD modeling is deployed for bluff-body analysis through CFX solver [32]. To simulate the wind-induced buckling, wind pressure data is imported from CFX solver to mechanical solver of eigen value buckling analysis [33].

2 Model Details

2.1 Bluff Body

The bluff-body modeling of open-top, ground-supported tank, for the CFD analysis is carried out based on the propositions of Architectural Institute of Japan (AIJ) [34] about the simulation of aerodynamic environment around structures. The tank-model H/D ratio is varied 4 times (0.25, 0.5, 0.75, and 1.0) by keeping tank's diameter 20 m and changing height from 5 to 20 m, with an interval of 5 m. The domain for virtual wind flow test has inlet, side faces, and ceiling at a distance of 5 times the major dimension (H or D) of the tank-model, so that the blockage ratio < 5% can be secured. Whereas, the outlet is 15 times the highest model dimension apart from the nearest model face to accommodate the trailing vortices at the model's wake and stop flow recirculation. All the faces of domain are free slip, except base, which is smooth,

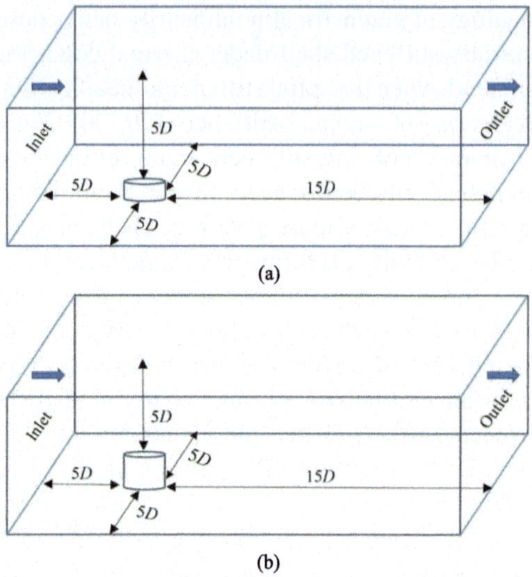

Fig. 1 Diagram of domain for tank H/D ratio **a** < 1.0, and **b** 1.0

to replicate the terrain condition similar to open terrain. The faces of the model are smooth. The diagram of the domain encasing the tank model is shown in Fig. 1.

2.2 *Stuctural Model*

The mechanical model of the tank for eigenvalue buckling analysis has fixed base. The top of the model is free and open. The tank's r/t ratio is varied from 1250 to 2000 with an interval of 250. Top ring-stiffener is attached to the model. The intermediate helical stiffener system is inspired from the structures of single stranded ribonucleic acid (RNA) and double stranded deoxyribonucleic acid (DNA). The helical stiffener is connected to the top ring-stiffener at the top end and a bottom ring-stiffener at the lowest end. The gauge length (L_g) of the helical stiffener is varied 4 times, which are $H/5$, $4H/15$, $2H/5$, and $4H/5$. The section moduli of top ring-stiffener (Z_{TR}) and intermediate helical-stiffener (Z_{IR}) are calculated following the empirical Eqs. (1) and (2) of American standard API 650 [31]. The bottom ring-stiffener is designed similar to the intermediate stiffener, as it is encompassing the tank circumference at the lower end and completing the lower circle of the helical stiffener. The design wind pressure (P_{wd}), basic wind velocity pressure (P_{wv}), and basic pressure (p) are expressed through Eqs. (3) to (5) from API 650 [31]. Where gust factor (G) is 0.85, design velocity (V) at 10 m of a gust with 3 s duration is 190 km/h, exposure coefficient (K_z) is 1.04 for velocity pressure, factor of topography (K_{zt}) is 1.0, factor of wind direction (K_d) is 0.95 for round tanks, structure's importance factor (I) is 1.0, the least yield strength (F_y) should be the smaller value between strength at highest temperature (in MPa) for working and 210 MPa. In case of single stranded helical stiffening formation, the stiffening spiral is attached to the external wall surface of the tank. When double stranded helical stiffening system is considered, one stiffening spiral is attached to the internal wall surface and another one is connected to the external surface of tank-wall. The diagram of the tank (H/D ratio 0.5) with dimensional details of the gauge length (L_g) of single and double helical stiffening system is showcased at Fig. 2.

$$Z_{TR} = \frac{6HD^2}{0.5F_y}\left(\frac{P_{wd}}{1.72}\right) \tag{1}$$

$$Z_{IR} = \frac{6H_{TI}D^2}{0.5F_y}\left(\frac{P_{wd}}{1.72}\right) \tag{2}$$

$$P_{wd} = P_{wv} + 0.24 \tag{3}$$

$$P_{wv} = p\left(\frac{V_b}{190}\right)^2 \tag{4}$$

$$p = 0.00256K_zK_{zt}K_dV^2IG \tag{5}$$

3 Numerical Model

3.1 CFD Analysis

Turbulence Model. Ansys CFX software is used for the turbulence modeling of aerodynamic environment [32]. Although some studies recommend Large Eddy Simulation (LES) turbulence model for simulation of wind environment around structures with complex geometry due to accurate formation of near wall vortices, this turbulence model needs high computational time and pricy advance system [35, 36]. The unsteady Reynolds Averaged Navier Stokes (URANS) model of turbulence is widely reported to exhibit considerably accurate and time-economic simulation performance of sub-grid to large-scale near surface vortices, with respect to LES [37–42]. Attracted by the advantages, the authors analyze three-dimensional (3D) URANS turbulence model by Reynolds decomposition of fluctuating and time-averaged quantities of continuity and momentum of wind flow through Eqs. (6) and (7) [43]. Due to the limitation of standard k-ε model in accurately replicating eddy-viscosity of incompressible Newtonian flow of wind, realizable k-ε model is applied for boundary layer separation with kinetic and dissipated energies of flow during turbulence, showcased in Eqs. (8) and (9) [44, 45]. Mean variation of momentum by the flow convection and unsteadiness is embodied by the isotropic stress due to mean pressure ($\overline{p}$), mean stress due to body force (f_i), Reynolds stress due to shifting speeds $u_i\prime$ and $u_j\prime$ along

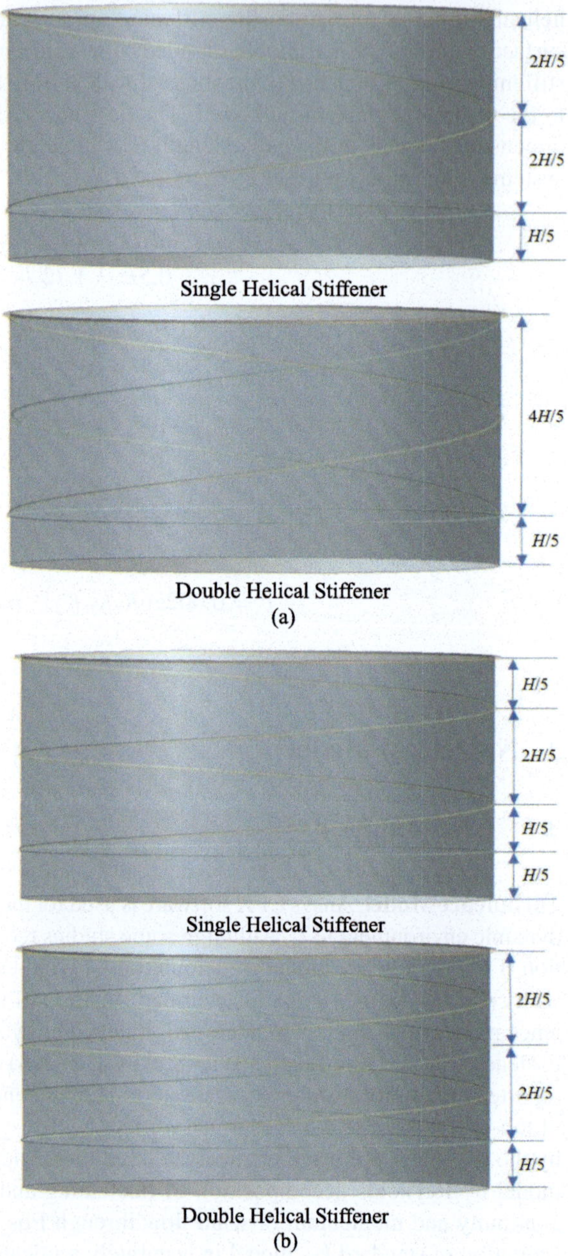

Fig. 2 Tank (H/D ratio 0.5) having 1 and 2 helical stiffeners with L_g **a** $4H/5$, **b** $2H/5$, **c** $4H/15$, and **d** $H/5$

Fig. 2 (continued)

Single Helical Stiffener

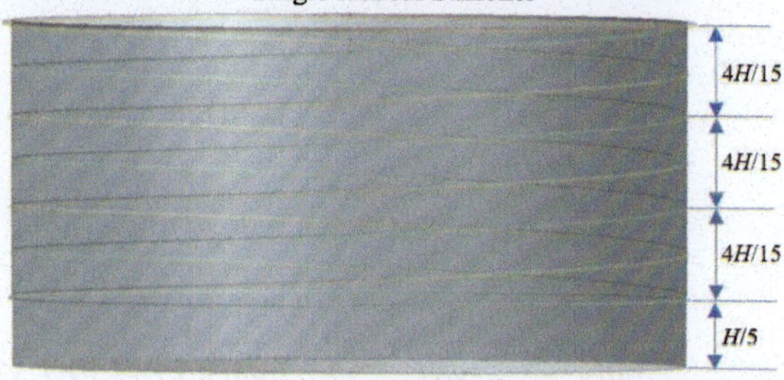

Double Helical Stiffener

(c)

Single Helical Stiffener

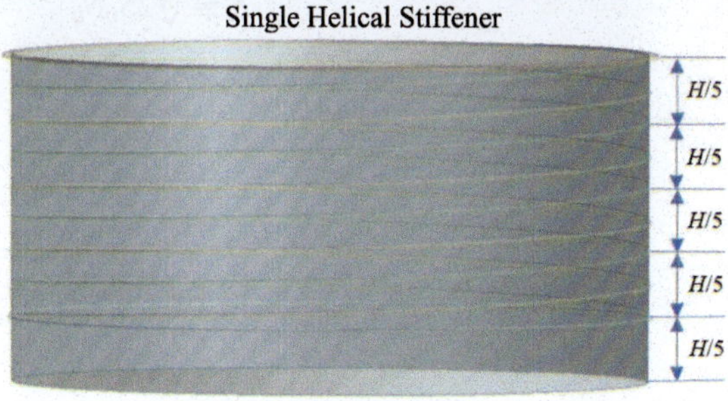

Double Helical Stiffener

(d)

ith and jth spatial co-ordinates x_i and x_j, and stress due to dynamic viscosity (μ). C_1 is described through Eq. (10). The kinetic energy (G_k) of average speed gradient, exhibited in Eq. (11), solved through modulus of strain tensor (S) and viscosity of flow turbulence (μ_t), which are expressed as Eqs. (12) and (13). Equations (14) to (18) express C_μ, A_s, $\tilde{\Omega}_{ij}$, ϕ, and $\tilde{S}$, , accordingly. Average rate of strain tensor (S_{ij}) is showcased as Eq. (19), where mean rotational tensor rate, air density, angular speed, kinematic, and dynamic viscosities are $\tilde{\Omega}_{ij}$, ρ, ω_k, v, and μ, respectively. u_i and u_j are mean speeds along x_i and x_j. A_0, σ_k, C_2, and σ_ε are equation constants, the value of which are 4.04, 1.0, 1.9, and 1.2, accordingly.

$$\frac{\partial u_i}{\partial x_i} = 0 \tag{6}$$

$$\frac{\partial}{\partial t}(\rho u_i) + \frac{\partial}{\partial x_j}(\rho u_i u_j) = -\frac{\partial \overline{p}}{\partial x_i} + \frac{\partial}{\partial x_j}(\mu + \mu_t)\left(\frac{\partial u_i}{\partial x_j} + \frac{\partial u_j}{\partial x_i}\right) \tag{7}$$

$$\frac{\partial}{\partial t}(\rho k) + \frac{\partial}{\partial x_j}(\rho k u_j) = \frac{\partial}{\partial x_j}\left[\left(\mu + \frac{\mu_t}{\sigma_k}\right)\left(\frac{\partial k}{\partial x_j}\right)\right] + G_k - \rho\varepsilon \tag{8}$$

$$\frac{\partial}{\partial t}(\rho\varepsilon) + \frac{\partial}{\partial x_i}(\rho\varepsilon u_i) = \frac{\partial}{\partial x_j}\left[\left(\mu + \frac{\mu_t}{\sigma_\varepsilon}\right)\left(\frac{\partial \varepsilon}{\partial x_j}\right)\right] + \rho C_1 S\varepsilon - \rho C_2\left(\frac{\varepsilon^2}{k + \sqrt{\nu\varepsilon}}\right) \tag{9}$$

$$C_1 = max\left[0.43, \frac{Sk/_\varepsilon}{Sk/_\varepsilon + 5}\right] \tag{10}$$

$$G_k = \mu_t S^2 \tag{11}$$

$$S \equiv \sqrt{2S_{ij}S_{ij}} \tag{12}$$

$$\mu_t = \rho C_\mu \frac{k^2}{\varepsilon} \tag{13}$$

$$C_\mu = \frac{1}{A_0 + A_s \frac{k\sqrt{S_{ij}S_{ij} + \tilde{\Omega}_{ij}\tilde{\Omega}_{ij}}}{\varepsilon}} \tag{14}$$

$$A_s = \sqrt{6}\cos\phi \tag{15}$$

$$\tilde{\Omega}_{ij} = \overline{\Omega}_{ij} - 3\varepsilon_{ijk}\omega_k \tag{16}$$

$$\phi = \frac{1}{3}\cos^{-1}\left(\sqrt{6}\frac{S_{ij}, S_{jk}, S_{ki}}{\tilde{S}^3}\right) \tag{17}$$

$$\tilde{S} = \sqrt{S_{ij} S_{ij}} \tag{18}$$

$$S_{ij} = \frac{1}{2}\left(\frac{\partial u_j}{\partial x_i} + \frac{\partial u_i}{\partial x_j}\right) \tag{19}$$

Setup of CFD Analysis and Pressure Coefficient Calculation. Iterative transient analysis of the time history of aerodynamic pressure is performed by applying second-order backward Euler method with 0.00001 target value of the root mean square (RMS) residual as the convergence criterion [46]. The time duration of the gust is considered as 3 s following the wind environment of Indian subcontinent as per IS-875, part 3 (2015) [47] and 0.001 s time step having 10 iterations at maximum is applied as per AIJ guidelines [34]. The general expressions of coefficients of wind pressure at the external (C_{pe}) and internal (C_{pi}) surfaces of the tank-wall, showed in Eqs. (20) and (21), are calculated according to the formulations followed by Portela and Godoy [48]. Where, P_{j_e} and P_{j_i} are pressure of wind at peripheral pressure taps on outside (j_e) and inside (j_i) faces of wall, which is commonly expressed as tapping point pressure P_j, in Eq. (22). P_{M_n} is the wind pressure measured from nth pressure tap having an angle of circumference $\Theta°$ at time T. The pressure (P_H) of free wind stream at the height H similar to the tank-height is provided in Eq. (23), according to CFD model of wind environment by Bairagi and Dalui [49]. The speed of wind at the similar height to the tank is U_H. The average external ($C_{pe_{mean}}$) and internal ($C_{pi_{mean}}$) coefficients of aerodynamic pressure are showcased as Eqs. (24) and (25). Here total sample number of time instances is expressed as N. Pressure taps are located at three levels of height on the tank-wall, which are 0.25H, 0.5H, and 0.75H.

$$C_{pi} = \frac{P_{j_i}}{P_H} \tag{20}$$

$$C_{pe} = \frac{P_{j_e}}{P_H} \tag{21}$$

$$P_j = P_{M_n}(\theta, T) \tag{22}$$

$$P_H = 0.5\rho U_H^2 \tag{23}$$

$$C_{pe_{mean}} = \frac{1}{N}\sum_{je=n-1}^{je=n} C_{pe}(T) \tag{24}$$

$$C_{pi_{mean}} = \frac{1}{N}\sum_{ji=n-1}^{ji=n} C_{pi}(T) \tag{25}$$

Validation with Experimental Literature by Mesh Sensitivity Study. Present CFD model is already validated through mesh sensitivity study with the experimentation of Portela and Godoy [50] in previously published study by the authors on wind environment simulation around an elevated cylindrical water reservoir [51]. The elevated water tank's *H*/*D* ratio (0.43) was similar to the experimental model (*H* 115.7 mm and *D* 269.2 mm) used by Portela and Godoy [50]. The basic velocity of wind was 64.8 m/s at 10 m, representing 19.8 m/s speed of wind at 116 mm inside the wind-tunnel of Portela and Godoy [50]. Both studies considered open terrain [50, 51]. The model and domain discretization in the present CFD is carried out with tetrahedral mesh as per the recommendations by Franke et al. [52] on CFD model discretization for aerodynamic simulation. The edge and corner mesh are inflated by using smallest sized element at these locations, to secure greater precision in separation of wind flow. In the previous study, the authors studied finer (FM), medium (MM), and coarse (CM) sizes of mesh, detailed in Table 1, to identify optimum mesh size for the present CFD model [51]. The discretized images of model and domain connection for this three mesh are shown in Fig. 3. For the validation purpose, the elevation of pressure tap at the elevated tank's wall was regarded as 0.6*H*, similar to the level of pressure tap on the experimental model of Portela and Godoy [50]. The circumferential distribution of coefficients of external wind pressure was compared with the same obtained from Portela and Godoy [50] in the previous study of Mukherjee et al. [51], which is presented in Fig. 4. The error in pressure coefficient values for FM, MM, and CM mesh with respect to Portela and Godoy [50] were 3.21%, 14.23%, and 36.63%, accordingly, which is exhibited in Fig. 5. The comparisons of three mesh types showcased the optimum accuracy of finer mesh (FM) and this particular mesh size is utilized for the discretization of the model and domain of the present CFD model, which is shown in Fig. 6.

Theoretical Validation by Studying Velocity and Turbulence Intensity Profile. According to the study by Sanyal and Dalui [53] on wind environment simulation following AIJ norms [34], power laws of wind velocity and turbulence intensity, showcased in Eqs. (26) and (27), are utilized for both CFD and theoretical model-based profiling of the wind flow speed and intensity of turbulence, which are presented in Fig. 7a and b. In these equations, wind speed (m/s) and turbulence intensity (%) at height z are U_z and I_z, accordingly. The velocity of wind at elevation z_{ref} is $U_{z_{ref}}$. . As this study is based on open terrain of Indian subcontinent, the power law coefficient (α) and gradient height (z_G) of atmospheric boundary layer are taken as 0.188 and 300 m, respectively following IS-875, part 3, 2015 [47] formulation and

Table 1 Details of different meshes

Mesh type	Element numbers	Node numbers	Element volume ratio	Edge length ratio	Skewness	Curvature angle
FM	77,606,245	16,753,678	$\leq$ 91.67	$\leq$ 80.43	$\leq$ 0.9	18°
MM	73,592,159	14,682,539	$\leq$ 96.32	$\leq$ 85.71	$\leq$ 0.95	24°
CM	57,937,626	12,323,547	$\leq$ 111.54	$\leq$ 107.92	$\leq$ 1.0	73°

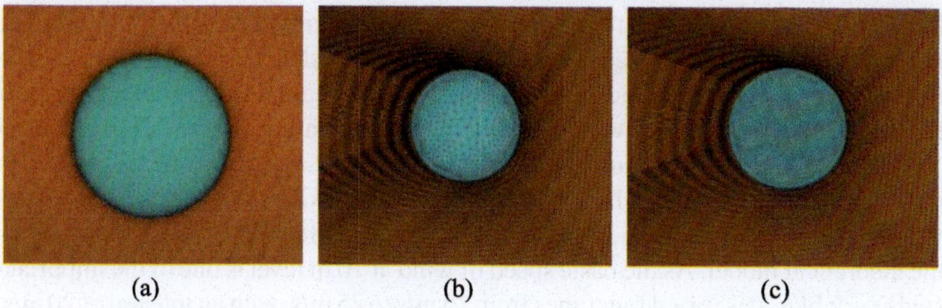

Fig. 3 Images of mesh **a** CM, **b** MM, and **c** FM

Fig. 4 Comparison of C_{pe} obtained from different mesh types with experimental study

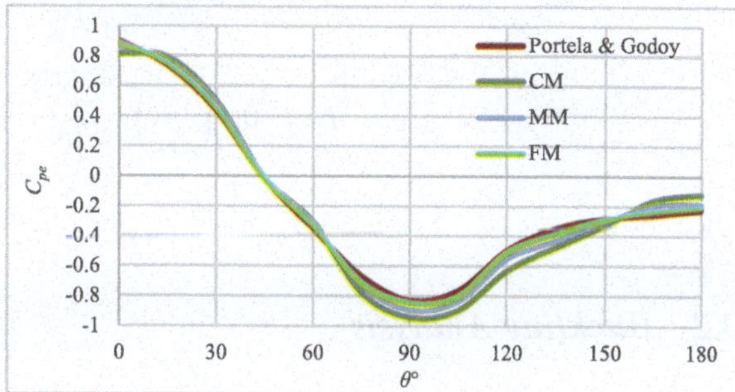

Fig. 5 Error comparison of different meshes with experimental study

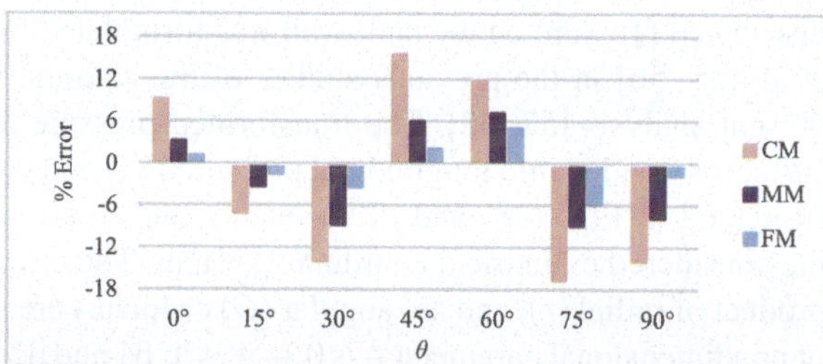

Fig. 6 Mesh of domain

IS-SP 64: 2001 [54] recommendation. The basic velocity of wind at 10 m and the intensity of turbulence at gradient height are considered as 50 m/s and 1%, respectively for the theoretical validation. From the turbulence intensity profile, it is clear that the error of CFD model with respect to theoretical model becomes significant (> 10%) beyond the height of tank, which will not affect the accuracy of wind stagnation pressure on the tank-wall, which is responsible for buckling. The velocity profile of wind, obtained from CFD model exhibits near perfect distribution in comparison to the theoretical model. As the basic speed of wind at 10 m level is one of the important study variables, it is varied four times from 15 m/s to 75 m/s, with an interval of 20 m/s.

$$U_z = U_{z_{ref}} \left(\frac{z}{z_{ref}} \right)^{\alpha} \tag{26}$$

$$I_z = 0.1 \left(\frac{z}{z_G} \right)^{-\alpha - 0.005} \tag{27}$$

3.2 *Buckling Analysis*

Analytical Model. Transformation of wind pressure as a nonuniform arbitrarily distributed pressure on the tank-wall was formulated following the analogy of Yang et al. [55, 56] in the previous studies of the authors, which is also utilized in the present analysis [57, 58]. The transformed pressure $P(z,\Phi)$ of wind on tank-wall surface is divided into longitudinal pressure $P(z)$ and circumferential pressure $P(\Phi)$, expressed in Eqs. (28) and (29) where z and Φ are axial and meridional axes of the considered cylindrical coordinate system. The circumferential ordinate (Φ), the product of radial (r), and arc angular (Θ) ordinates are shown in Eq. (30). The range of nondimensional parameter $\in$ is $0 \leq \in \leq 1$. β1 and β2 are the constants of equation. As the stagnation pressure of wind is reported as the most important load for wind-induced tank buckling, the wind velocity pressure is regarded to be the pressure of reference ($P_{z_{ref}}$) [59]. Wind-induced buckling stress (P_B) and radial deformation (δ_{r_B}) are solved through Eqs. (31) and (32). P_U and δ_{r_U} are uniform pressure and its response as radial deformation. δ_{r_η} and P_η are the radial deformation and the stress causing the response at ηth function of load, which summarizes surface pressure variation. The nondimensional longitudinal axis (ζ) is represented as the ratio of z and H. The load multiplier of buckling (λ) is calculated as Eq. (33). As present study is based on eigenvalue buckling analysis of helically stiffened, ground-supported, cylindrical, steel tank, induced by wind, the governing equation of wind-induced buckling is provided as Eq. (34) [33]. In the buckling equation, tangent and stress stiffening matrices are K_T and S. λ_m is the load multiplier of buckling and ψ_m is the eigen vector of mode m.

$$\Phi = r\theta \tag{28}$$

Fig. 7 Distributions of **a** velocity, and **b** intensity of turbulence with the elevation

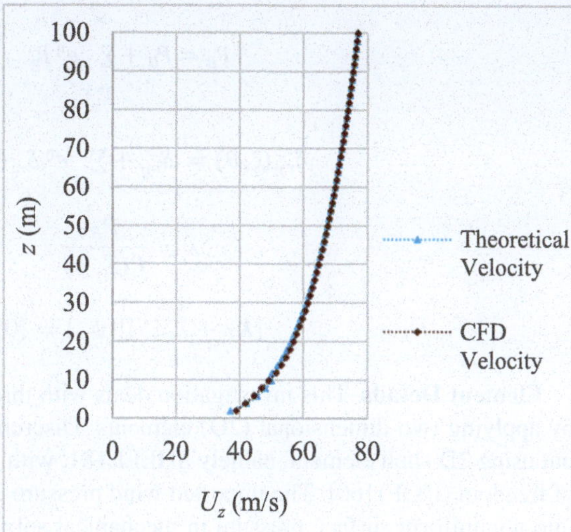

(a)

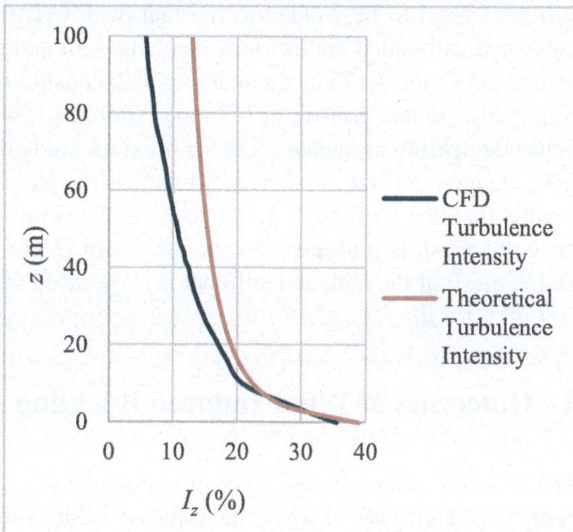

(b)

$$P(z) = P_{Z_{ref}}\left[1 + \in \left(\frac{Z}{H} + \frac{1}{2}\right)\right] \tag{29}$$

$$P(\theta) = P_{Z_{ref}}[1 + \beta_1 + \in \beta_1(\cos\theta - 1) + \ominus\beta_2(\cos 2\theta)] \tag{30}$$

$$P_B = P_U + \sum_{\eta=1}^{\infty} \epsilon^{\eta} P_{\eta} \tag{31}$$

$$\delta_{r_B}(\zeta,\theta) = \delta_{r_U} + \sum_{\eta=1}^{\infty} \epsilon^{\eta} \delta_{r_{\eta}}(\zeta,\theta) \tag{32}$$

$$\lambda = \frac{P_B}{P(Z,\phi)} \tag{33}$$

$$[K_T + \lambda_m[S]]\{\psi_m\} = \{0\} \tag{34}$$

Element Details. This investigation deals with three-dimensional (3D) analyses by applying two-dimensional (2D) elements. Discretization of tank-wall is carried out using 2D shell element, namely SHELL181, with four nodes having six degrees of freedom (DOF) [60]. The imported wind pressure data from CFX is transformed into nonuniform surface pressure in mechanical solver of Ansys workbench using surface enveloping 2D element SURF154 having four to eight nodes with three DOF [61]. The stiffener is discretized with 2D BEAM188 element with two nodes having 6 DOF [60]. Following the international standards of vertical steel tank design, the stiffeners need to be welded to the tank-wall [31, 62]. The welded connection is replicated as bonded contact between tank-wall and stiffener in this study with the help of 2D CONTA175 element, along with considering Lagrangian method of non-penetrable contact constraint [63–66]. Similar to finite element analysis of wind-induced buckling of tank-shell in the previous study of the authors, Mukherjee et al. [58], the element size is taken as 50 mm, as the element size need to be 1256 times smaller than the shell circumference as per Sun et al. [67]. Among all the tanks, the image of mesh is presented for the tank with *H*/*D* ratio 0.5 in Fig. 8. The contact region mesh of the tank and stiffener is showcased in Fig. 9.

4 Outcomes of Wind-Induced Buckling Analyses

The structure under study in this investigation is a cylindrical, open-top, ground-supported steel-tank, of which the dimensions are similar to the studies by Uematsu et al. [17]. The distribution of pressure coefficients of wind on the outside and inside faces of the tank's wall is calculated from the measured pressure from CFD analyses, which is provided in Fig. 10a to d, subsequently for four different cases of *H*/*D* ratio of the tank ranging from 0.25 to 1.0. The outside positive pressure and suction shows increasing pattern with the increments of *H*/*D* and tapping elevation (z) to tank height (H) ratio. The inside wind suction subsequently increases and decreases with the increase in *H*/*D* and *z*/*H* ratios. The mechanism responsible for the wind-induced buckling of tank-shell is showcased here through Fig. 11, which is explained according to the authors previous studies [57, 58]. The wind pressure distribution

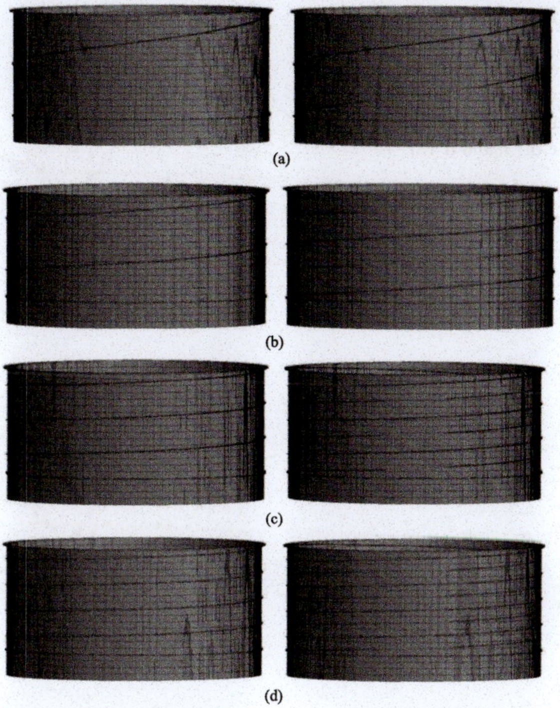

Fig. 8 Mesh of tank (H/D 0.5) having 1 and 2 helical stiffeners with L_g **a** $4H/5$, **b** $2H/5$, **c** $4H/15$, and **d** $H/5$

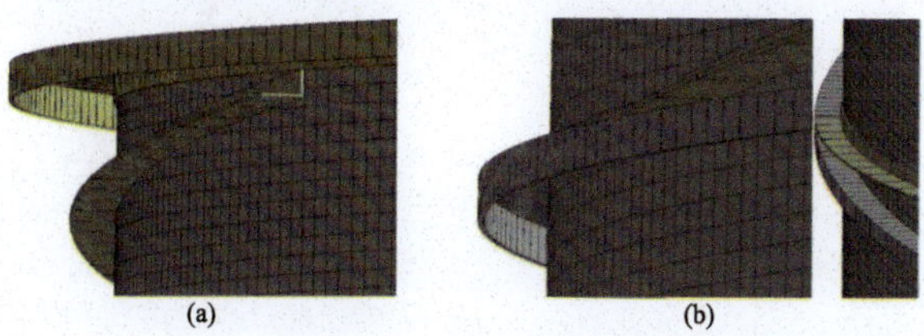

Fig. 9 The mesh for connection of tank-wall and helical stiffener with **a** top and **b** bottom ring-stiffener

and consequent buckling wave formation on tank-wall are presented in Fig. 11a and b, respectively. Due to the frontal interaction with the wind velocity–pressure, high amount of positive pressure gets concentrated at the stagnation zone of the tank's windward area. Suction is noticed at the other portions of the tank other than frontal area, which is highest at the sides perpendicular to the direction of the wind. This characteristic of aerodynamic pressure distribution on the tank-wall, generates membrane stresses circulation starting from the leeward face through two opposite side faces to meet at the frontal region of the tank-wall, which causes buckling waves to occur, especially at the windward arc of the tank's circumference for short ($H/D \leq 1.0$) tank.

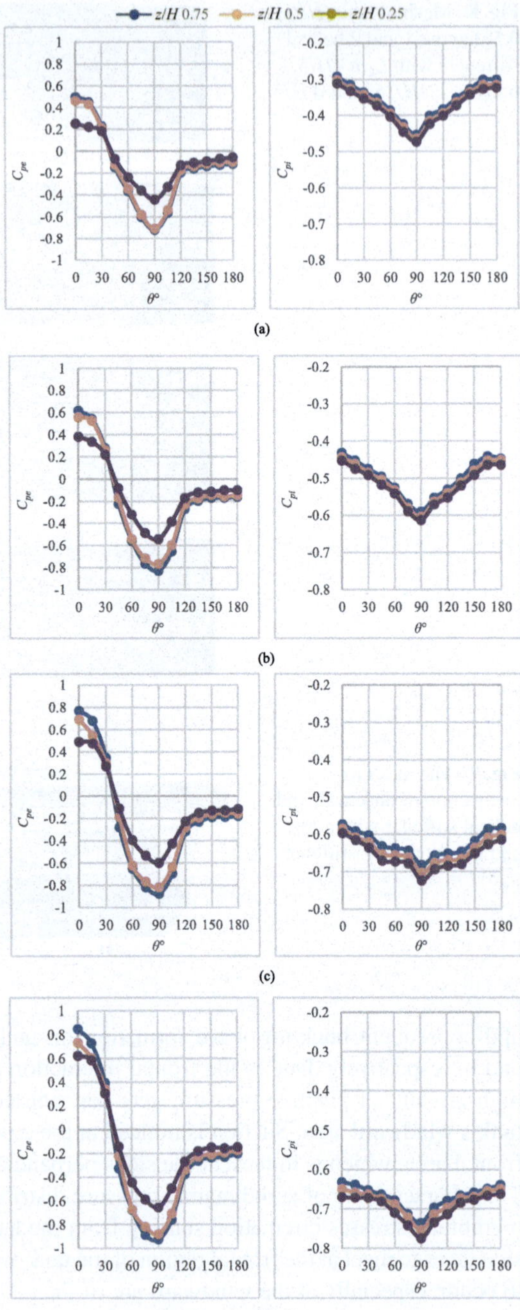

Fig. 10 Mean coefficients of aerodynamic pressure on exterior and interior wall-surfaces for open-top tank *H*/*D* ratio **a** 0.25, **b** 0.5, **c** 0.75, and **d** 1.0

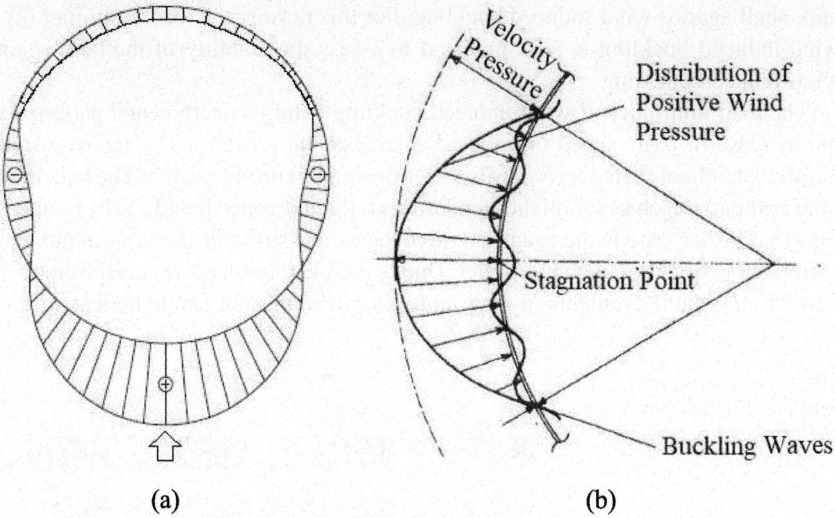

Fig. 11 Diagrams for the patterns of **a** pressure of wind, and **b** wind-induced buckling wave

The pressure of wind on the external and internal surfaces of the open-top tank's wall combinedly produces the membrane stress, which is responsible for the occurrence of circumferential buckling waves at the wall. The deformation of the tank-wall due to wind-induced buckling showcases the extent of the damage of the wall, which is expressed non-dimensionally in Fig. 12, as, normalized radial deformation ($\delta_{r_{BN}}$) contour. The normalization is carried out with respect to the maximum value of radial deformation. The radial deformation of any portion of the tank-wall, respectively towards outside and inside from the initial circumferential position, is regarded as positive and negative radial deformation. The velocity pressure of wind at the stagnation zone tends to increase with the height, which results in the higher concentration of membrane forces at the upper portion, resulting in the buckling wave. The extent of normalized radial deformation due to wind-induced buckling gradually reduces with the increase in the number of helical stiffener (N_{HS}) and the decrease in the gauge length (L_g). From the comparison of the contour plots between single and double helically stiffened open-roof tanks, it is clear that dual helical stiffener provides greater stiffness to the tank-wall against wind-induced buckling for similar gauge length. Also, it is noticed that the increasing slenderness (H/D) ratio contributes to the higher amount of radial deformation. In case of single helical stiffener attachment, the circumferential wave number increases with the decrease in gauge length. Whereas, decrease in the gauge length of double helical stiffener up to $2H/5$, increases the wave number. If the reduction in gauge length of dual helical stiffener is $< 2H/5$, circumferential waves become less in number and occur near the base of the tank, creating elephant's foot formation. Although radial deformation and wave number can be useful for assessing the wind-induced buckling damage to the tank-wall, it is not dependable for understanding the stability of a tank-shell against wind-induced buckling. For this reason, the load multiplier (λ) of wind-induced buckling is also analyzed to assess the stability of the tank against wind-induced buckling.

The load multiplier of wind-induced buckling exhibits decremental pattern with the increase in basic speed of wind, slenderness and r/t ratio. The increase in the number of helical stiffener (N_{HS}) also rises the load multiplier value. The load multiplier comparison chart for all the slenderness ratio and gauge lengths are provided in Fig. 13. The decrease in the gauge length of the helical stiffener also contributes to the increment in buckling load multiplier. Due to the increase in basic speed from 15 m/s to 35 m/s, the decrements in load multiplier of tank with single helical stiffener

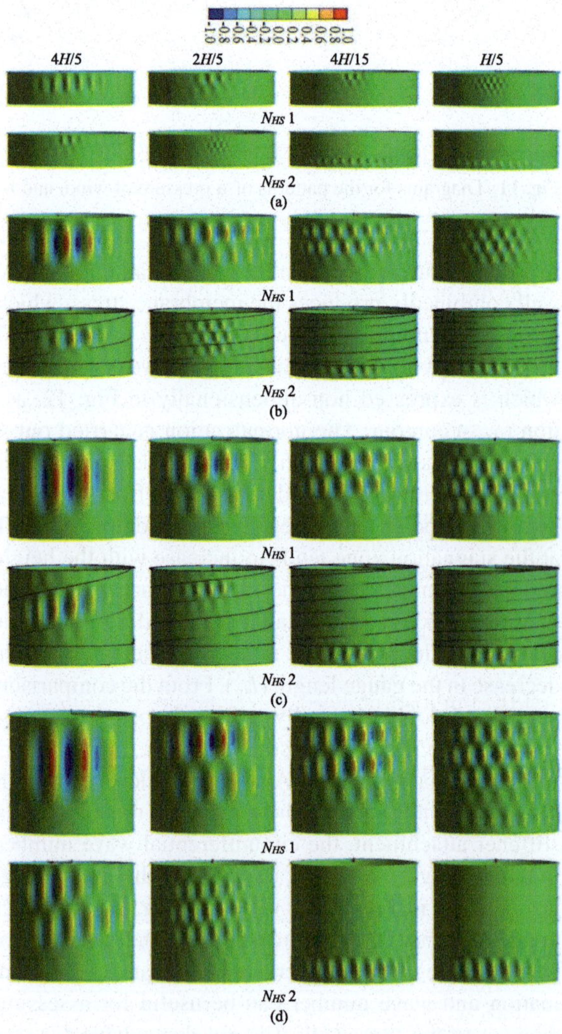

Fig. 12 $\delta_{r_{BN}}$ contour of helically stiffened tank for H/D Ratio **a** 1.0, **b** 0.75, **c** 0.5, and **d** 0.25

are 3.28%, 3.92%, 4.14%, and 16.1% for H/D ratio from 0.25 to 1.0, respectively. The rise in velocity from 35 m/s to 55 m/s reduces the multiplier of buckling load of single helically stiffened tank by 6.1%, 11%, 7.83%, and 6.69%, accordingly for H/D ratio from 0.25 to 1.0. The load multiplier of wind-induced buckling of the tank strengthened with single helical stiffener is lessened by 10.02%, 16.82%, 11.71%, and 7.86%, subsequently for H/D ratio from 0.25 to 1.0, when the basic wind speed rises from 55 m/s to 75 m/s. In case of double helical stiffener attachment, drops of 1.89%, 2.63%, 5.57%, and 7.36% are noticed in the multiplier of buckling load of open-top tank, respectively for H/D ratio from 0.25 to 1.0 due to the basic wind velocity rise from 15 m/s to 35 m/s. Increment of velocity from 35 m/s to 55 m/s intensifies the reductions in wind-induced buckling load multiplier by 5.78%, 6.44%, 8.13%, and 5.51% for H/D ratio from 0.25 to 1.0 of open-top tank with dual helical stiffener, accordingly. Further basic velocity of wind surge from 55 m/s to 75 m/s against tank with double helical stiffener having H/D ratio of 0.25 up to 1.0, subsequently experience downfall of load multiplier by 9.99%, 8.1%, 8.69%, and 11.32%. When the r/t ratio increases by 250, the average decrement of single helically stiffened open-roof tank's load multiplier of wind-induced buckling is 31.59%. For the rise in r/t ratio of 250 in case of dual helically stiffened tank with H/D ratio 0.25, the buckling load multiplier decreases respectively by 5.56%, 6.58%, 23.25%, and 32.1%, when the gauge length of helical stiffening system surges from $H/5$ to $4H/5$. In case of double helically stiffened tank with H/D ratio from 0.5 to 1.0, the load multiplier's mean decrease is 32.1% for the increment of r/t ratio by 250. With the drop in gauge length by $H/5$, the average increments of single helically stiffened tank's load multiplier of wind-induced buckling are 133.19%, 55.92%, 47.24%, and 41.19%, accordingly for H/D ratio from 0.25 to 1.0. The decrements of gauge length from $4H/5$ to $2H/5$, $2H/5$ to $4H/15$, and $4H/15$ to $H/5$ cause the increments in load multiplier of dual helically stiffened tank with H/D ratio 0.25 by 456.57%, 261.76%, and 21.15%, accordingly. Decreasing changes of double helical stiffening system's gauge length from $4H/5$ to $2H/5$ and $2H/5$ to $4H/15$ create rise in buckling load multiplier by 107.29% and 22.4% respectively for tank having H/D ratio 0.5, whereas, multiplier increments are 50.63% and 38.07% accordingly, when double helically stiffened tank's H/D ratio is 0.75. For both of the cases of H/D ratio 0.5 and 0.75, the decrease of gauge length from $4H/15$ to $H/5$ exhibits no significant rise (<1.0%) in load multiplier of wind-induced buckling of double helically stiffened tank. In case of double helically stiffened tank with H/D ratio 1.0, reductions in gauge length from $4H/5$ to $2H/5$, $2H/5$ to $4H/15$, and $4H/15$ to $H/5$ show surge in buckling load multiplier by 68.66%, 9.52%, and 7.89%, subsequently. The gradual increase in slenderness ratio from 0.25 to 0.5, 0.5 to 0.75, and 0.75 to 1.0 reduces the load multiplier of wind-induced buckling of single helically stiffened tank by 68.81%, 32.04%, and 19.68%, respectively, whereas, for tanks with dual helical stiffener the load multiplier decrements are 83.73%, 34.18%, and 25.04%, subsequently.

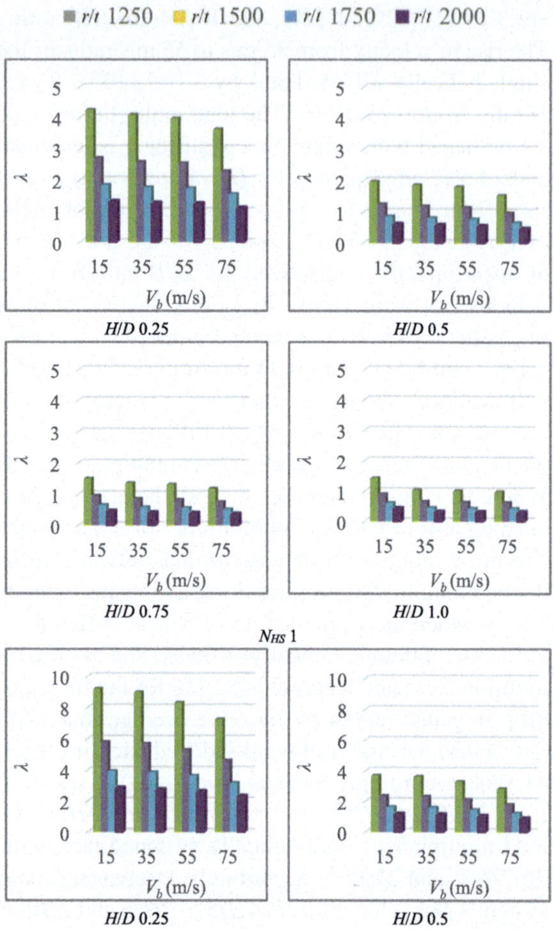

Fig. 13 Buckling load multiplier for gauge length **a** $4H/5$, **b** $2H/5$, **c** $4H/15$, and **d** $H/5$

5 Conclusive Remarks

The conclusive remarks on the applicability of helical stiffener for rendering stability to empty, open-top, cylindrical, ground-supported, steel tank against wind-induced buckling are discussed in this section based on the safe stability criterion ($\lambda \geq 3$) following API 650 [31] and Zeybek et al. [20, 21], depending on Euler's theory of buckling, peak gust speed region according to IS 875, part 3, 2015 [47], and ensuring economy by less material consumption in the form of higher r/t ratio of tank-wall. The buckling load multiplier (λ) value < 1 and $= 1$, respectively, signify the instability and critical stability. The low stability condition is denoted by the load multiplier range $1 < \lambda < 3$. The conclusions are as follows:

Fig. 13 (continued)

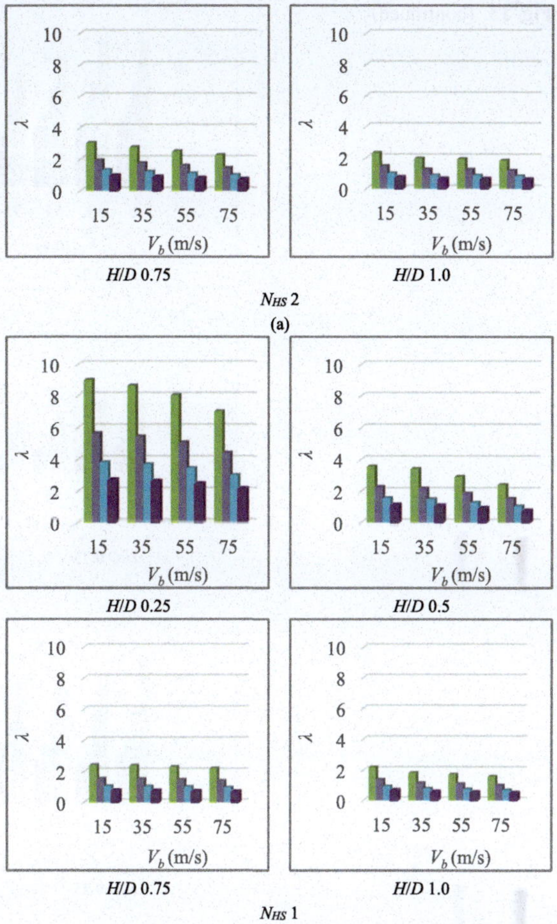

Fig. 13 (continued)

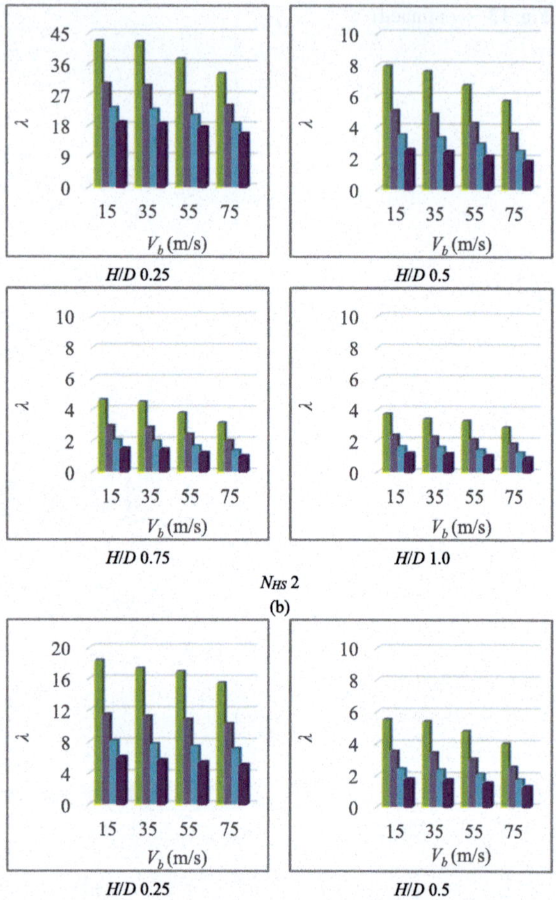

Fig. 13 (continued)

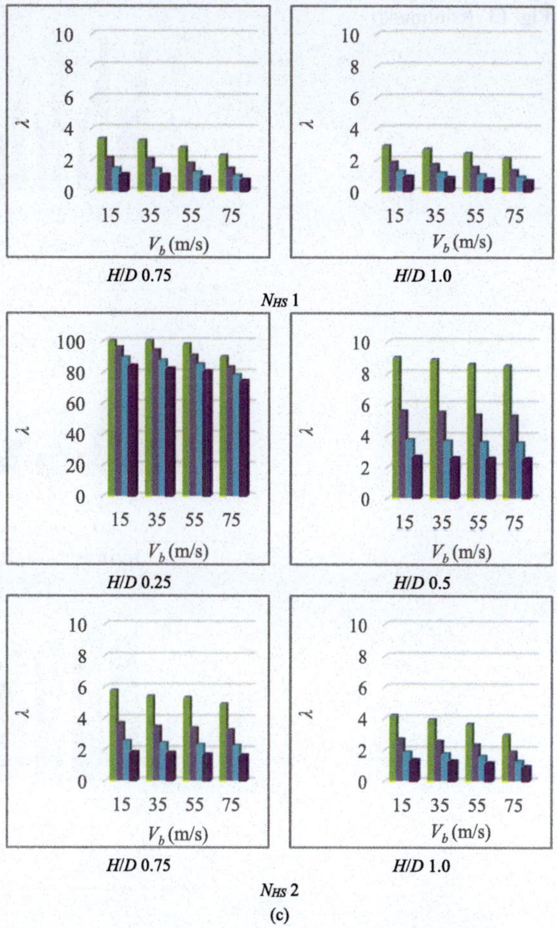

Fig. 13 (continued)

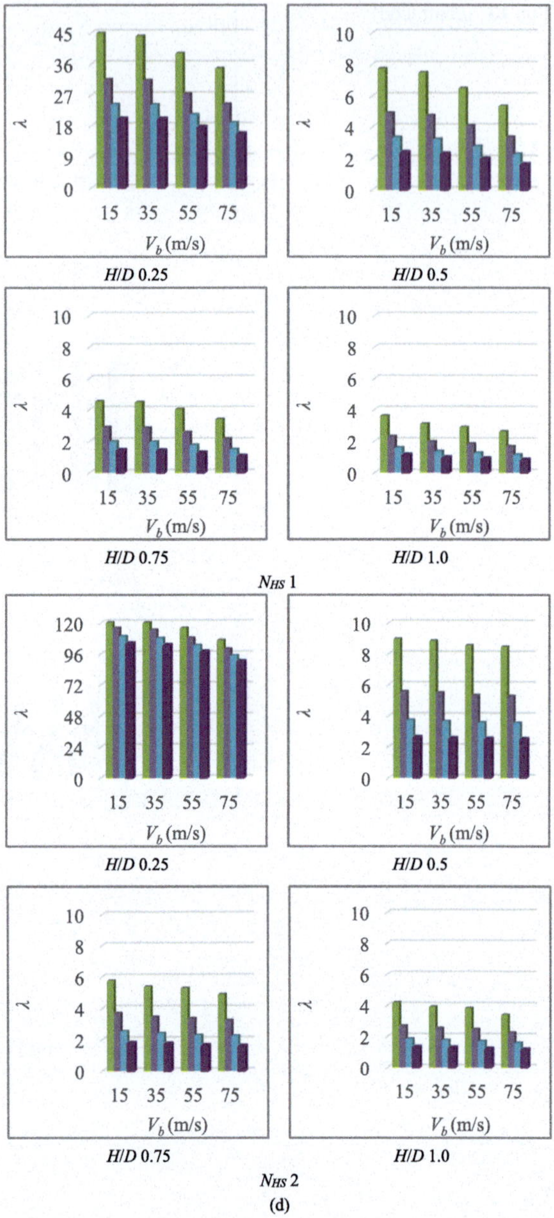

- Gauge length of 4*H*/5 is applicable safely and economically at the wind zone having basic speed up to 55 m/s, when *r*/*t* ratio is 1250 and 1750, respectively for tank (*H*/*D* 0.25) with single and double helical stiffener. In case of tank's *H*/*D* ratio 0.5 and double stiffener is attached, economic safety can be maintained with *r*/*t* ratio 1250, against wind-induced buckling, up to 55 m/s wind zone. For the *H*/*D* ratio of 0.75, only economic low stability is observed for *r*/*t* ratio 1250 and 1500 up to 55 m/s wind zone, accordingly for single and double helical stiffener. Double helical stiffener and *r*/*t* ratio 1500 produce only economic critical stability in open-top tank with *H*/*D* ratio 1.0 up to 55 m/s wind zone.
- In case of 2*H*/5 gauge length, safe and economic stability of single helically stiffened tank is observed for *r*/*t* ratio 1750 and 1250, respectively for *H*/*D* ratio 0.25 and 0.5, up to 55 m/s wind zone. Subsequently, economic low and critical stability is monitored up to 55 m/s wind zone, for *r*/*t* ratio 1500 of tank with single helical stiffener, when *H*/*D* ratio is 0.75 and 1.0. In case of dual helical stiffener safety with economy is secured respectively for *r*/*t* ratio 2000, 1500, and 1250 of tank with *H*/*D* ratio 0.25, 0.5, both 0.75 and 1.0, up to 55 m/s wind zone.
- For the gauge length of 4*H*/15, the safe stability along with material economy is noticed up to 55 m/s wind zone, for *r*/*t* ratio 1500 and 1750 respectively when single and double helical stiffener is applied to tank with *H*/*D* ratio 0.5. Single helically stiffened tank with *H*/*D* ratio 0.75 and 1 accordingly showcases economic low and critical stability for *r*/*t* ratio 1750 up to 55 m/s wind zone. Double helically stiffened tank having *H*/*D* ratio 0.75 and 1.0 exhibits safety and economy up to 55 m/s wind zone, for *r*/*t* ratio 1500 and 1250, respectively. Tank with *H*/*D* ratio 0.25 is completely safe for both single and double helical stiffener up to 55 m/s wind zone.
- When gauge length is *H*/5, the economic safety is monitored up to 55 m/s for single helically stiffened tank's *H*/*D* ratio 0.5 and 0.75, when the *r*/*t* ratio is 1500 and 1250, accordingly. In case of single helical stiffener applied to tank with *H*/*D* ratio 1.0, stability with economy is observed for *r*/*t* ratio 1250, up to 33 m/s wind zone. When double helical stiffener is attached to tank with *H*/*D* ratios 0.5, 0.75, and 1.0, this exhibits safe buckling stability along with material economy up to 55 m/s wind zone, respectively for *r*/*t* ratios 1750, 1500, and 1250. When tank's *H*/*D* ratio is 0.25, it is absolutely stable if most economic *r*/*t* ratio 2000 is provided for both cases of single and dual helical stiffener.

References

1. Jerath, S., & Sadid, H. (1985). Buckling of orthotropic cylinders due to wind load. *Journal of Engineering Mechanics, 111*(05), 610–622, ASCE. https://doi.org/10.1061/(ASCE)0733-9399(1985)111:5(610)
2. Krajcinovic, D. (1970). Semimembrane analysis of cylindrical shells subjected to wind loading. *Journal of Applied Mechanics, 37*(04), 995–1001, ASME. https://doi.org/10.1115/1.3408729

3. Pecknold, D. A. (1989). Load transfer mechanisms in wind-loaded cylinders. *Journal of Engineering Mechanics, 115*(11), 2353–2367, ASCE. https://doi.org/10.1061/(ASCE)0733-9399(1989)115:11(2353)
4. Pircher, M. (2004). Medium-length thin-walled cylinder under wind loading—a case study. *Journal of Structural Engineering, 130*(12), 2062–2069, ASCE. https://doi.org/10.1061/(ASCE)0733-9445(2004)130:12(2062)
5. Uematsu, Y., & Uchiyama, K. (1985). Deflection and buckling behavior of thin, circular cylindrical shells under wind loads. *Journal of Wind Engineering and Industrial Aerodynamics, 18*(03), 245–261, Elsevier. https://doi.org/10.1016/0167-6105(85)90084-4
6. Briassoulis, D., & Pecknold, D. A. (1986). Behaviour of empty steel grain silos under wind loading: Part 1: The stiffened cylindrical shell. *Engineering Structures, 08*(04), 260–275, Elsevier. https://doi.org/10.1016/0141-0296(86)90034-9
7. Chen, L., & Rotter, J. M. (2012). Buckling of anchored cylindrical shells of uniform thickness under wind load. *Engineering Structures, 41*, 199–208, Elsevier. https://doi.org/10.1016/j.engstruct.2012.03.046
8. Maleki, S., & Mehretehran, A. M. (2018). 3D wind buckling analysis of long steel corrugated silos with vertical stiffeners. *Engineering Failure Analysis, 90*, 156–167, Elsevier. https://doi.org/10.1016/j.engfailanal.2018.03.031
9. Uchiyama, K., Uematsu, Y., & Orimo, T. (1987). Experiments on the deflection and buckling behavior of ring-stiffened cylindrical shells under wind pressure. *Journal of Wind Engineering and Industrial Aerodynamics, 26*(02), 195–211, Elsevier. https://doi.org/10.1016/0167-6105(87)90044-4
10. Schmidt, H., Binder, B., & Lange, H. (1998). Postbuckling strength design of open thin-walled cylindrical tanks under wind load. *Thin-walled Structures, 31*(01–03), 203–220, Elsevier. https://doi.org/10.1016/S0263-8231(98)00009-3
11. Maleki, S., & Mehretehran, A. M. (2019). 3D wind buckling analysis of steel silos with stepped walls. *Thin-Walled Structures, 142*, 236–261, Elsevier. https://doi.org/10.1016/j.tws.2019.05.007
12. Burgos, C. A., Batista-Aberu, J. C., Calabró, H. D., Jaca, R. C., & Godoy, L. A. (2015). Buckling estimates for oil storage tanks: Effect of simplified modeling of the roof and wind girder. *Thin-Walled Structures, 91*, 29–37, Elsevier. https://doi.org/10.1016/j.tws.2015.02.006
13. Resinger, F., & Greiner, R. (1982). Buckling of wind loaded cylindrical shells—application to unstiffened and ring-stiffened tanks. In: *Buckling of Shells*, pp. 305–331, Springer, Berlin. https://doi.org/10.1007/978-3-642-49334-8_10
14. Schneider, W., & Zahlten, W. (2004). Load-bearing behaviour and structural analysis of slender ring-stiffened cylindrical shells under quasi-static wind load. *Journal of Constructional Steel Research, 60*(01), 125–146, Elsevier. https://doi.org/10.1016/j.jcsr.2003.08.002
15. Azzuni, E., & Guzey, S. (2017). Stability of open top cylindrical steel storage tanks: Design of top wind girder. *Journal of Pressure Vessel Technology, 139*(03), 031207–031217, ASME. https://doi.org/10.1115/1.4035507
16. Sun, T., Azzuni, E., & Guzey, S. (2018). Stability of open-topped storage tanks with top stiffener and one intermediate stiffener subject to wind loading. *Journal of Pressure Vessel Technology, 140* (01), 011204–011217, ASME. https://doi.org/10.1115/1.4038723
17. Uematsu, Y., Yamaguchi, T., & Yasunaga, J. (2018). Effects of wind girders on the buckling of open-topped storage tanks under quasi-static wind loading. *Thin-walled Structures, 124*, 01–12, Elsevier. https://doi.org/10.1016/j.tws.2017.11.044
18. Bu, F., & Qian, C. (2015). A rational design approach of intermediate wind girders on large storage tanks. *Thin-walled Structures, 92*, 76–81, Elsevier. https://doi.org/10.1016/j.tws.2015.02.024
19. Bu, F., & Qian, C. (2016). On the rational design of the top wind girder of large storage tanks. *Thin-walled Structures, 99*, 91–96, Elsevier. https://doi.org/10.1016/j.tws.2015.10.014
20. Zeybek, Ö., Topkaya, C., & Rotter, J. M. (2019). Stress resultants for wind girders in open-top cylindrical steel tanks. *Engineering Structures, 196*, 109347, Elsevier. https://doi.org/10.1016/j.engstruct.2019.109347

21. Zeybek, Ö., Topkaya, C., & Rotter, J. M. (2022). Stiffness requirements for wind girders in open-top cylindrical steel tanks. *Thin-Walled Structures, 176*, 109353, Elsevier. https://doi.org/10.1016/j.tws.2022.109353
22. Słowiński, K., Piekarczyk, M., & Dybel, P. (2023). Effect of simplified wind girder modelling on MNA-LBA analysis of open steel tanks. *International Journal of Pressure Vessels and Piping, 203*, 104941, Elsevier. https://doi.org/10.1016/j.ijpvp.2023.104941
23. Shao-Wen, Y. (1980). Buckling of cylindrical shells with spiral stiffeners under uniform compression and torsion. *Computers & Structures, 11*(06), 587–595, Elsevier. https://doi.org/10.1016/0045-7949(80)90065-6
24. Shaterzadeh, A., & Foroutan, K. (2016). Post-buckling of cylindrical shells with spiral stiffeners under elastic foundation. *Structural Engineering and Mechanics, 60*(04), 615–631, Techno-Press. https://doi.org/10.12989/sem.2016.60.4.615
25. Shaterzadeh, A., Foroutan, K., & Ahmadi, H. (2019). Nonlinear static and dynamic thermal buckling analysis of spiral stiffened functionally graded cylindrical shells with elastic foundation. *International Journal of Applied Mechanics, 11*(01), 1950005, World Scientific Publishing. https://doi.org/10.1142/S1758825119500054
26. Ahmadi, H., & Foroutan, K. (2019). Nonlinear vibration of stiffened multilayer FG cylindrical shells with spiral stiffeners rested on damping and elastic foundation in thermal environment. *Thin-Walled Structures, 145*, 106388, Elsevier. https://doi.org/10.1016/j.tws.2019.106388
27. Ahmadi, H., & Foroutan, K. (2019). Nonlinear primary resonance of spiral stiffened functionally graded cylindrical shells with damping force using the method of multiple scales. *Thin-Walled Structures, 135*, 33–44, Elsevier. https://doi.org/10.1016/j.tws.2018.10.028
28. Nam, V. H., Phuong, N. T., Van Minh, K., & Hieu, P. T. (2018). Nonlinear thermo-mechanical buckling and post-buckling of multilayer FGM cylindrical shell reinforced by spiral stiffeners surrounded by elastic foundation subjected to torsional loads. *European Journal of Mechanics-A/Solids, 72*, 393–406, Elsevier. https://doi.org/10.1016/j.euromechsol.2018.06.005
29. Tu, T. M., Thai, D. K., Hoan, P. V., & Hoa, L. K. (2022). Nonlinear behavior of FG porous cylindrical sandwich shells reinforced by spiral stiffeners under torsional load including thermal effect. *Mechanics of Advanced Materials and Structures, 29*(27), 5860–5875, Taylor & Francis. https://doi.org/10.1080/15376494.2021.1967530
30. Foroutan, K., & Dai, L. (2024). Nonlinear torsional stability and buckling of spiral stiffened sigmoid FG thin circular cylindrical shells. *European Journal of Mechanics-A/Solids, 106*, 105328, Elsevier. https://doi.org/10.1016/j.euromechsol.2024.105328
31. API Standard 650. (2020). Welded tanks for oil storage. Washington: American Petroleum Institute.
32. Khan, A. K., Shah, T. R., Khosa, A. A., & Ali, H. M. (2023). Evaluation of wind load effects on solar panel support frame: A numerical study. *Engineering Analysis with Boundary Elements, 153*, 88–101, Elsevier. https://doi.org/10.1016/j.enganabound.2023.05.016
33. Baiz, P. M., & Aliabadi, M. H. (2007). Buckling analysis of shear deformable shallow shells by the boundary element method. *Engineering Analysis with Boundary Elements, 31*(04), 361–372, Elsevier. https://doi.org/10.1016/j.enganabound.2006.07.008
34. AIJ Guidelines. (2004). Recommendations for loads on buildings. Tokyo: Architectural Institute of Japan.
35. Sun, X., Li, W., Huang, Q., Zhang, J., & Sun, C. (2020). Large eddy simulations of wind loads on an external floating-roof tank. *Engineering Applications of Computational Fluid Mechanics, 14*(01), 422–435, Taylor & Francis. https://doi.org/10.1080/19942060.2020.1718757
36. Blocken, B. (2018). LES over RANS in building simulation for outdoor and indoor applications: A foregone conclusion? *Building Simulation, 11*(05), 821–870. Springer. https://doi.org/10.1007/s12273-018-0459-3
37. Pasley, H., & Clark, C. (2000). Computational fluid dynamics study of flow around floating-roof oil storage tanks. *Journal of Wind Engineering and Industrial Aerodynamics, 86*(01), 37–54, Elsevier. https://doi.org/10.1016/S0167-6105(99)00138-5
38. Kuroda, S., Uejima, H., Ishida, K., Yoshida, S., Shiratori, M., Sekine, K., Tsuchida, T., & Iwata, K. (2012). Simulation for a floating roof behavior of cylindrical storage tank due to

wind load: part 1—CFD analysis. In: *Proceedings of the ASME 2012 pressure vessels and piping conference (PVP 2012)* (Vol. 4) Fluid-Structure Interaction, pp. 125–137. American Society of Mechanical Engineers. https://doi.org/10.1115/PVP2012-78447
39. Hillewaere, J., Degroote, J., Lombaert, G., Vierendeels, J., & Degrande, G. (2013). Computational aspects of simulating wind induced ovalling vibrations in silo groups. *Journal of Computational and Applied Mathematics, 246*, 161–173, Elsevier. https://doi.org/10.1016/j.cam.2012.06.033
40. Hillewaere, J., Degroote, J., Lombaert, G., Vierendeels, J., & Degrande, G. (2015). Wind-structure interaction simulations of ovalling vibrations in silo groups. *Journal of Fluids and Structures, 59*, 328–350, Elsevier. https://doi.org/10.1016/j.jfluidstructs.2015.09.013
41. Liu, Q., Zhao, Y., Cai, S., & Dong, S. (2020). Wind loads and wind-resistant behaviour of large cylindrical tanks in square-arrangement group. Part 2: CFD simulation and finite element analysis. *Wind and Structures, 31*(06), 495–508, Techno-Press. https://doi.org/10.12989/was.2020.31.6.495
42. Sun, W., Wang, Z., Feng, J., & Gu, W. (2022). Surface wind pressure distribution of molten-salt power tower by CFD analysis. *Arabian Journal for Science and Engineering, 47*(10), 12497–12507, Springer. https://doi.org/10.1007/s13369-021-06501-x
43. Tominaga, Y. (2015). Flow around a high-rise building using steady and unsteady RANS CFD: Effect of large-scale fluctuations on the velocity statistics. *Journal of Wind Engineering and Industrial Aerodynamics, 142*, 93–103, Elsevier. https://doi.org/10.1016/j.jweia.2015.03.013
44. Bairagi, A. K., & Dalui, S. K. (2021). Wind environment around the setback building models. *Building Simulation, 14*(05), 1525–1541, Springer. https://doi.org/10.1007/s12273-020-0758-3
45. Hassan, S., Molla, M. M., Nag, P., Akhter, N., & Khan, A. (2022). Unsteady RANS simulation of wind flow around a building shape obstacle. *Building Simulation, 15*, 291–312, Springer. https://doi.org/10.1007/s12273-021-0785-8
46. CFX-Solver. (2006). Ansys theory guide. Release II.
47. Indian Standard Code IS 875, Part 3—2015. (Reaffirmed 2020). Code of practice for design loads (other than earthquake loads), for building and structures, part 3—wind loads. Bureau of Indian Standards, New Delhi
48. Portela, G., & Godoy, L. A. (2007). Wind pressure and buckling of grouped steel tank. *Wind and Structures, 10*(01), 23–44, Techno-Press. https://doi.org/10.12989/was.2007.10.1.023
49. Bairagi, A. K., & Dalui, S. K. (2021). Prediction of pressure coefficient on setback building by artificial neural networks. *Canadian Journal of Civil Engineering, 48*(10), 1364–1385, Canadian Science Publishing. https://doi.org/10.1139/cjce-2020-0100
50. Portela, G., & Godoy, L. A. (2005). Wind pressure and buckling of cylindrical steel tanks with a conical roof. *Journal of Constructional Steel Research, 61*(06), 786–807, Elsevier. https://doi.org/10.1016/j.jcsr.2004.11.002
51. Mukherjee, S., Reddy, J. S. K., & Singha Roy, D. K. (2024). Prediction of wind-pressure coefficients on circular elevated-tanks through the novel CFD-ANN model for various building-interference circumstances. *International Journal of Computational Fluid Dynamics, 37*(05), 367–394, Taylor & Francis. https://doi.org/10.1080/10618562.2024.2325709
52. Franke, J., Hellsten, A., Schlunzen, K. H., & Carissimo, B. (2011). The COST 732 best practice guideline for CFD simulation of flows in the urban environment: a summary. *International Journal of Environment and Pollution, 44*(01–04), 419–427, Inderscience Publishers. https://doi.org/10.1504/IJEP.2011.038443
53. Sanyal, P., & Dalui, S. K. (2022). Forecasting of aerodynamic coefficients of tri-axially symmetrical Y plan shaped tall building based on CFD data trained ANN. *Journal of Building Engineering, 47*, 103889, Elsevier. https://doi.org/10.1016/j.jobe.2021.103889
54. Indian Standard Code IS- SP 64. (2001). Explanatory Handbook on Indian Standard Code of Practice for Design Loads (other than Earthquake Loads) for Buildings and Structures, Part 3—Wind Loads. New Delhi: Bureau of Indian standards.
55. Yang, L., Luo, Y., Qiu, T., Zheng, H., & Zeng, P. A. (2022). Novel analytical study on the buckling of cylindrical shells subjected to arbitrarily distributed external pressure. *European*

Journal of Mechanics/A Solids, 91, 104406, Elsevier. https://doi.org/10.1016/j.euromechsol.2021.104406
56. Yang, L. (2023). Analytical investigation on the buckling of cylindrical shells under combined non-uniform axial compression and external pressure. *Proceedings of the Institution of Mechanical Engineers, Part C: Journal of Mechanical Engineering Science, 237*(24), 6009–6019, Sage. https://doi.org/10.1177/09544062231168727
57. Mukherjee, S., & Singha Roy, D. K. (2024). Wind-structure interaction mechanism of circular storage shells-a critical review. *Mechanics of Advanced Materials and Structures, 31*(30), 13284–13307, Taylor & Francis. https://doi.org/10.1080/15376494.2024.2337852
58. Mukherjee, S., Mondal, U., & Singha Roy, D. K. (2024). Proposition of economic ring-stiffener cross-section to meet the requirements of wind-buckling stability of open-top cylindrical tank. *Asian Journal of Civil Engineering, 25*, 5089–5107, Springer. https://doi.org/10.1007/s42107-024-01100-1
59. Yasunaga, J., & Uematsu, Y. (2020). Dynamic buckling of cylindrical storage tanks under fluctuating wind loading. *Thin-Walled Structures, 150*, 106677, Elsevier. https://doi.org/10.1016/j.tws.2020.106677
60. Song, Z. J., Xu, T. L., Ren, X. H., Xu, R. J., & Li, X. B. (2024). Load-end shortening curve of stiffener elements with beam-column buckling considering steel grades based on empirical formulae of critical stress. *Ocean Engineering, 306*, 118115, Elsevier. https://doi.org/10.1016/j.oceaneng.2024.118115
61. Lv, H., Zhang, H., Zhu, Z. Y., Dong, S. L., & Xie, X. (2024). Structural morphologyand dynamic characteristics analysis of drum-shaped honeycomb-type iii cable dome with quad-strut layout. *Advanced Steel Construction—an International Journal, 20*(01), 81–92, The Hong Kong Institute of Steel Construction. https://doi.org/10.18057/IJASC.2024.20.1.9
62. Indian Standard Code IS 803-1976 (Reaffirmed 2017). Indian standard code of practice for design, fabrication and erection of vertical mild steel cylindrical welded oil storage tanks. Bureau of Indian Standards, New Delhi, India.
63. Chen, S., Deng, Y., & Liang, X. (2024). Finite element analysis of birdcage buckling of umbilical cable with damaged sheath. *Ocean Engineering, 304*, 117777, Elsevier. https://doi.org/10.1016/j.oceaneng.2024.117777
64. Zeng, J., Yang, Y., Ma, H., Yang, Y., & Fan, C.:(2023). Dynamic modelling strategy of a shaft-disk-blade coupling system integrating beam and shell theories. *International Journal of Mechanics and Materials in Design, 20*(01), 107–127, Taylor & Francis. https://doi.org/10.1007/s10999-023-09664-7
65. Gerosa, F. M., & Marsden, A. L. (2024). A mechanically consistent unified formulation for fluid-porous-structure-contact interaction. *Computer Methods in Applied Mechanics and Engineering, 425*, 116942, Elsevier. https://doi.org/10.1016/j.cma.2024.116942
66. Moherdaui, T. F., Neto, A. G., & Wriggers, P. (2024). A second-order penalty-based node-to-segment contact using the virtual element method. *Finite Elements in Analysis and Design, 237*, 104183, Elsevier. https://doi.org/10.1016/j.finel.2024.104183
67. Sun, B., Ma, D., Gao, L., He, M., Peng, Z., Li, X., & Wang, W. (2022). Wind buckling analysis of a large-scale open-topped steel tank with harmonic settlement-induced imperfection. *Buildings, 12*(11), 1973, MDPI. https://doi.org/10.3390/buildings12111973

Seismic Performance Evaluation of Different Structural Shapes Using MCDM-Based Analysis

Dabanshanskhem Mawblei, Borun Nandeibam, Sunepsongla Ozukum, Bhargavi Nanda Phukan, Mrinal Kanti Sen, and Supran Chakravarty

Abstract Understanding how building shapes influence their response to earthquakes is key for structural engineers aiming to design safer, more resilient structures. This study explores the seismic behavior of buildings with different plan configurations—regular and irregular—under earthquake-induced lateral forces. Using SAP2000 software, three models were analyzed: a rectangular shape, a C-shape, and an I-shape. The study focused on how story displacement changes along the x- and y-axes as building height increases. The analysis followed the Linear Static Analysis method (also known as Equivalent Static Analysis) in compliance with Indian Standard codes, which consider various loading conditions. To compare the seismic performance of the three models, the Best Worst Method was used to assign weightages to each, culminating in a combined response curve that reflects their relative behavior under seismic forces. The results showed that buildings with irregular shapes, like the C-shaped and I-shaped models, experience significantly larger story displacements compared to the rectangular model. This highlights their greater vulnerability to earthquakes. The findings emphasize that the geometry of a building's plan plays a crucial role in its seismic performance. By shedding light on the impact of irregularities in building plans, this study underscores the importance of optimizing configurations to improve earthquake resistance. These insights provide structural engineers with valuable guidance for designing safer buildings that perform better during seismic events.

Keywords Seismic response · Regular and irregular buildings · Story displacement · Best worst method · Earthquake-resistant design

---

D. Mawblei · B. Nandeibam · S. Ozukum · B. N. Phukan · M. K. Sen (✉) · S. Chakravarty
Department of Civil Engineering, Assam Don Bosco University, Azara, Guwahati, India
e-mail: mksen88@gmail.com; mrinal.sen@dbuniversity.ac.in

P. Danish et al. (eds.), *Innovative Technologies for Smart and Sustainable Infrastructure*,
https://doi.org/10.1007/978-3-031-92801-7_4

1 Introduction

For millions of years, earthquakes have occurred due to tectonic shifts within the Earth's crust, often causing significant destruction to infrastructure and buildings [5]. These seismic events pose a serious threat to human life and property, making it essential for civil engineers to thoroughly understand their impact on structures. To mitigate seismic forces, engineers utilize various structural elements. Shear walls, typically constructed from reinforced concrete or masonry, provide vertical rigidity and help transfer lateral forces safely to the foundation [3]. Bracing systems, incorporating diagonal steel members, offer additional reinforcement and enhance overall stability [10]. Another effective approach involves dampers, which absorb and dissipate seismic energy, thereby reducing vibrations within a structure [9]. Different types of dampers, including viscous, friction, and tuned mass dampers, contribute to minimizing dynamic responses, improving both safety and comfort. Meanwhile, base isolation techniques involve placing flexible components, such as rubber bearings or sliding plates, between a building and its foundation, allowing independent movement and reducing seismic impact [2].

The results suggest that pseudo-elastic behavior is mainly influenced by structural strength rather than ductility. Although bracing and outrigger systems effectively reduce lateral drift and second-order effects, they also induce higher floor accelerations, which can be a safety concern. Furthermore, the study reveals that response spectrum analysis (RSA) tends to yield conservative estimates when assessing outrigger systems, potentially compromising structural safety as building height increases. Considering seismic forces is imperative when designing high-rise buildings, whether symmetrical or asymmetrical, especially in earthquake-prone regions. Architectural designs often introduce irregularities in both plan and elevation, making it challenging to ensure stability in unconventional structures. Uneven distribution of mass, stiffness, and strength increases a structure's susceptibility to collapse when exposed to unexpected dynamic forces.

This study conducts a comparative analysis of 3D asymmetrical frames along both single and multiple directions while maintaining a consistent floor area for all models. Each structural frame is designed with equal floor space and a uniform number of stories. Seismic analysis is performed using the Response Spectrum Method, with STAAD.Pro employed to evaluate building responses. Key structural parameters, including horizontal displacement, torsion, and drift, are analyzed across different models. The findings suggest that symmetrical structures generally exhibit better seismic performance compared to asymmetrical designs.

1.1 Literature Review

Farhan and Bommisetty [4] conducted research on the seismic response of buildings with both regular and irregular plan configurations. Both linear static and dynamic analyses were performed to evaluate how regular and irregular buildings behave under seismic loading. A comparative assessment of different heights and zones revealed that irregular structures experience greater displacement than regular ones. Haque et al. [6] investigated the effects of static and dynamic forces on various RCC building frames with different geometries, including W-shaped, L-shaped, rectangular, and square designs. The study compared the maximum displacement responses of different structural shapes under static and dynamic loads. Findings demonstrated that irregular buildings exhibited larger displacements than regular structures, indicating their greater susceptibility to earthquake-induced forces.

Kabir et al. [7] explored how the shape of multi-story buildings, maintaining identical weight, influences their response to seismic and wind loads. The study concluded that if overall mass remains relatively constant, structural responses are comparable. However, L-shaped and C-shaped buildings were found to be more vulnerable to seismic and wind forces than regular-shaped structures. Ankon [1] examined seismic behavior in five 15-story buildings, where two models had regular shapes (square and rectangular) and three had irregular configurations. Results indicated that irregular buildings exhibited longer periods, increased drift, and greater displacement, making them more prone to earthquake-related damage.

Syriac [13] focused on the seismic performance of G+11 buildings with regular and vertically irregular configurations. Using Staad Pro for modeling, ten different structures were analyzed under seismic loads. Findings revealed that lateral displacement was higher in irregular buildings compared to their regular counterparts. The study also highlighted that among all tested models, U-shaped buildings displayed the highest displacement, while stepped-regular structures showed the least. Overall, results confirmed that regular buildings generally performed better than irregular ones under seismic conditions.

Previous studies have extensively analyzed the seismic behavior of regular and irregular structures using software such as ETABS and Staad Pro. These studies focused on various structural configurations across different seismic zones, including rectangular, L-shaped, C-shaped, and W-shaped buildings. However, most of the research has concentrated on seismic variation in different shapes but none of these studies evaluate the weightage of various shapes. Additionally, while previous studies have compared displacement and drift in different irregular configurations, there has been little focus on I-shaped structures. This research aims to address these gaps by modeling and analyzing G+5 regular and irregular structures, including rectangular, C-shaped, and I-shaped buildings, in SAP-2000 and evaluating their seismic response under linear static analysis and also merging the curves of all the models. The objectives of this research are as follows: (i) To model G+5 regular and irregular structures in SAP-2000, including rectangular, C-shaped, and I-shaped configurations, (ii) To compare storey displacement patterns among different structural configurations, and (iii) To evaluate weightage of each model by using multi criteria decision model making (MCDM) tool. This study will provide insights into how different structural forms respond to seismic forces in low-rise buildings, contributing to a more comprehensive understanding of earthquake-resistant design.

2 Methodology and Structural Modelling

The framework of the study is illustrated in Fig. 1. Initially, various works related to seismic hazards were reviewed. Following this, three models were developed: M1—a rectangular section, M2—a C-shaped section, and M3—an I-shaped section. Subsequently, seismic analysis was conducted using Linear Static analysis, during which the storey displacement of all the developed models was calculated along the X and Y axes. Afterward, the storey displacement curves were generated concerning storey height for all the models. Six curves were developed, representing displacement with respect to height for the rectangular, C-shaped, and I-shaped models in both the X and Y axes. These curves were then combined into two graphs, each containing three curves (one for each shape). Next, an MCDM tool was employed to determine the weightage of each model. Based on the assigned weightage, the curves were merged. A combined curve for the X and Y axes was then developed. The BWM tool was utilized to determine the weightage in this process.

2.1 Structural Modeling

The details of size, reinforcement and materials of various structural elements for all developed models are shown in annexure I. The table provides a detailed description

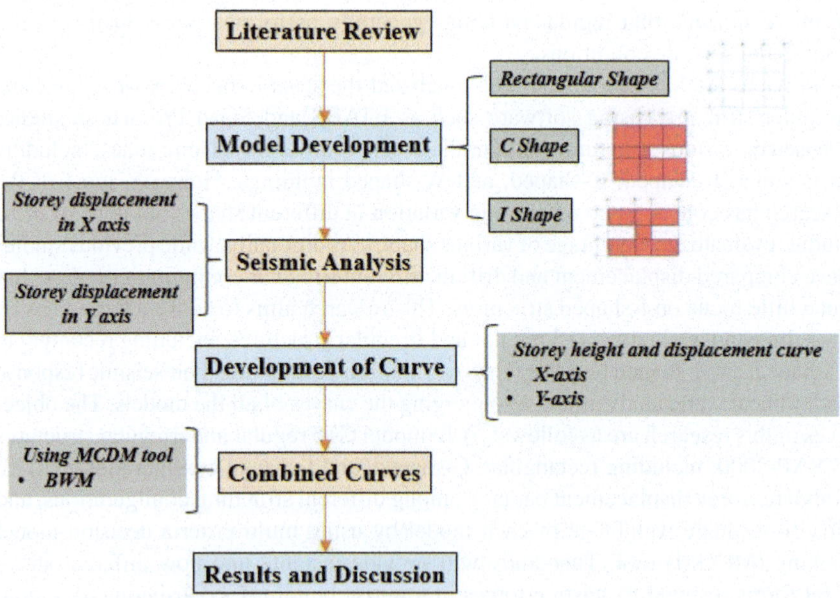

Fig. 1 Framework of the study

of various structural elements considered in the study. The structure has a plan size of 10 m × 10 m with a floor height of 3 m and comprises G+5 storeys. The beam dimensions are 300 mm × 600 mm, while the column size is 400 mm × 400 mm, ensuring adequate load-bearing capacity. The slab thickness is 125 mm, and M30-grade concrete is used, which offers good strength and durability.

Regarding material properties, the unit weight of brick is 19.2 kN/m^3, and the unit weight of concrete is 25 kN/m^3. The structure is located in Seismic Zone V, which is the most seismically active zone in India, with a seismic zone factor of 0.36. A response reduction factor of 5 and an importance factor of 1 are considered in the seismic design. The damping ratio is 5%, and the soil type is classified as Type II, indicating medium-strength soil conditions.

The building frame is an Ordinary Moment Resisting Frame (OMRF), which provides moderate resistance to lateral forces. The live load on floors is taken as 2 kN/m^2, while the roof live load is 1.5 kN/m^2. Additional loads include a floor finish load of 1 kN/m^2 and a partition load of 1 kN/m^2. The reinforcement details for columns specify HYSD415 and Mild250 steel. The column reinforcement configuration is rectangular, with lateral reinforcement provided in the form of ties, spaced at 25 mm intervals. The clear cover for longitudinal reinforcement in columns is 40 mm, and each column has 3 bars in both the X and Y directions. Similarly, the beam reinforcement consists of HYSD415 and Mild250 steel, with a clear cover of 75 mm to protect against corrosion and fire damage. These specifications ensure structural integrity, safety, and durability under applied loads and seismic forces. After defining all the specifications of all structural elements, the models (rectangular, C, and I shapes) are prepared by using SAP2000 [11]. The developed models are shown in Figs. 2, 3 and 4.

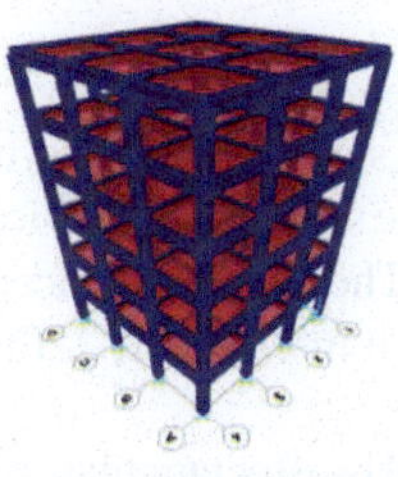

(a) 3D view of rectangular shape

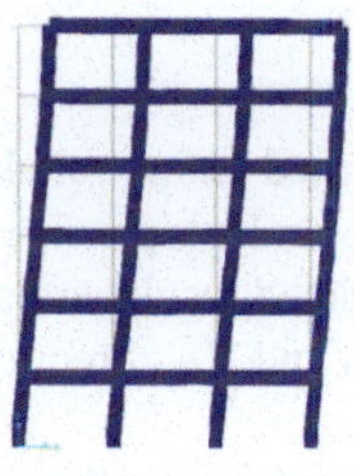

(b) Displacement in the X axis

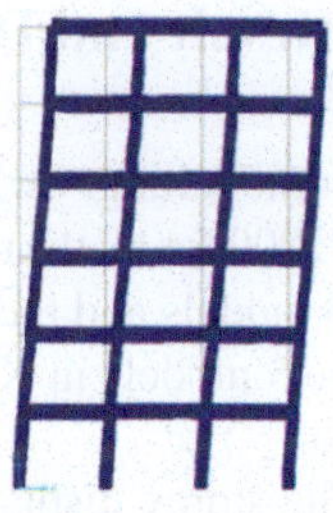

(c) Displacement in the Y axis

Fig. 2 Modelling of rectangular shape

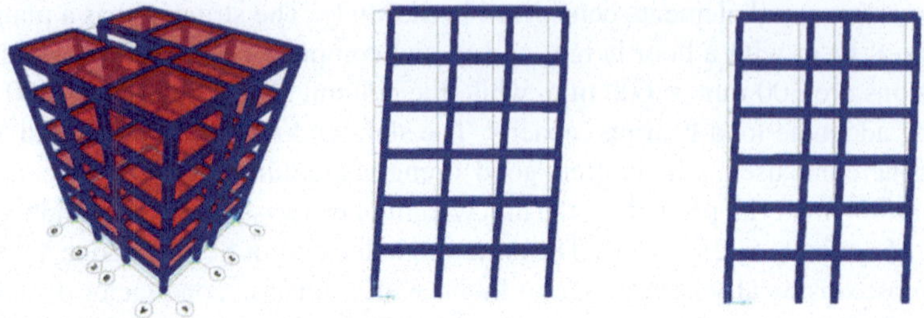

(a) 3D view of C-shape (b) Displacement in the X axis (c) Displacement in the Y axis

Fig. 3 Modelling of C shape

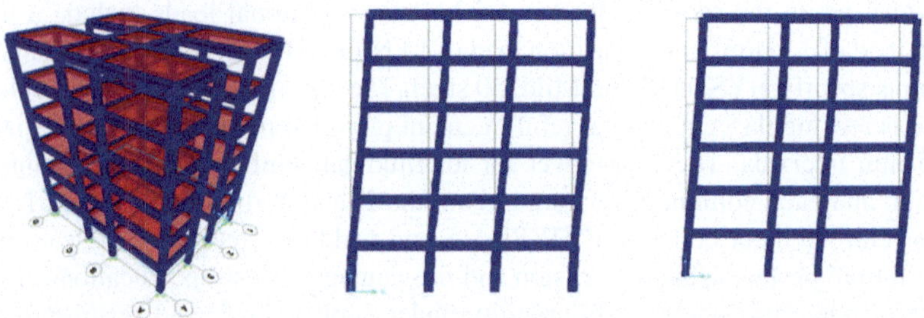

(a) 3D view of I shape (b) Displacement in the X axis (c) Displacement in the Y axis

Fig. 4 Modelling of I shape

3 Result and Discussion

After modelling, we have performed seismic analysis (linear static analysis) by using SAP2000 for all the models. The analysis provides us with the structural behaviour of the 3 models and then the calculated results are compared. The storey displacement of the 3 models in X—direction and Y—direction are shown in Tables 1, 2, Figs. 5 and 6.

The storey displacement values for the three models (M1—Rectangular, M2—C-Shaped, and M3—I-Shaped) exhibit an increasing trend with height, which is expected due to the cumulative effect of lateral seismic forces. However, differences in displacement values among the models indicate the influence of sectional geometry on structural behavior. At the lowest storey height (3 m), all three models show minimal displacement, with M2 having the highest displacement, followed by M3 and M1. As the height increases, the differences in displacement values become more noticeable. For instance, at 18 m, M3 exhibits the highest displacement, while M2 has the least displacement among the three.

Table 1 Displacement of the 3 models in X direction

Storey height (in meters)	Storey displacement in mm		
	M1	M2	M3
3	0.000816	0.000882	0.000835
6	0.00193	0.00199	0.00198
9	0.00303	0.00307	0.00311
12	0.00404	0.00404	0.00413
15	0.00485	0.00481	0.00494
18	0.00536	0.00528	0.00544

Table 2 Displacement of the 3 models in the Y direction

Storey height (in meters)	Storey displacement in mm		
	M1	M2	M3
3	0.000817	0.000887	0.000887
6	0.00214	0.00273	0.00233
9	0.00344	0.00456	0.00375
12	0.00463	0.00625	0.00505
15	0.00558	0.00762	0.00609
18	0.00617	0.00851	0.00673

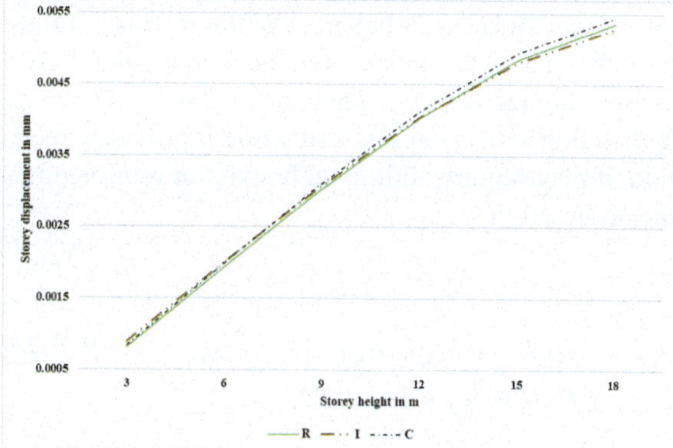

Fig. 5 Displacement of the 3 models in X direction

These variations can be attributed to the torsional stiffness and moment of inertia of each section. The rectangular section (M1) has a relatively uniform stiffness distribution, making it more stable. The C-shaped section (M2), being asymmetric, may experience higher torsional effects, but its shape also provides some resistance against lateral forces. The I-shaped section (M3), which has a higher moment of inertia along one axis but is relatively flexible on the other, shows slightly higher displacement at greater heights.

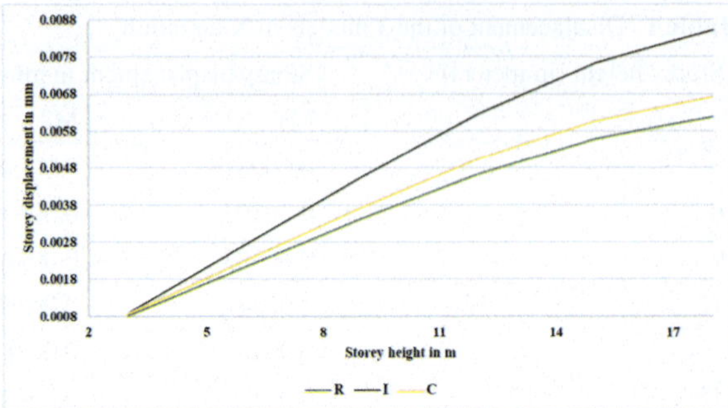

Fig. 6 Displacement of the 3 models in the Y direction

The storey displacement values in both the X and Y directions exhibit an increasing trend with height, which is expected due to the accumulation of lateral forces acting on the structure. However, the displacement behavior differs between the two directions and across the three models, highlighting the influence of sectional geometry on lateral stability. In the Y-direction, the displacement values are significantly higher, especially for M2. At 18 m, M2 shows the highest displacement, while M1 has the least, and M3 falls in between. This suggests that the C-shaped model experiences greater flexibility in the Y-direction, possibly due to a lower moment of inertia along this axis, making it more susceptible to lateral drift. M1 (Rectangular) displays consistent performance in both directions, with the least displacement in the Y-direction, making it a stable choice for resisting lateral forces. M2 shows the least displacement in the X-direction but the highest in the Y-direction, indicating directional stiffness imbalance. While it resists forces well in the X-direction, it is more flexible in the Y-direction, making it potentially vulnerable to lateral drift. M3 exhibits higher displacement in the X-direction but remains relatively balanced in the Y-direction. This suggests that while it provides good lateral stiffness in one direction, it may require additional bracing or reinforcement to improve performance in the X-direction.

3.1 Model Weightage Assessment and Integration of Developed Curves

After determining the storey displacement for each model concerning storey height in both the X and Y axes, we assessed the weightage of each model based on the displacement values. The Best–Worst Method (BWM), a multi-criteria decision-making (MCDM) approach developed by Jafar Rezaei in 2015 [8, 12], was employed for this evaluation. In this study, M1 was selected as the best criterion for both the X and Y axes, while M2 and M3 were considered the worst criteria for their respective axes. Using the BWM tool, the weightage of each model was determined, with the evaluated values presented in Table 3.

Table 3 Evaluated weightage of all models

Models	Weightage	
	X-axis	Y-axis
M1	0.7604	0.6706
M2	0.1563	0.0588
M3	0.0833	0.2706

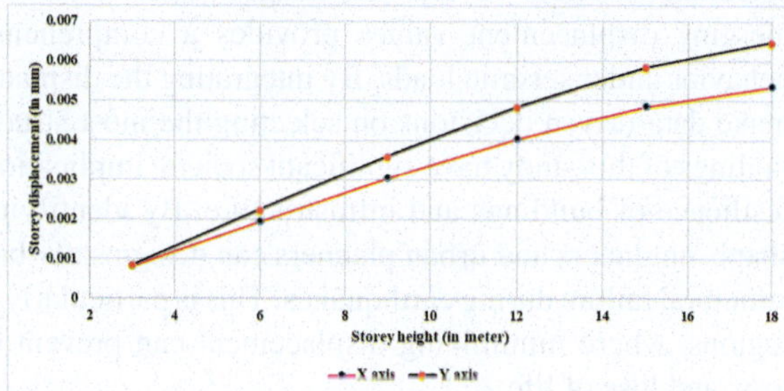

Fig. 7 Combine curve of storey displacement against storey height in X and Y axis

Based on the evaluated weightage using the BWM, Model M1 exhibits the highest weightage in both the X and Y directions, with values of 0.7604 and 0.6706, respectively. This indicates that M1 demonstrates superior structural stability and lower displacement, making it the most favorable option among the three models. Model M2 has a significantly lower weightage, particularly in the Y-axis, suggesting that it experiences higher displacement and lower stiffness in that direction. In the X-axis, its weightage is still considerably lower than M1, reinforcing its relatively weaker performance. Model M3 shows moderate performance, with a weightage of 0.0833 in the X-direction and 0.2706 in the Y-direction. While it performs better than M2 in the Y-axis, it remains less stable than M1 overall.

After determining the weightage, each model's weightage was multiplied with its respective output to calculate the combined effect of storey displacement against storey height. A combined storey displacement curve was then developed for the X and Y axes for all models as shown in Fig. 7.

4 Conclusions

This study aimed to evaluate the seismic performance of three structural models—M1 (Rectangular), M2 (C-Shaped), and M3 (I-Shaped)—by analyzing their storey displacement under seismic loads. Using Linear Static Analysis, the displacement of each model was determined along the X and Y axes at different storey heights. The displacement trends indicated that M1 (Rectangular) exhibited the least deformation, making it the most stable structure. In contrast, M2 (C-Shaped) and M3 (I-Shaped) experienced higher displacements, suggesting greater flexibility and lower lateral stiffness. To quantitatively assess the structural efficiency, the Best–Worst Method (BWM) was employed as a Multi-Criteria Decision-Making (MCDM) tool. The weightage calculations confirmed that M1 was the most effective model, with the highest weightage values (0.7604 in the X-axis and 0.6706 in the Y-axis). The combined storey displacement curves further validated these findings, demonstrating that the rectangular model performed best in resisting lateral forces.

The process of merging the storey displacement values using weightage-based calculations offers several advantages. Instead of evaluating models separately, merging displacement values provides a comprehensive overview of structural behavior under seismic loads. By integrating the displacement effects, engineers can make data-driven decisions on selecting the most structurally efficient design. The findings of this study have significant societal implications for improving the seismic resilience of buildings and infrastructure. By identifying the most stable structural shape, engineers and urban planners can design safer buildings, reducing the risk of structural failure during earthquakes. This is particularly crucial for earthquake-prone regions, where minimizing displacement can prevent structural damage, financial loss, and loss of life.

Acknowledgements All the authors acknowledge Assam Don Bosco University for providing the laboratory facilities.

Competing Interests The authors have no conflicts of interest to declare that are relevant to the content of this chapter.

Appendix 1: Details of Various Structural Elements

SI no	Parameters	Description
1	Plan size	10 m × 10 m
2	Floor height	3 m
3	Number of storeys	G + 5
4	Beam size	300 mm × 600 mm
5	Column size	400 mm × 400 mm
6	Slab thickness	125 mm
7	Grade of concrete	M30
8	Unit weight of brick	19.2 kN/m^3
9	Unit weight of concrete	25 kN/m^3
10	Seismic zone	V
11	Seismic zone factor	0.36
12	Response reduction factor	5
13	Importance factor	1

(continued)

(continued)

SI no	Parameters	Description
14	Damping ratio	5%
15	Soil type	II
16	Type of building frame	Ordinary moment resistance frame
17	Live load	2 kN/m^2
18	Roof live load	1.5 kN/m^2
19	Floor finish	1 kN/m^2
20	Partition load	1 kN/m^2
21	Rebar material (column)	HYSD415 and Mild250
22	Column reinforcement configuration	Rectangular
23	Lateral reinforcement (column)	Ties
24	Lateral reinforcement spacing (column)	25 mm
25	Clear cover longitudinal (column)	40 mm
26	Number of bars in X-direction (column)	3
27	Number of bars in Y-direction (column)	3
28	Rebar material (beam)	HYSD415 and Mild250
29	Clear cover (beam)	75 mm

References

1. Ankon, S. B. (2020). A comparative study of behavior of multi-storied regular and irregular buildings under static and dynamic loading. *Open Journal of Civil Engineering, 10*(04), 337.
2. Casciati, F., Magonette, G., & Marazzi, F. (2006). *Technology of semiactive devices and applications in vibration mitigation*. John Wiley & Sons.
3. El-Dakhakhni, W., & Ashour, A. (2017). Seismic response of reinforced-concrete masonry shear-wall components and systems: State of the art. *Journal of Structural Engineering, 143*(9), 03117001.
4. Farhan, M. A., & Bommisetty, J. (2019). Seismic analysis of multistoried RCC buildings regular and irregular in plan. *International Journal of Engineering Research & Technology (IJERT), 8*(11).
5. Ghose, M. K. (2011). Disaster, risk and vulnerability due to earthquakes and designing of seismic resistant structure for mitigation. *Applied Disaster Research, 61*.
6. Haque, M., Ray, S., Chakraborty, A., Elias, M., & Alam, I. (2016). Seismic performance analysis of RCC multi-storied buildings with plan irregularity. *American Journal of Civil Engineering, 4*(3), 68–73.
7. Kabir, R., Sen, D., & Islam, M. (2015). Response of multi-storey regular and irregular buildings of identical weight under static and dynamic loading in context of Bangladesh. *International Journal of Civil & Structural Engineering, 5*(3), 252–260.
8. Kheybari, S., Rezaie, F. M., & Rezaei, J. (2019). Measuring the importance of decision-making criteria in biofuel production technology selection. *IEEE Transactions on Engineering Management, 68*(2), 483–497.
9. Lu, Z., Wang, Z., Zhou, Y., & Lu, X. (2018). Nonlinear dissipative devices in structural vibration control: A review. *Journal of Sound and Vibration, 423*, 18–49.

10. Oh, S., Kurama, Y. C., Mohle, J., Polster, L., Manning, M., & Weldon, B. D. (2023). A novel reinforced-concrete buckling-restrained brace for precast concrete lateral-load-resisting frames. *PCI Journal, 68*(3), 95–116.
11. Poluraju, P., & Rao, N. (2011). Pushover analysis of reinforced concrete frame structure using SAP 2000. *International Journal of Earth Sciences and Engineering, 4*(6), 684–690.
12. Rezaei, J. (2015). Best-worst multi-criteria decision-making method. *Omega, 53*, 49–57.
13. Syriac, T. (2021). Comparative study on the behaviour of a multi-storeyed building for regular & irregular plan configuration. *International Journal of Innovative Research in Technology.*

Smart Infrastructure

Wireless Fidelity (Wi-Fi) Attenuation Value in Identifying the Chloride Concentration in Concrete for Sustainable Health Monitoring

Md. Fahim Al Mamun, Suvash Chandra Paul, Noor Md. Sadiqul Hasan, and M. Shariful Islam

Abstract Chloride-induced corrosion of reinforcing steel in concrete structures poses a significant threat, undermining the durability of the structure and leading to deterioration and costly repairs. Traditional methods of detecting chlorides are highly labor-intensive and time-consuming. This paper explores a non-destructive testing (NDT) technique that employs Wi-Fi signals to detect chloride ion concentrations in concrete for sustainable health monitoring. Samples with chloride concentrations of 0%, 5%, 10%, and 15% were prepared using both direct and indirect application methods for three different concrete strength grades. The specimens were placed inside Faraday's shield, where the signal attenuation of Wi-Fi was measured. The analysis revealed a direct correlation between signal attenuation, chloride content, and concrete strength. This demonstrates the efficacy of the NDT method for early chloride detection. The technique is straightforward and efficient for monitoring the health of structural components, ultimately enhancing the performance and durability of reinforced concrete structures.

Keywords Wireless fidelity (Wi-Fi) · Attenuation · Chloride solution · Non-destructive test (NDT) · Machine learning technique · Concrete grades · Health monitoring

Md. F. A. Mamun · S. C. Paul (✉) · N. Md. S. Hasan
Department of Civil Engineering, International University of Business Agriculture and Technology, Dhaka, Bangladesh
e-mail: suvashpl@iubat.edu

Md. F. A. Mamun
e-mail: fahimalmamun.ro@iubat.edu

N. Md. S. Hasan
e-mail: nmshasan@iubat.edu

M. S. Islam (✉)
Department of Engineering, South Carolina State University, Orangeburg, SC, USA
e-mail: mislam1@scsu.edu

P. Danish et al. (eds.), *Innovative Technologies for Smart and Sustainable Infrastructure*,
https://doi.org/10.1007/978-3-031-92801-7_5

1 Introduction

The sustainable health monitoring is essential for the integrity of reinforced concrete (RC) structures as concrete materials may degrade over time due to various natural hazards. One such RC degradation is chloride-induced corrosion of steel reinforcement caused by breaking the protected layer of steel in concrete [1]. Typically, carbonation induced corrosion of reinforcing steel is a slow process, while the chloride induced corrosion is relatively faster than that of the carbonation [2]. Chloride-induced reinforcement corrosion has been claimed to cause an alarming worldwide durability problem that deteriorates RC structures [3]. Infrastructures such as buildings, bridges, dams, tunnels, and port constructions are usually affected by the reinforcing steel corrosion, particularly, in tidal zones as well as coastal locations [4].

The longevity of coastal concrete structures is relatively lower than anticipated operation lifetime due to distinctive salts in seawater [5]. Lower concrete cover depth is prone to a faster corrosion in RC structures as the structure's longevity or service life is frequently determined by the concrete cover depth [6, 7]. Various properties of material and environmental factors play significant role in the gradual complex process of chloride penetration in the RC structures [8]. The chloride concentration threshold (typically, 0.15% of cement weight) at which active corrosion of steel reinforcement begins, is influenced by several variables such as cement type, cement's C_3A concentration, mix proportions, pH levels, temperature, relative humidity, steel surface conditions, and chloride penetration sources [9]. Typically, chloride ion penetration leads to the onset of corrosion for steel rebars by breaking the passive layer of the steel, which ultimately causes the rebars to corrode, crack, and spall of concrete [10, 11].

Generally, X-ray diffraction (XRD) or other chemical testing methods are utilized in detection of chloride ions in concrete, which require skills and time [12]. Researchers have used non-destructive test (NDT) methods to detect chloride in concrete using radio frequency, electric resistivity, and ultrasonic detection [13]. These methods assess the durability of RC structures, considering different concrete mixes. Some researchers used radio frequencies (ranges from 0.1 to 40.1 MHz) in detecting the permittivity, this indicates that corrosion can be studied by the ability of storing electric energy in an electric field in the reinforced concrete materials [14]. The study depicts that dielectric constant values can be detected by radio frequency for different concentration levels of NaCl in concrete samples. Similarly, ultrasound was used in another study for identifying the presence of chloride ions in cement paste samples [15]. It was observed that the ultrasonic pulse velocity (UPV) increases with NaCl concentration. The electrical resistivity test also examined concrete resistance against chloride penetration [16]. From the discussion above, it is evident that researchers used various advanced techniques for identifying the presence of chloride contents in concretes, while easily available low-cost NDT techniques, such as Wi-Fi signals have been explored in one of these author's previous study [11].

Nowadays, Wi-fi signals are being used at an increasing rate for a variety of tasks, including traffic monitoring [17], metal detection [18], breathing signal monitoring [19], and moisture and humidity estimation [20]. Wi-Fi signals have also been utilized to monitor health and safety in commercial buildings and civil engineering industries [21]. Similarly, machine learning technique (MLT) has been a widely accepted method in recent years that utilizes algorithms to solve various complicated civil engineering challenges, such as surface chloride concentration detection, rate of penetration detection, diffusion coefficient estimation, and for assessing chloride migration sensitivity for various durability concern of concrete materials [22]. Previous researchers predicted surface chloride concentration using MLT applying the same data source for concrete materials that were exposed to marine environments [23]. Machine learning and deep learning can also be a suitable technique in obtaining the optimum mixing ingredients, identifying the certain environmental conditions and exposure time along with the prediction of strength of concrete materials [23, 24].

To increase the serviceability and/or sustainable health monitoring of RC structures, current study adopted a rapid NDT setup that utilized a Wireless Fidelity (Wi-Fi) signal to detect chloride contents in concrete. Concrete samples with known chloride contents (0, 5, 10, 15%), three different thicknesses (25 mm, 50 mm, and 75 mm), and three different grades were produced for the experiment. The specimens were placed in a Faraday's shield, where a Wi-Fi signal was transmitted through one face and received from the opposite face of the samples by means of attenuation. The response of the samples was estimated based on the difference between two attenuation values by the analyzer software, while three different MLT were applied for predicting chloride percentage of concrete samples towards sustainable health monitoring.

2 Research Significance

The degradation of RC structures caused by the chloride-induced corrosion is a serious durability concern which can lead to significant economic challenges because of the large amount of maintenance cost. Timely detection of chlorides is crucial for mitigating deterioration and preventing structural failure. NDT methods are optimal for chloride detection without compromising structural integrity. Additionally, existing methods for sample collection, such as chemical testing, XRD, and XRF, etc. are costly and time-consuming. Therefore, the present study offers a simplified approach utilizing Wi-Fi signals to detect chloride ions content in RC structures to reduce issues associated with chloride-induced corrosion.

Table 1 The current NDT test set-up employed the following device with their specifications

Name of devices	Quantity	Device specification
Laptop	2	• Intel core i7 (10th gen) • 8 GB RAM (DDR4)
Wi-fi adapter	2	• 5dBi High gain antenna • Dual band 867 + 400 Mbps speed • Frequency 2.4 of 5.0 GHz
Software	1	• Microsoft Wi-Fi analyzer

3 Research Methodology

3.1 Device and Software Specification

A Faraday shield, two computers, and two adapters were utilized in the investigation to measure the amount of chloride ions in concrete samples. An Ethernet connection was made between a laptop and a router having a frequency of 2.4 GHz through the passage of 10 Mbps Wi-Fi signal. A 5 decibels (dBi) high gain antenna and adaptor were attached to one laptop to transfer the Wi-Fi frequency to another. Microsoft Wi-Fi analyzer software was used to transform data attenuation in dB from the web signal. Both laptops' built-in adapters were turned off so that data could only be received and sent via external adapters. During the data transfer process, to prevent any signal frequency loss, the Faraday shield was employed that was similar to one of these author's previous study [11]. The device specifications for the current NDT test set-up have been provided in Table 1.

3.2 Detection Procedure of Chloride and Preparation of Samples

In the current test set-up, the concrete samples were covered by a Faraday shield, thus the Wi-Fi signal transmitter can transmit the signals through the specimens without losing the data. The Wi-Fi analyzer software was used to convert the frequency attenuation into dB after obtaining the signals by the receiver. Three structural grades (M20, M30, and M40) contributed 216 samples in total as summarized in Table 2, where the NaCl was applied in two distinct groups. Group A was directly induced chloride (DIC) samples, which refers to incorporating NaCl into the concrete specimens during the casting process, similar to one of these author's previous studies [11]. Following the casting of initial 108 samples with DIC, these were subjected to a water tank for curing of up to 28 days. Each batch encompassed 36 specimens, differentiated by three distinct grades and four varying concentrations of NaCl, specifically 0%, 5%, 10%, and 15%, respectively. Conversely, in group B, 108 samples were cast without incorporating chloride. Three different grades and 36 specimens similarly categorized these specimens. Nonetheless, these samples underwent curing in water mixed with chloride (WMC) concentrations of 0%, 5%, 10%, and 15%, respectively. While one of these authors' previous study investigated the same technique with DIC samples [11], the present study observed the samples with both DIC and WMC in order to obtain wide range of acceptability for this NDT technique for sustainable health monitoring. In terms of the attenuation data, at least 10 data points were collected from every specimen to ensure in-depth coverage of almost the whole specimen surface area. The final data for each specimen was the mean value derived from these ten data points. The technique used to detect chloride in multiple batches and grades of concrete specimens is shown in Fig. 1a, and the concrete samples are displayed in Fig. 1b. A Faraday Shield, which can function as an enclosure and block electromagnetic fields, was applied to shield the specimens from any potential external electromagnetic fields during the testing phase. They were placed in the core of the space between two transmitters [25]. Additionally, it guarantees that Wi-Fi signals are not interfered with by either the transmitter or the receiver.

Table 2 Summary of physical specifications for specimens

Application of NaCl	Grade	Dimension (mm)	Number of samples
Group A: NaCl mixed in samples and Group B: samples cured in NaCl solution	M20	175 × 75 × 25	72
	M30	175 × 75 × 50	72
	M40	175 × 75 × 75	72

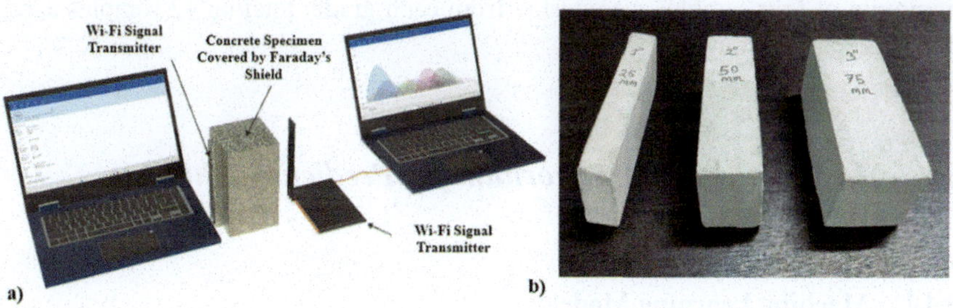

Fig. 1 Test set-up and specimens: **a** 3 Dimension view of the setup of detecting chloride, and **b** Concrete samples used in this study

Table 3 Concrete mixing compositions for different grades in (kg/m^3)

Grade	Cement	Sand	Coarse aggregates	Water
M20	393	663	1244	177
M30	665	561	1052	299
M40	1080	304	900	460

3.3 *Sample Preparation of Concrete for Compression and Chloride Test*

Three different grades of concrete were used to cast the specimens for a detailed experimental analysis; Table 3 lists the material compositions for each grade. CEM II 42.5N grade-cement was used having a specific gravity of 3.15 as the binder of the concrete matrix. Concrete was produced using locally collected fine aggregate, particularly river sand with a fineness modulus of 2.5. The concrete mixture contained coarse aggregates, which were designed as stone chips no larger than 19 mm. The water-to-cement (w/c) ratio was kept constant for every mix at 0.45. The typical composition of 100 g of NaCl is approximately 60.66 g of chlorine (Cl) and 39.34 g of sodium (Na) [26]. Fresh concrete from group A was poured into molds after chloride induction, and it was left to set for 24 h at room temperature in the laboratory before curing for 28 days in water at 21 ± 2 °C and 60 ± 5% relative humidity. In identical weather conditions, the freshly mixed concrete samples from group B that did not directly contain chloride were also poured into molds. To evaluate and/or verify the compressive strength for different grade of concrete (M20 = 20 MPa, M30 = 30 MPa, M40 = 40 MPa), cubic samples with a size of 75 mm were prepared for each grade without directly inclusion of NaCl or cured with NaCl mixed water. A minimum of four samples were tested from each grade, totaling 12 samples across three grades at 28 days.

3.4 *Machine Learning Algorithm for Predicting Chloride Percentage*

3.4.1 Machine Learning Models

Machine learning models are usually used in various fields due to their predictive accuracy and analytical performance. Similarly, the current investigation applied three independent machine learning algorithms: the Random Forest (RF), Extreme Gradient Boosting (XGB), and Convolutional Neural Networks (CNN) models.

Random Forest (RF) Model

The RF model, introduced by Breiman (2001) [27], is a powerful ensemble learning method based on bagging and random feature selection, widely used for regression and classification tasks. Its robustness against overfitting and ability to handle high-dimensional data has made it a popular choice in predictive modeling [28]. Unlike traditional decision trees, RF combines multiple tree predictions, averaging their outputs in regression problems, which improves generalization performance. For this study, 1030 data points from 108 samples were analyzed in each group. Concrete grades, volumes, and attenuations comprise the input data, while the chloride percentage depicts the output data. Similarly, there were separate input variables (X) and target variables (Y) of datasets as the numerical and categorical input data had grade, attenuation, as well as volume columns. The target variable was the anticipated percentage of chloride, while the dataset was divided into training and testing datasets: 80% was training and 20% was model assessment with a random arbitrary state of 44.

Extreme Gradient Boosting (XGB) Model

One of the most effective and scalable applications of gradient boosting algorithms is the XGB model, which was first presented by Chen and Guestrin [29]. It optimizes computational speed and model performance through regularization techniques, advanced tree pruning, and parallelization. The model's hyperparameters include objectives, learning_rate, max_depth, n_estimators, were essential for maximizing the predictive performance while avoiding overfitting. Techniques like early stopping and cross-validation are often employed to refine the model further [30]. The present study used XGB to predict the chloride percentage in materials based on input features such as volume, attenuation, and grade. With a dataset split into training (80%) and testing (20%), and hyperparameters including random_state = 42, n_estimators = 100, max_depth = 3, and learning_rate = 0.2, the model was tailored to balance learning efficiency and generalization.

Convolutional Neural Networks (CNN) Model

CNNs i.e., convolutional neural networks, were first developed for image recognition applications (LeCun et al. 1998) [31] and have emerged as a versatile tool for predictive modeling across diverse domains, including materials science and engineering. Recently, studies have demonstrated CNNs' ability to capture spatial dependencies and nonlinear patterns in corrosion analysis [32]. Using volume, attenuation, and grade as input features, a CNN model was obtained for this investigation to predict the chloride percentage. To ensure adequate evaluation, the collection of data was split into training (80%) and testing (20%) sections. Note that convolutional layers enhance the ability to learn feature interactions across input dimensions, while dropout size 0.2, epochs 64, and batch size 32 normalization layers mitigate overfitting and improve generalization.

Performance Analysis of the Model

The coefficient of determination (R^2), mean squared error (MSE), mean absolute error (MAE), and root-mean-squared error (RSME) were assessed by the proposed models among the metrics [33]. These statistical measurements were calculated using the separate mathematical equations from (1) to (4).

$$MSE = \frac{1}{n}\sum_{i=1}^{n}(y_i - \hat{y}_i)^2 \tag{1}$$

$$MAE = \frac{1}{n}\sum_{i=1}^{n}|yi - \hat{y}i| \tag{2}$$

$$RMSE = \sqrt{\frac{1}{n}\sum_{i=1}^{n}(y_i - \hat{y}_i)^2} \tag{3}$$

$$R^2 = 1 - \frac{\sum_{i=1}^{n}(y_i - \hat{y}_i)^2}{\sum_{i=1}^{n}(y_i - \bar{y})^2} \tag{4}$$

In this case, n represents the number of observations, $\hat{y}i$ represents the value predicted by the model, yi represents the actual value, and $\bar{y}$ represents the mean actual value.

4 Results and Discussion

The ultimate compressive strength at 28 days for three different types of concrete were approximately 18 MPa, 28 MPa, and 42 MPa for Grade M20, M30, and M40, correspondingly. This indicates the obtained compressive strength were similar to the target strength for three different grades of concrete. Typically, addition of NaCl into concrete can temporarily improve its strength due to acting as an accelerant at its initial stage of hydration, while the strength may decrease in later age along with durability concerns [34]. The authors in the current study also have a plan to conduct further study to better understand the influence of NaCl on the DIC and WMC samples in the mechanical and durability properties.

4.1 Attenuation Values of Concrete for Different Grades

Figure 2 reveals how the attenuation values increase with increasing NaCl concentrations for different grade's sample with varying thickness. The graph further exhibits difference in attenuation values between DIC and WMC samples. The DIC samples exhibit more attenuation value in all grades, thicknesses, and NaCl concentrations than that of the WMC. Generally, the variability in attenuation can be attributed to the methodology employed for chloride induction into the samples. Group A was subjected to direct induction, thereby ensuring precise NaCl concentration, in contrast to group B samples, which underwent curing through the mixing of NaCl with water. It is understandable that larger quantity of NaCl concentration may not permeate into the body of group B samples as the samples were cured in NaCl mixed water, likely the system will take more time to ingress into it. Overall, the thickest (75 mm) DIC sample (M40) had the highest attenuation value, whereas the slightest (25 mm) WMC sample had the lowest attenuation values for all grades of concrete. The thickest WMC samples (75 mm) even had lower attenuation values compared to the slightest DIC samples (25 mm) for all grades of concrete, while the attenuation values and thicknesses were observed to vary throughout all grades.

For DIC samples of grade M40 with a thickness of 25 mm and NaCl concentrations of 0%, 5%, 10%, and 15%, the attenuation values increased by approximately 24%, 18%, 27%, and 15%, respectively, compared to grade M20. This trend was similar for the concrete samples with a thickness of 50 mm, and overall, these results are consistent with one of these author's previous studies [11]. Similarly, for 75 mm thick samples, attenuation values increased by around 61%, 40%, 33%, and 34% at corresponding NaCl concentrations for grade M40 with respect to M20 grade of concrete. For WMC samples of grade M40 with a thickness of 75 mm and NaCl concentrations of 0%, 5%, 10%, and 15%, the attenuation values enhanced by about 30%, 8%, 5%, and 4%, respectively, in comparison with grade M20. Based on the results, is seen that the attenuation values increased with the increasing thickness of concrete, which was expected as the lower number of signals passed through the thick concrete media.

Figure 2 further demonstrates the attenuation values increased with the increasing content of chloride irrespective of application techniques e.g., DIC or WMC. This indicates, the current NDT technique is fully capable of measuring the presence and/or level of chloride ions in the concrete for sustainable health monitoring of RC infrastructures. The chemical composition of the concrete mix and the frequency of application are two other variables that may influence the attenuation levels. In some other studies, microwave techniques are employed to detect chloride. According to a report, the dielectric characteristics and microwave reflections of concrete, including its transmission and reflection parameters, constants of dielectric, and loss factors, helped identifying the presence of chloride in concrete [35]. Generally, the dielectric constants and loss factors associated with concrete exhibit an upward trend in

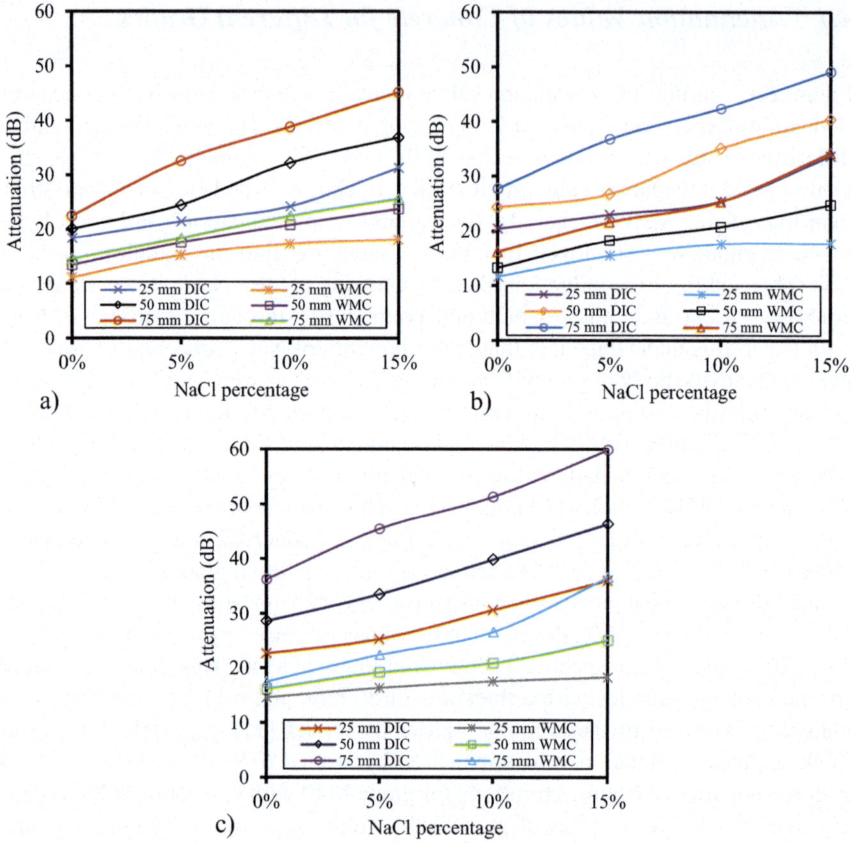

Fig. 2 Attenuation versus NaCl concentration in different grades of concrete: **a** M20, **b** M30, **c** M40

response to increasing chloride concentrations. Microwave measurements demonstrate sensitivity to chloride levels; however, they are additionally affected by variations in temperature and moisture content presented in the concrete matrix. These variables, as illustrated in Fig. 3, which depicts the relationship between attenuation and NaCl concentrations across various grades of concrete samples, must be meticulously accounted for during the on-site evaluation of chloride content [36]. The proportional relationship between the molarity of NaCl and signal-to-noise ratio (SNR) was demonstrated by the nuclear magnetic resonance (NMR) analysis by the previous studies [37]. In that study, the rate of SNR in the model identically grows as the proportion of NaCl increases [37], which is also consistent with the current study as the attenuation value increases with the increasing percentage of NaCl. Typically, chloride content exceeding 0.15% by weight of cement is considered the threshold level per ACI 318 [39]. Therefore, likely a further study is required to obtain a suitable correlation between Wi-Fi attenuation and chloride threshold levels to better estimate the remaining service life of the RC structures for sustainable health monitoring.

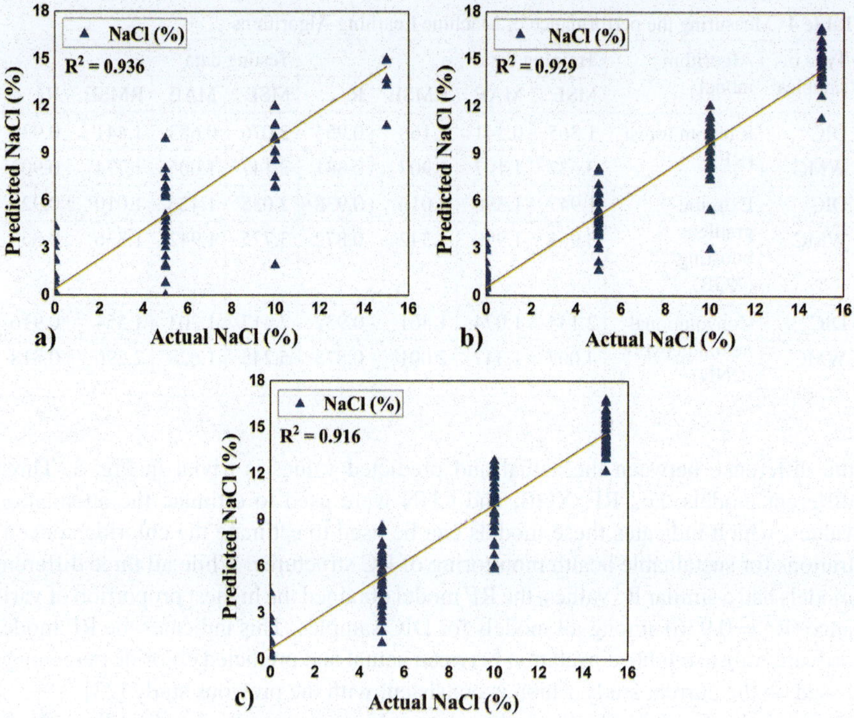

Fig. 3 Relationship between actual and predicted NaCl percentage of direct induced chloride (DIC) samples with different prediction model: **a** RF, **b** XGB, and **c** CNN

4.2 *Prediction of Chloride Using Machine Learning Algorithms*

The efficacy parameters of the proposed strategies are shown in Table 4. The R^2 was calculated for each of the two samples i.e., the training and testing datasets. According to the training data, the R^2 values of DIC for the three distinct algorithms (RF, XGB, and CNN models) were 0.955, 0.938, and 0.932. Similarly, the R^2 values of WMC samples were 0.881, 0.872, and 0.873, consequently. Conversely, the testing data's R^2 DIC values are 0.936, 0.929, and 0.916, to be precise. The model indicates that the CNN, XGB, and RF models have WMC R^2 values of 0.818, 0.868, and 0.903, correspondingly. MSE, RMSE, and MAE are also shown for testing and training data.

Figure 3 illustrates the relationship between actual and estimated NaCl percentages for DIC. For this portion of the testing dataset, the diagram makes it easier to visually evaluate how closely the model's predictions match the actual value. The actual percentage results are extracted directly from the measurement data; the estimated percentage value produced by a prediction model for a similar dataset. These results represent the model's prediction of the target variable. It is noteworthy that

Table 4 Measuring the performance of Machine Learning Algorithms

Type of samples	Algorithm model	Training data				Testing data			
		MSE	MAE	RMSE	R^2	MSE	MAE	RMSE	R^2
DIC	Random forest (RF)	1.365	0.541	1.168	0.955	2.076	0.683	1.441	0.936
WMC		3.637	1.107	1.907	0.881	3.147	1.005	1.774	0.903
DIC	Extreme gradient boosting (XGB)	1.947	1.395	1.016	0.938	2.035	1.426	1.010	0.929
WMC		3.961	1.990	1.342	0.872	3.775	1.942	1.336	0.868
DIC	Convolutional neural network (CNN)	2.134	1.074	1.461	0.932	2.417	1.203	1.554	0.916
WMC		4.007	1.347	2.001	0.873	5.245	1.628	2.290	0.818

the difference between the actual and predicted values is trivial in Fig. 3. Three different models, i.e., RF, XGB, and CNN were used to estimate the attenuation values, which indicates these models can be used to estimate the chloride concentrations for sustainable health monitoring of RC structures. While all three different models had a similar R^2 values, the RF model obtained the highest proportion of variance ($R^2 = 0.936$) among all models for DIC samples. This indicates the RF model provides most suitable correlation between actual and predicted chloride percentage based on the current study which is consistent with the previous study [23].

Similarly, Fig. 4 demonstrates the relationship between actual and predicted NaCl percentage for WMC concrete samples. Here, the prediction data were evaluated similar way to the DIC samples. Likely, further research is required to improve the model as shown in Figs. 3 and 4 by adding other characteristics, including porosity, aggregates, and cement types. The visual analysis shows that the RF model with highest R^2 values (0.90) predicts the chloride percentage precisely for both WMC and DIC data. The RF model also displays the regression analysis more accurately than the other models; however, the different models are represented closer to the RF model.

To ensure the durability of buildings and structures in coastal locations, several researchers employed the RF model to forecast the chloride resistance in concrete [38]. Machine learning has been used in many studies to measure longevity and service life in the construction and healthcare sectors. However, the RF model was proven more accurate and precise in identifying surface chloride in marine concrete in coastal locations despite different models being utilized in separate investigations [23].

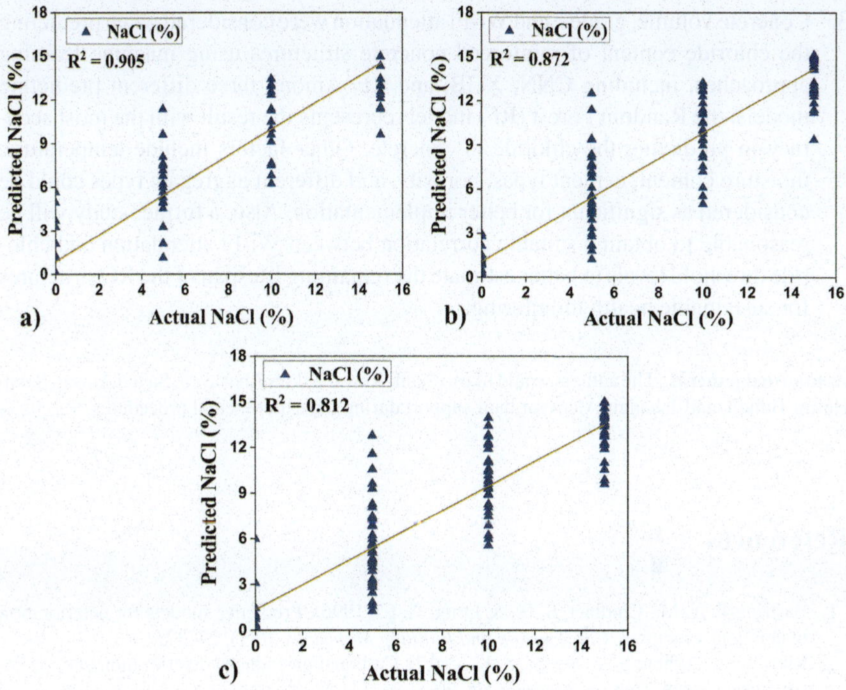

Fig. 4 Relationship between actual and predicted NaCl percentage of water-mixed chloride (WMC) samples with different prediction models: **a** RF, **b** XGB, and **c** CNN

5 Conclusions

The results presented in this research can stimulate the correlation between Wi-Fi attenuation and concrete's chloride content as Wi-Fi signals can be utilized as a novel way to identify chloride content in concrete. This study suggested an approach for monitoring and determining the degree of corrosion of RC structures based on the threshold level of chloride content. The following results are derived from the current investigation:

1. Concerning the detection of NaCl in concrete, the attenuation value can be measured using non-destructive techniques such as Wi-Fi signals. As the NaCl ratio in the concrete sample increases, the attenuation value increases accordingly. The thicker the structure and the higher the concrete grade, the greater the value of attenuation.
2. The attenuation value could differ depending on how the chloride particles penetrate the concrete structures. This also explains why two different approaches of direct-induced (DIC) and water-mixed chloride (WMC) are adopted in this study, and the innovative approach of Wi-Fi has demonstrated the ability to detect different attenuation values for different chloride induction methods.

3. Concrete volume, grades, and Wi-Fi attenuation were considered when predicting the chloride content of reinforced concrete structures using machine learning approaches, including CNN, XGB, and RF. Among three different predicting models, the Random Forest (RF) model represents the result with the most accuracy in predicting the chloride in concrete. Other factors include temperature, moisture content, cement types, porosity, and different aggregate types could be considered as significant for better implementation. Also, a further study will be reasonable to obtain a suitable correlation between Wi-Fi attenuation and chloride threshold levels to better estimate the remaining lifespan of the RC structures for sustainable health monitoring.

Acknowledgements The authors would like to thank Anwar Uddin Abusayed, Sayed Anwar, Sakib Hasan Tumon, and Jahangir Alam for their support during the experimental procedures.

References

1. Basheer, P. A. M., Chidiact, S. E., & Long, A. E. (1996). Predictive models for deterioration of concrete structures. *Construction and Building Materials, 10*(1), 27–37.
2. Rizvi, S. S., Akhtar, S., & Verma, S. K. (2017). Carbonation-induced deterioration of concrete structures. *Indian Concrete Journal, 65*, 70.
3. Alexander, M., & Beushausen, H. (2019). Durability, service life prediction, and modelling for reinforced concrete structures–review and critique. *Cement and Concrete Research, 122*, 17–29.
4. Melchers, R. E., & Li, C. Q. (2006). Phenomenological modeling of reinforcement corrosion in marine environments. *ACI Materials Journal, 103*(1), 25.
5. Val, D. V., & Stewart, M. G. (2003). Life-cycle cost analysis of reinforced concrete structures in marine environments. *Structural Safety, 25*(4), 343–362.
6. Costa, A., & Appleton, J. (2002). Case studies of concrete deterioration in a marine environment in Portugal. *Cement and Concrete Composites, 24*(1), 169–179.
7. Balayssac, J. P., Detriche, C. H., & Grandet, J. (1993). Interêt de l'essai d'absorption d'eau pour la caracterisation du béton d'enrobage. *Materials and Structure, 26*, 226–230.
8. Gharaibeh, E. S., Hanai, T., Xi, Y., & Frangopol, D. M. (2000). Effects of uncertainties on chloride penetration in saturated concrete. In: *8th ASCE specialty conference on probabilistic mechanics and structural reliability* (pp. 241–244). Notre Dame, Indiana.
9. Soroka, I. (1993). *Concrete in hot environments*. CRC Press.
10. Güneyisi, E., Gesoğlu, M., Karaboğa, F., & Mermerdaş, K. (2013). Corrosion behavior of reinforcing steel embedded in chloride-contaminated concretes with and without metakaolin. *Composites Part B: Engineering, 45*(1), 1288–1295.
11. Paul, S. C., Al Mamun, F., Hasan, N. M. S., Shaun, F. J., Kong, S. Y., Wong, L. S., & Babafemi, A. J. (2024). Chloride detection in concrete using wireless fidelity (Wi-Fi) signal. *Innovative Infrastructure Solutions, 9*(3), 60.
12. Ribeiro, M. S., Lage, J., & Gonçalves, A. (2019). Chloride assessment in structures: Influences of sampling and test location. *Cement and Concrete Research, 117*, 82–90.
13. Polder, R. B., & Peelen, W. H. (2002). Characterization of chloride transport and reinforcement corrosion in concrete under cyclic wetting and drying by electrical resistivity. *Cement and Concrete Composites, 24*(5), 427–435.
14. Al-Qadi, I. L., Haddad, R. H., & Riad, S. M. (1997). Detection of chlorides in concrete using low radio frequencies. *Journal of Materials in Civil Engineering, 9*(1), 29–34.

15. Ramírez-Ortíz, A. E., Castellanos, F., & Cano-Barrita, P. F. D. J. (2018). Ultrasonic detection of chloride ions and chloride binding in Portland cement pastes. *International Journal of Concrete Structures and Materials, 12*(1), 1–12.
16. Basheer, P. A. M., Gilleece, P. R. V., Long, A. E., & Mc Carter, W. J. (2002). Monitoring electrical resistance of concretes containing alternative cementitious materials to assess their resistance to chloride penetration. *Cement and Concrete Composites, 24*(5), 437–449.
17. Won, M., Zhang, S., & Son, S. H. (2017). WiTraffic: Low-cost and non-intrusive traffic monitoring system using Wi-Fi. In: *26th International conference on computer communication and networks (ICCCN)* (pp. 1–9). IEEE.
18. Hanif, A., Chughtai, M. S., Qureshi, A. A., Aleem, A., Munir, F., Tahir, M., & Uppal, M. (2018). Non-obtrusive detection of concealed metallic objects using commodity Wi-Fi radios. In: *IEEE global communications conference (GLOBECOM)* (pp. 1–6). IEEE.
19. Wang, X., Yang, C., & Mao, S. (2017). PhaseBeat: Exploiting CSI phase data for vital sign monitoring with commodity Wi-Fi devices. In: *IEEE 37th international conference on distributed computing systems (ICDCS)* (pp. 1230–1239). IEEE.
20. Zhang, X., Ruby, R., Long, J., Wang, L., Ming, Z., & Wu, K. (2016). WiHumidity: A novel CSI-based humidity measurement system. In: *International conference on smart computing and communication* (vol. 1, pp. 537–547). Springer.
21. Zabielski, J., & Srokosz, P. (2020). Monitoring of structural safety of buildings using wireless network of mems sensors. *Buildings, 10*(11), 193.
22. Taffese, W. Z., & Espinosa-Leal, L. (2022). A machine learning method for predicting the chloride migration coefficient of concrete. *Construction and Building Materials, 348*, Article 128566.
23. Cai, R., Han, T., Liao, W., Huang, J., Li, D., Kumar, A., & Ma, H. (2020). Prediction of surface chloride concentration of marine concrete using ensemble machine learning. *Cement and Concrete Research, 136*, Article 106164.
24. Deng, F., He, Y., Zhou, S., Yu, Y., Cheng, H., & Wu, X. (2018). Compressive strength prediction of recycled concrete based on deep learning. *Construction and Building Materials, 175*, 562–569.
25. Wang, Q., Chen, C., & Lu, N.: Electromagnetic and Ohmic heating analyses for Faraday shield in large and powerful RF ion sources. In: *7th international symposium on advances in electrical, electronics, and computer engineering* (vol. 12294, pp. 291–295). SPIE.
26. Gkoutos, G. V., & Hoehndorf, R. (2012). Ontology-based cross-species integration and analysis of Saccharomyces cerevisiae phenotypes. *Journal of Biomedical Semantics, 3*(2), S6.
27. Breiman, L. (2001). Random forests. *Machine Learning, 45*, 5–32.
28. Liaw, A., & Wiener, M. (2002). Classification and regression by random forest. *R News, 2*(3), 18–22.
29. Chen, T., & Guestrin, C. (2016). Xgboost: A scalable tree boosting system. In: 22nd ACM SIGKDD international conference on knowledge discovery and data mining (pp. 785–794). ACM.
30. Wang, C. C., Kuo, P. H., & Chen, G. Y. (2022). Machine learning prediction of turning precision using optimized xgboost model. *Applied Sciences, 12*(15), 7739.
31. LeCun, Y., Bottou, L., Bengio, Y., & Haffner, P. (1998/2022). Gradient-based learning applied to document recognition. *Proceedings of the IEEE, 86*(11), 2278–2324.
32. Atha, D. J., & Jahanshahi, M. R. (2018). Evaluation of deep learning approaches based on convolutional neural networks for corrosion detection. *Structural Health Monitoring, 17*(5), 1110–1128.
33. Sattari, M. T., Farkhondeh, A., & Patrick Abraham, J. (2018). Estimation of sodium adsorption ratio indicator using data mining methods: a case study in Urmia Lake basin, Iran. *Environmental Science and Pollution Research, 25*, 4776–4786.
34. Dimri, A., Varshney, J. K., Verma, V. K., & Gupta, S. (2015). A review on strength of concrete in seawater. *International Journal of Engineering Research and Technology, 4*, 844–847.
35. Jamil, M., Hassan, M. K., Al-Mattarneh, H. M. A., & Zain, M. F. M. (2013). Concrete dielectric properties investigation using microwave nondestructive techniques. *Materials and Structures, 46*, 77–87.

36. Al-Mattarneh, H. (2016). Determination of chloride content in concrete using near-and far-field microwave non-destructive methods. *Corrosion Science, 105*, 133–140.
37. Yun, H., Patton, M. E., Garrett Jr, J. H., Fedder, G. K., Frederick, K. M., Hsu, J. J., Sides, P. J., et al. (2004). Detection of free chloride in concrete by NMR. *Cement and Concrete Research, 34*(3), 379–390.
38. Taffese, W. Z., & Espinosa-Leal, L. (2022). Prediction of chloride resistance level of concrete using machine learning for durability and service life assessment of building structures. *Journal of Building Engineering., 60*, Article 105146.
39. ACI Committee. (2008). *Building code requirements for structural concrete (ACI 318–08) and commentary*. American Concrete Institute.

Smart Cities and Sustainability Issues: A SWOT Analysis Approach

Aamir Ahmad Teeli, Saima Nazir, and Ch. Sankar Rao

Abstract Smart cities are connected urban environments which do have a potential to improve efficiency, ensure quality of life for their citizens and enhance sustainability by integrating cutting edge technologies. Smart cities are at the forefront of urban development, utilizing state-of-the-art technologies to solve pressing environmental issues along with enhancing the quality of life for citizens. This article explores the complex relationship that exists between sustainability and smart city projects, explaining how both programs have the potential to transform urban living. Through the examination of technical frameworks like Internet of Things (IoT), big data analytics and artificial intelligence, we reveal the revolutionary potential of smart cities in terms of improving environmental sustainability and resource management. We also examine the challenges which are there in the way of fully realizing these advantages, including as data privacy concerns, technology interoperability issues, and the requirement for strong legal frameworks.

Keywords Smart cities · Technological development · Sustainability issues

1 Introduction

Cities are growing at a rate never seen before due to the global population's continued urbanization, which puts a tremendous amount of strain on the world's infrastructure and resources. This presents major challenges for quality of life, environmental sustainability, and resource management [37]. Conventional urban management strategies frequently cannot keep up with the demands of growing populations, which

A. A. Teeli (✉) · Ch. S. Rao
Department of Economics, Central University of Tamil Nadu, Thiruvarur, India
e-mail: imamireconomics@gmail.com

Ch. S. Rao
e-mail: Sankaraochiral@cutn.ac.in

S. Nazir
Department of Commerce, University of Kashmir, Srinagar, J&K, India

P. Danish et al. (eds.), *Innovative Technologies for Smart and Sustainable Infrastructure*,
https://doi.org/10.1007/978-3-031-92801-7_6

leads to problems like traffic jams, pollution of the air and water, waste management issues, and social injustices. As a result, there is pressing need for more resilient, efficient, and sustainable urban systems. Smart cities, by utilizing cutting-edge ICT, aim to create interconnected urban environments where data-driven decision-making improves sustainability and efficiency [9, 11].

Smart cities go beyond simple technological integration; they are comprehensive approaches to urban development that put sustainability, inclusivity, and resilience first [39, 61]. Smart cities can optimize a number of urban functions, such as energy consumption, transportation, water management, public safety, and health services, by utilizing a variety of technologies like artificial intelligence (AI), the IoT, and big data analytics [23, 24, 38, 51, 59]. These technologies enable for real-time monitoring and management of urban infrastructure, which enables proactive and adaptive responses to emerging challenges. Nevertheless, major obstacles must be looked into, in order to ensure smart city initiatives to be implemented successfully. These obstacles include data privacy concerns, technological interoperability [18]. In this regard current study intends to investigate how smart cities can be associated to sustainability issues and try to point out some ways to overcome these barriers.

2 Review of Literature

Numerous studies have emphasized the rapid urbanization phenomenon, also known as the "Century of the City," as a distinctive feature of the twenty-first century [43]. Projections suggest that share of people living in cities will increase to roughly around 70% by 2050 [27]. This extraordinary change is posing serious problems in the social, infrastructure, and environmental spheres and calling for creative solutions. A promising framework to address these complex urban issues is smart cities, which makes use of information and communication technologies (ICTs) [2, 5].

Smart City initiatives revolve around the integration of technologies, including big data analytics, machine learning, artificial intelligence, and the (IoT) Internet of Things. These technologies make it easier to monitor and review urban dynamics in real-time, giving managers and planners of cities crucial information to improve urban governance [63]. Urban environments are becoming increasingly permeable with Internet of Things (IoT) sensors, making it possible to gather enormous amounts of data about public safety, energy consumption, traffic patterns, and air quality. This robust data ecosystem facilitates the making of well-informed decisions [20, 66].

Big data analytics have played a crucial role in Smart Cities since it makes it possible to derive useful insights derived from the data that is gathered. According to studies [15, 33], the use of predictive analytics in traffic management can greatly reduce congestion and enhance the effectiveness of public transportation systems. By predicting traffic patterns and recommending alternate routes, these analytics can maximize urban mobility while reducing their negative effects on the environment. Smart City initiatives have a lot to offer the healthcare industry in addition to transportation. Thanks to wearable technology and real-time health monitoring systems which allow for continuous tracking of patients' vital signs, which enables prompt medical interventions and individualized healthcare delivery [42]. By encouraging preventive care, these systems not only enhance patient outcomes but also lessen the strain on the healthcare system [3].

An additional crucial component of development of Smart Cities is efficient energy management. Urban environments are experiencing a surge in energy demand as a result of economic activity and population growth. Better resource allocation, consumption forecasting, and enhancing renewable energy sources are ensured by smart grids and energy-efficient technologies [7, 59, 62]. Cities can lessen the effects of climate change, improve sustainability, and save operating expenses by implementing data-driven approaches to energy management. The utilization of advanced data analytics to support environmental monitoring is essential for comprehending the dynamics of urban ecosystems. In order to promote healthier urban living conditions, cities can address problems like noise pollution, urban heat islands, and air and water quality with the aid of environmental data analysis [66]. Such monitoring also helps ensure that environmental laws are followed and provide information for decisions made in urban planning that support sustainable development [22].

Urban environments require high level of safety and security. Smart Cities use artificial intelligence (AI) and machine learning (ML) does create predictive models which can lead to threats of public safety and crime. Studies have shown that use of machine learning algorithms can improve the precision of crime predictions, allowing law enforcement organizations to more efficiently allocate resources and create focused intervention plans [19, 21, 27]. Additionally, AI-driven surveillance systems can keep an eye on public areas, improving overall public safety and assisting in the prompt detection of incidents [4].

The use of these cutting-edge technologies is not without its difficulties, though. Data privacy, algorithmic bias, and consequences of automated decision-making are ethical issues that are becoming more and more prominent [63]. In order to fully benefit from Smart City technologies, cities must prioritize accountability, transparency, and inclusivity when establishing strong frameworks for data governance [6]. Maintaining this equilibrium is crucial to make sure that technological developments don't exacerbate already-existing disparities but rather support fair urban development.

3 Technological Framework

The technological foundation of smart city initiatives is an amalgam of state-of-the-art technology with the goal of transforming urban life. This framework consists primarily of an advanced integration of big data analytics, artificial intelligence (AI), and the Internet of Things (IoT) [6, 48]. Cities can now gather, process, and use enormous volumes of real-time data, which facilitates proactive decision-making, efficient resource use, and improved service delivery. Future generations can live in cities that are more resilient, sustainable, and livable, thanks to the use of these technologies.

3.1 Internet of Things (IoT)

One revolutionary technology that is essential for the creation of smart cities is the Internet of Things (IoT). It enables smooth data transfer between the digital and physical realms by linking a wide range of gadgets and sensors to the internet. Real-time monitoring and control of urban infrastructure is made possible by this connectivity, which improves responsiveness and efficiency across a range of industries. Applications of IoT in smart cities include streamlining waste management procedures, boosting transportation networks, and maximizing energy consumption and water management [8, 28, 41]. Cities may benefit from increased operational transparency, automated control, and predictive maintenance by integrating IoT, which will ultimately make urban areas more sustainable and habitable. Urban infrastructure may be monitored and controlled in real time, thanks to IoT. Examples of applications are:

Smart Grids

In smart cities, smart grids signal a paradigm shift in how energy is distributed and used. These grids facilitate monitoring and optimization of distribution of electricity through the integration of IoT, allowing for dynamic adjustments which need to be made in response to fluctuations in demand and supply of renewable energy sources [43]. As a result, the electrical grid is more reliable, energy efficiency is increased, and transmission losses are decreased. Furthermore, by cleverly controlling the intermittent nature of renewable energy sources like solar and wind power, smart grids make it easier to integrate them. Cities may lower carbon emissions, increase energy system sustainability, and build resilience to shocks by implementing smart grids [35].

Water Management

Urban sustainability requires effective water management, and IoT-enabled technologies present hitherto unseen chances for conservation and optimization. Real-time, continuous monitoring of water pressure, flow rates, and quality is made possible by IoT sensors embedded throughout water infrastructure [50]. By enabling early detection of leaks, broken pipes, and other problems, this data helps utilities save money on operating expenses and minimize water loss. In addition, IoT-driven water management systems can enhance urban green space irrigation strategies, guaranteeing effective water usage and fostering robust vegetation [31]. Smart cities may increase water conservation efforts, strengthen infrastructure resilience, and lessen the effects of water scarcity by utilizing IoT in water management.

Waste Management

Another area where IoT technologies are revolutionizing sustainability and efficiency is waste management. Real-time waste level monitoring in dumpsters and bins is made possible by automated waste collection systems that are fitted with Internet of Things sensors [49]. Waste management agencies can limit greenhouse gas emissions related to waste transportation, optimize collection routes, and save fuel usage with the help of this data. Additionally, sorting and recycling technologies enabled by the Internet of Things improve recycling processes' efficiency by raising diversion rates and lowering the amount of garbage dumped in landfills. Smart cities may significantly lessen their environmental effect, advance the ideas of the circular economy, and transition to a zero-waste future by implementing IoT in trash management.

3.2 Artificial Intelligence (AI)

The foundation of smart city efforts is artificial intelligence (AI), which provides sophisticated decision-making and prediction capabilities that spur efficiency and creativity. By analyzing vast amounts of data to identify patterns, predict trends, and automate challenging tasks, AI technologies are transforming how cities operate and address problems. There are numerous beneficial uses of AI with regards to smart cities: these include energy optimization, public safety, traffic control, and environmental monitoring [40]. Cities may use AI to lower operating costs, identify problems before they arise, and enhance the standard of living in cities [38]. AI integration strengthens urban systems' ability to respond in real time while also facilitating long-term strategic planning, which increases cities' adaptability and resilience to environmental changes and growing urbanization. AI improves prediction and decision-making skills [1]. It is crucial to highlight important uses of AI in the context of smart cities.

Traffic Management

Artificial intelligence (AI)-powered algorithms for controlling traffic use real-time data and predictive analytics to enhance the flow of traffic and reduce traffic jams, revolutionizing urban mobility. Artificial intelligence (AI) has the ability to employ complex algorithms to evaluate traffic patterns, identify areas of congestion, and automatically modify the signals for traffic to reduce bottlenecks and improve the efficiency of transportation networks [29]. Moreover, AI-powered systems for traffic control provide intelligent traffic signal coordination and adaptive routing, guaranteeing commuters' reduced journey times and improved traffic flow. Artificial intelligence (AI)-based traffic management systems not only improve urban living conditions but also cut carbon emissions and improve urban air quality by decreasing traffic and trip delays [58, 60].

Energy Management

Artificial Intelligence (AI) is an essential element of energy management because it helps maximize energy consumption, increase efficiency, and promote sustainability. AI algorithms analyze past consumption data, weather forecasts, along with other variables to anticipate patterns in demand for energy and optimize energy distribution in real-time [14]. This enables utilities in making adjustments to supply levels, installing energy storage devices, as well as participating in demand-response initiatives to efficiently trade-off between demand and supply. Furthermore, control systems for energy, which are powered by AI facilitate the grid's integration of clean energy sources by forecasting generation outputs and coordinating their deployment alongside traditional energy sources [54]. Finally, by optimizing renewable energy utilization and minimizing waste, AI improves the resilience and sustainability of energy management systems.

Public Safety

AI-powered public safety systems improve emergency response times and urban security by providing sophisticated analysis, decision assistance, and surveillance. Massive volumes of data are processed by AI algorithms from sensors, social media feeds, surveillance cameras, and other sources in order to identify possible dangers, spot unusual activity, and immediately notify authorities. Law enforcement organizations can also foresee crime hotspots, allocate resources efficiently, and proactively handle new security threats, thanks to predictive modeling driven by AI. Artificial intelligence (AI) algorithms are capable of analyzing incoming data streams in emergency response scenarios in order to prioritize response efforts, evaluate the severity of incidents, and best allocate resources for first responders. Cities can increase situational awareness, expedite response times, and ensure the protection and security of their residents by utilizing AI in public safety [38].

3.3 Big Data Analytics

The emergence of smart cities depends heavily on the use of big data analytics, which provides profound understanding of tendencies and patterns that guide urban management and planning. Big data analytics helps cities make more effective and efficient decisions by processing and analyzing vast volumes of information from multiple urban sources. With the use of this technology, forecasting and resource allocation become more precise as patterns in areas like population increase, transportation flow, energy consumption, and environmental effect are identified. There is increased utilization of big data analytics to manage utilities more effectively, monitor environmental conditions, improve emergency response times, and optimize public transit routes in smart cities. Cities may create sustainable and eco-friendly surroundings for their residents by using data intelligently to generate more strategic and adaptive approaches to urban concerns [12]. By revealing patterns and trends, big data analytics aid in the management and development of cities. Examples of applications are:

Urban Planning

Urban planning is revolutionized by big data analytics, which offer priceless insights into zoning, infrastructure development, and land use management decision-making processes. Planners can more efficiently identify regions of growth and development, prioritize investments, and distribute resources by examining data on population demographics, socioeconomic trends, and urban mobility patterns. Additionally, planners can evaluate how planned developments would affect air quality, traffic congestion, and community well-being thanks to big data analytics, which makes for more sustainable and well-informed planning choices [47]. Furthermore, planners can produce more resilient and habitable communities for future generations by using data-driven simulations and modeling tools that enable them to envision multiple scenarios and forecast the long-term effects of various development methods.

Resource Allocation

To maximize the distribution of public resources and provide citizens with effective services, resource allocation is crucial. Big data analytics, which offer insights into demand patterns, service usage trends, and infrastructure performance metrics, are essential to this process [55]. Through the analysis of data from several sources, including social services, utilities, and public transit systems, city officials are able to pinpoint inefficiencies, optimize service delivery procedures, and distribute resources according to current demand and priorities. Moreover, cities can apply focused interventions to address inequalities in service access and enhance results for marginalized communities through decision-making based on data. By employing big data analytics to inform strategic resource allocation, cities may lower costs, improve service quality, and advance social fairness and inclusion.

Environmental Monitoring

To track and mitigate the impact of human activity on the ecosystem and public health, environmental monitoring is crucial. By compiling and evaluating data from sensors, satellite imaging, and other sources, big data analytics helps environmental monitoring initiatives by tracking pollution hotspots, evaluating ecological health indicators, and monitoring air and water quality [32, 46]. Environmental authorities may detect new hazards, rank conservation initiatives, and create evidence-based regulations to save ecosystems and public health by examining patterns in pollution levels, biodiversity, and habitat degradation. Authorities may also react quickly to environmental catastrophes like natural disasters, industrial mishaps, and oil spills by using big data analytics-powered real-time surveillance as well as alert mechanisms [46]. This reduces damage and protects people and ecosystems. Urban areas may reduce environmental hazards, encourage sustainability, and save natural resources for future generations by using proactive environmental monitoring backed by big data analytics.

4 Sustainability Issues

Urban landscapes face substantial problems in their long-term survival due to sustainability issues, which calls for creative ways to lessen their effects and promote resilience. The issues of resource depletion, environmental degradation, social injustices, and economic disparities highlight how urgent it is to implement sustainable methods in urban growth [16, 65]. In order to successfully handle the diverse nature of sustainability concerns facing cities, we analyze their ramifications and look at how smart city projects might provide revolutionary solutions.

4.1 Environmental Sustainability

Smart cities are setting the standard in utilizing state-of-the-art technologies and creative approaches to reduce their negative environmental effects. To cut down on carbon emissions, conserve resources, and improve ecological resilience, these urban settings make use of advanced infrastructure and environmentally friendly solutions. By integrating clean sources of energy, putting automated transportation networks in place, and streamlining waste management practices, smart cities aim to create environmentally friendly urban environments [34, 49, 52]. A proactive strategy to environmental stewardship is fostered by the deployment of technologies like IoT, AI, and big data analytics, which enable real-time tracking and adaptive management of natural resources. By addressing these issues, smart cities ensure a healthier and more sustainable future for their citizens while also addressing the short-term environmental problems caused by urbanization. Smart city initiatives seek to reduce environmental impact through many ways.

Energy Efficiency

Sustainable urban growth is predicated on energy efficiency, and smart networks are essential for optimizing energy use and cutting carbon emissions. IoT, AI, and big data analytics are just a few of the state-of-the-art technologies that smart grids use to track, control, and enhance transmission of energy instantaneously. Cities can reduce their reliance on fossil fuels and emit less greenhouse gases by employing smart grid technology to incorporate energy from sustainable sources such as wind and solar energy into the grid [36]. Moreover, demand-response programs facilitated through modern smart grid systems enable utilities to reward customers for moving their energy use to off-peak times, when renewable energy is most plentiful [30]. Furthermore smart grids enable the integration of intermittent green energy sources while boosting the stability and resilience of the grid through the more effective management of energy storage devices. Cities can significantly reduce energy use, save utility costs for citizens and companies, and build a more resilient and sustainable energy infrastructure by implementing smart grid technologies.

Emission Reduction

Intelligent transportation systems present creative ways to reduce emissions and enhance urban air quality. Intelligent transportation systems (ITSs) help cities minimize car emissions, manage traffic flow, and ease traffic congestion by utilizing AI, IoT, and big data analytics [67]. Algorithms for advanced traffic control system reduce idle time and fuel consumption by applying congestion pricing schemes, improving signal timing, and dynamically modifying traffic signals through the analysis of real-time traffic data. In order to further reduce emissions from conventional combustion engine cars, intelligent transportation systems also make it easier for electric vehicles (EVs) and other alternative forms of transportation like bicycles and scooters to be deployed [67]. Additionally, by assisting vehicles in finding parking spots more quickly, smart parking management systems lessen traffic jams and the pollution that comes with driving around looking for a spot. By implementing intelligent transportation systems, cities may significantly lower their emissions of greenhouse gases, enhance air quality, and provide more sustainable and livable metropolitan environments for their citizens.

Waste Management

Sustainable waste management in smart cities requires improved recycling and waste-to-energy procedures. Cities may maximize resource recovery and reduce landfill trash by optimizing waste collecting, sorting, and processing processes through the use of state-of-the-art technology such as IoT, AI, and robotics. By installing Internet of Things (IoT) sensors in waste bins, waste management organizations may keep track of fill levels in real-time, optimize waste fetching routes, cut fuel costs, and minimize emissions related to waste transportation [64]. Furthermore, recyclable items are identified and separated from mixed waste streams by AI-powered sorting devices using machine learning algorithms, which raises recycling rates and lowers contamination. Moreover, communities can turn organic waste into renewable energy by using technologies that transform waste into energy like anaerobic decomposition and incineration with energy recovery, which minimizes dependency on fossil fuels and lowers the release of greenhouse gases. Cities can reduce landfill waste significantly, lessen their environmental effect, and support a more circular and sustainable economy by implementing modern waste management systems.

4.2 *Social Sustainability*

An essential element of the development of smart cities is social sustainability, which focuses on building inclusive urban environments where all citizens have fair access to the advantages of cutting-edge technologies. A variety of issues must be addressed in order to ensure social sustainability, including closing the digital divide, encouraging community involvement, and advancing social fairness [45, 57]. Ensuring equitable accessibility to digital infrastructure along with literacy programs is crucial in order to enable all demographic groups to take part in and reap the benefits of smart city efforts. Furthermore, fostering community ownership and trust requires including residents in decision-making processes through open government and inclusive platforms. Smart cities may improve everyone's quality of life, lessen inequality, and create strong, unified communities that prosper in an increasingly technologically advanced urban environment by putting social sustainability first. Ensuring equitable access to the benefits of smart cities is crucial. Among the difficulties are:

Digital Divide

Addressing the digital divide is essential to ensuring that all residents have equitable access to the benefits of smart city initiatives. Inequalities in literacy and access to technology can amplify already-existing ones, limiting prospects for social and economic inclusion [56]. Cities must undertake focused interventions, like building out broadband infrastructure, supplying reasonably priced internet access, and providing digital literacy training programs, in order to close the digital gap [45]. Furthermore, programs to make computers, tablets, and smart phones accessible can enable marginalized populations to take full advantage of the digital economy and the vital services that smart city platforms provide. Cities may foster more equal and inclusive settings where all citizens have the chance to prosper in the digital age by giving priority to addressing the digital divide.

Public Participation

To ensure that smart city programs represent the requirements and preferences of the communities they serve, it is imperative that individuals be included in the decision-making process. Residents are empowered to take part in the planning, execution, and assessment of smart city projects when there is meaningful public engagement, which promotes openness, accountability, and trust between the community and the local government. Through online forums, smart phone applications, and virtual town hall meetings, cities may use technology to encourage public participation [57]. This gives citizens an opportunity to voice opinions, share problems, and collaborate to resolve crucial urban problems. Furthermore, by allowing citizens to directly allocate funds to activities that address community objectives, programs like participatory budgeting promote civic engagement and a sense of ownership. To attain more inclusive and environmentally friendly urban development, communities may make sure that smart city programs are responsive to the varied necessities and ambitions of their citizens by placing a high priority on public engagement.

Inclusivity

Promoting social justice and cohesiveness requires making sure that smart city programs serve all demographic groups. Access to technology is just one aspect of inclusivity; other aspects include opportunities and resources being distributed fairly among all communities. To guarantee that no group is left behind, cities must proactively address obstacles to inclusivity, such as linguistic hurdles, cultural differences, and socioeconomic gaps. This calls for focused outreach initiatives, programming that take cultural sensitivity into account, and policies that give marginalized and vulnerable communities' needs top priority. Cities can also encourage collaborations with grassroots projects, advocacy groups, and community-based organizations to strengthen the voices of marginalized people and guarantee that their viewpoints are considered during the process of making decisions [25]. Cities may create more cohesive, dynamic, and resilient communities where all individuals have the opportunity to succeed through making inclusiveness a fundamental tenet.

Inclusion and fair access to smart city innovations are key to social sustainability. It will need focused efforts to give marginalized groups access to technology and education if we are to close the digital divide [25]. Platforms for public engagement make sure that the opinions of citizens are heard, and laws need to be made to ensure that everyone may benefit from smart city initiatives.

4.3 Economic Sustainability

In order to maintain long-term survival, smart cities must both promote ecological principles and achieve economic growth. Economic development policies that encourage investment, innovation, and job creation while limiting adverse effects on the environment and society must be carefully considered in order to maintain this delicate balance. By fostering innovation ecosystems that encourage entrepreneurship, research, and development in emerging technologies, smart cities promote economic growth [17].To guarantee that all citizens have access to economic possibilities, these cities also place a high priority on the creation of jobs in industries like advanced manufacturing, sustainable energy, and information technology. In order to reduce their environmental impact and increase resilience, smart cities also incorporate sustainable practices including resource efficiency, the establishment of green infrastructural setup, and the application of circular economy concepts. Smart cities may build vibrant urban economies that are socially inclusive, environmentally conscious, and economically robust by adopting both sustainability and economic growth goals. Sustainable development and economic growth must coexist in smart cities.

Innovation Ecosystems

Smart cities must support ecosystems of creative thinking in order to strike a balance between sustainability and economic growth. Cities may drive technology innovations that address sustainability concerns and generate economic growth by fostering an atmosphere that encourages creativity, collaboration, and entrepreneurship. Research facilities, co-working spaces that foster knowledge sharing and promote businesses, incubators, and accelerators are important parts of innovation ecosystems [53]. Furthermore, programs like technology parks and innovation districts offer real locations where industry, government, and academics may work together to develop and market creative ideas. Cities can draw talent, encourage investment, and accelerate the creation of innovations that advance sustainability and boost economic prosperity by fostering innovation ecosystems.

Job Creation

For all citizens of smart cities to gain from economic progress, new jobs in technology-driven businesses must be created. As smart technologies develop further, there will be a need for qualified professionals in data science, renewable energy, sophisticated manufacturing, and information technology. By funding workforce development projects, vocational training programs, as well as educational collaborations that equip citizens with the know-how required to thrive in the digital economy, smart cities may promote job creation [17]. Furthermore, communities can develop a diverse and resilient workforce that fosters innovation and economic growth by implementing talent attraction and retention strategies like industry alliances, internship programs, and talent retention incentives. In order to promote equitable prosperity and lower unemployment, smart cities should prioritize the creation of jobs in technology-driven businesses while advancing sustainability goals.

Sustainable Investment

In order to achieve economic resilience and sustainability, smart cities must make sure that their investments in smart technologies pay off over time. In keeping with the triple bottom line of people, planet, and profit, sustainable investment methods give priority to initiatives that produce favorable social, environmental, and financial returns over the long run. Sustainable investment techniques can be adopted by smart cities through the incorporation of environmental, social, and governance (ESG) considerations into the process of decision-making [26]. Projects are assessed on the basis of their capacity to generate enduring value for stakeholders. Furthermore, the public and private sectors can share risks and gains while raising finance for sustainable infrastructure projects through public–private partnerships and cutting-edge financing techniques like impact investing and green bonds. In order to ensure that economic growth is socially, ecologically, and economically resilient—and to facilitate more prosperous and sustainable urban future—smart cities should adopt sustainable investment principles. Establishing an atmosphere that encourages entrepreneurship and innovation is essential to economic sustainability in smart cities [13]. In addition to guaranteeing that funds invested in smart infrastructure produce economic advantages over the long run, this promotes employment creation in developing technological sectors. Together with rapid technical breakthroughs, policymakers must give priority to long-term economic viability.

5 Conclusion

Through the application of state-of-the-art technology, smart cities provide a revolutionary approach to urban planning that tackles urgent sustainability issues. Cities may enhance the quality of life for their residents, upgrade the management of resources, and reduce their negative environmental effects by integrating cutting-edge solutions like IoT, AI, and big data analytics [10, 47]. However, the adoption of inclusive policies and successful stakeholder participation are prerequisites for realizing these benefits. Smart cities have the power to open the door for a more egalitarian and sustainable urban future by placing a high priority on sustainability, equity, and resilience.

Smart cities present a glimmer of hope for developing cities that are technologically cutting edge, environmentally sustainable and socially inclusive in the face of increasing urbanization and environmental pressures. Policies, business executives, and the general public must work together to embrace innovation, promote teamwork, and place a high priority on sustainability in order to advance smart urbanization. Cities may steer towards a sustainable and happier future for everybody by taking into account the insights gained from prosperous smart city endeavors and tackling the recognized obstacles.

6 Policy Implications

The expansion of smart cities is heavily reliant on policy consequences, making strong frameworks necessary to effectively steer their development. To ensure the safe and effective deployment of technology, it is important to create clear norms and standards. Encouraging cooperation between the public and commercial sectors stimulates investment and innovation, which advances smart city projects. Securing sufficient funds through different funding channels is equally crucial for sustaining long-term development initiatives and large-scale projects. The course of smart cities is largely determined by policymakers, whose efforts in putting laws into place will also affect how successful and long-lasting initiatives involving urban innovation are.

References

1. Agrawal, A., Gans, J., & Goldfarb, A. (2018). Prediction, judgment, and complexity: A theory of decision-making and artificial intelligence. In *The economics of artificial intelligence: An agenda* (pp. 89–110). University of Chicago Press.
2. Aguilera, U., Peña, O., Belmonte, O., & López-de-Ipiña, D. (2017). Citizen-centric data services for smarter cities. *Future Generation Computer Systems, 76*, 234–247.
3. Al Nuaimi, E., Al Neyadi, H., Mohamed, N., & Al-Jaroodi, J. (2015). Applications of big data to smart cities. *Journal of Internet Services and Applications, 6*(1), 1–15.
4. Ali, Z., Chaudhry, S. A., Ramzan, M. S., & Al-Turjman, F. (2020). Securing smart city surveillance: A lightweight authentication mechanism for unmanned vehicles. *IEEE Access, 8*, 43711–43724.
5. Alifi, M. R., & Supangkat, S. H. (2016). Information extraction for traffic congestion in social network: Case study: Bekasi city. In *2016 International conference on ICT For smart society (ICISS)* (pp. 53–58). IEEE.

6. Allam, Z., & Dhunny, Z. A. (2019). On big data, artificial intelligence and smart cities. *Cities, 89*, 80–91.
7. Altomare, A., Cesario, E., & Vinci, A. (2019). Data analytics for energy-efficient clouds: Design, implementation and evaluation. *International Journal of Parallel, Emergent and Distributed Systems, 34*(6), 690–705.
8. Alzahrani, A. I., Chauhdary, S. H., & Alshdadi, A. A. (2023). Internet of things (IoT)-based wastewater management in smart cities. *Electronics, 12*(12), 2590.
9. Anthopoulos, L. G. (2015). Understanding the smart city domain: A literature review. In: *Transforming city governments for successful smart cities* (pp. 9–21).
10. Asopa, P., Purohit, P., Nadikattu, R. R., & Whig, P. (2021). Reducing carbon footprint for sustainable development of smart cities using IoT. In: *2021 Third international conference on intelligent communication technologies and virtual mobile networks (ICICV)* (pp. 361–367). IEEE.
11. Bakici, T., Almirall, E., & Wareham, J. (2013). A smart city initiative: The case of Barcelona. *Journal of the Knowledge Economy, 4*(2), 135.
12. Bassoo, V., Ramnarain-Seetohul, V., Hurbungs, V., Fowdur, T. P., & Beeharry, Y. (2018). Big data analytics for smart cities. In: *Internet of things and big data analytics toward next-generation intelligence* (pp. 359–379).
13. Bătăgan, L. (2011). Smart cities and sustainability models. *Informatica Economică, 15*(3), 80–87.
14. Bedi, P., Goyal, S. B., Rajawat, A. S., Shaw, R. N., & Ghosh, A. (2022). Application of AI/IoT for smart renewable energy management in smart cities. In: *AI and IoT for smart city applications* (pp. 115–138).
15. Bejan, A. I., Gibbens, R. J., Evans, D., Beresford, A. R., Bacon, J., & Friday, A. (2010). Statistical modelling and analysis of sparse bus probe data in urban areas. In *13th International IEEE conference on intelligent transportation systems* (pp. 1256–1263). IEEE.
16. Beura, D. (2017). Depletion of water bodies due to urbanisation and its management. *International Education and Research Journal, 3*(6).
17. Bonte, D. (2018). *Role of smart cities for economic development* (pp. 1–16). ABI Research.
18. Braun, T., Fung, B. C., Iqbal, F., & Shah, B. (2018). Security and privacy challenges in smart cities. *Sustainable Cities and Society, 39*, 499–507.
19. Catlett, C., Cesario, E., Talia, D., & Vinci, A. (2019). Spatio-temporal crime predictions in smart cities: A data-driven approach and experiments. *Pervasive and Mobile Computing, 53*, 62–74.
20. Cesario, E. (2019). Big data analysis for smart city applications. In: *Encyclopedia of big data technologies* (pp. 1–6). Springer.
21. Cesario, E. (2023). Big data analytics and smart cities: Applications, challenges, and opportunities. *Frontiers in Big Data, 6*, 1149402.
22. Cesario, E., Uchubilo, P. I., Vinci, A., & Zhu, X. (2022). Multi-density urban hotspots detection in smart cities: A data-driven approach and experiments. *Pervasive and Mobile Computing, 86*, Article 101687.
23. Chau, K. W. (2006). A review on integration of artificial intelligence into water quality modelling. *Marine Pollution Bulletin, 52*(7), 726–733.
24. Chui, K. T., Lytras, M. D., & Visvizi, A. (2018). Energy sustainability in smart cities: Artificial intelligence, smart monitoring, and optimization of energy consumption. *Energies, 11*(11), 2869.
25. Colding, J., Nilsson, C., & Sjöberg, S. (2024). Smart cities for all? Bridging digital divides for socially sustainable and inclusive cities. *Smart Cities, 7*(3), 1044–1059.
26. de Bem Machado, A., Rodrigues dos Santos, J., Richter, M. F., & Sousa, M. J. (2021). Smart cities: Building sustainable cities. In: *Green Technological innovation for sustainable smart societies: Post pandemic era* (pp. 1–19).
27. Diro, A. A., & Chilamkurti, N. (2018). Distributed attack detection scheme using deep learning approach for internet of things. *Future Generation Computer Systems, 82*, 761–768.

28. Ejaz, W., Naeem, M., Shahid, A., Anpalagan, A., & Jo, M. (2017). Efficient energy management for the internet of things in smart cities. *IEEE Communications Magazine, 55*(1), 84–91.
29. Englund, C., Aksoy, E. E., Alonso-Fernandez, F., Cooney, M. D., Pashami, S., & Åstrand, B. (2021). AI perspectives in smart cities and communities to enable road vehicle automation and smart traffic control. *Smart Cities, 4*(2), 783–802.
30. Gellings, C. W., & Samotyj, M. (2013). Smart grid as advanced technology enabler of demand response. *Energy Efficiency, 6*, 685–694.
31. Gimpel, H., Graf-Drasch, V., Hawlitschek, F., & Neumeier, K. (2021). Designing smart and sustainable irrigation: A case study. *Journal of Cleaner Production, 315*, Article 128048.
32. Hamza, M. A., Shaiba, H., Marzouk, R., Alhindi, A., Asiri, M. M., Yaseen, I., Rizwanullah, M., et al. (2022). Big data analytics with artificial intelligence enabled environmental air pollution monitoring framework. *Computers, Materials & Continua, 73*(2).
33. Herrera, J. C., Work, D. B., Herring, R., Ban, X. J., Jacobson, Q., & Bayen, A. M. (2010). Evaluation of traffic data obtained via GPS-enabled mobile phones: The mobile century field experiment. *Transportation Research Part C: Emerging Technologies, 18*(4), 568–583.
34. Iqbal, R. (2019). Intelligent transportation systems in smart cities. In: *Sustainable city logistics planning: Methods and applications* (Vol. 1, pp. 167–196). Nova Science Publishers, Inc.
35. Kabeyi, M. J. B., & Olanrewaju, O. A. (2022). The use of smart grids in the energy transition. In *2022 30th Southern African universities power engineering conference (SAUPEC)* (pp. 1–8). IEEE.
36. Kalair, A., Abas, N., Saleem, M. S., Kalair, A. R., & Khan, N. (2021). Role of energy storage systems in energy transition from fossil fuels to renewables. *Energy Storage, 3*(1), Article e135.
37. Kalhor, K., & Mahdisoltani, M. (2015). Urbanization and its effects on the environment and society along with sustainable development. In *Third international symposium on environmental and water resources engineering, Tehran, Iran* (Vol. 2, No. 3, pp. 1–12).
38. Mahor, V., Rawat, R., Kumar, A., Garg, B., & Pachlasiya, K. (2023). IoT and artificial intelligence techniques for public safety and security. In *Smart urban computing applications* (pp. 111–126). River Publishers.
39. Martin, C. J., Evans, J., & Karvonen, A. (2018). Smart and sustainable? Five tensions in the visions and practices of the smart-sustainable city in Europe and North America. *Technological Forecasting and Social Change, 133*, 269–278.
40. McMillan, L., & Varga, L. (2022). A review of the use of artificial intelligence methods in infrastructure systems. *Engineering Applications of Artificial Intelligence, 116*, Article 105472.
41. Milić, D. C., Tolić, I. H., & Peko, M. (2020). Internet of things (IoT) solutions in smart transportation management. *Business Logistics in Modern Management.*
42. Muhammed, T., Mehmood, R., Albeshri, A., & Katib, I. (2018). UbeHealth: A personalized ubiquitous cloud and edge-enabled networked healthcare system for smart cities. *IEEE Access, 6*, 32258–32285.
43. Neffati, O. S., Sengan, S., Thangavelu, K. D., Kumar, S. D., Setiawan, R., Elangovan, M., Velayutham, P., et al. (2021). Migrating from traditional grid to smart grid in smart cities promoted in developing country. *Sustainable Energy Technologies and Assessments, 45*, Article 101125.
44. O'Dwyer, E., Pan, I., Acha, S., & Shah, N. (2019). Smart energy systems for sustainable smart cities: Current developments, trends and future directions. *Applied Energy, 237*, 581–597.
45. Povetkina, N. A., Zueva, A. S., & Kopina, A. A. (2023). Reducing the digital divide as a mechanism to ensure sustainable economic and social development. *Sustainable development risks and risk management: A systemic view from the positions of economics and law* (pp. 27–31). Springer International Publishing.
46. Rahman, M. S., Di, L., & Esraz-Ul-Zannat, M. (2017). The role of big data in disaster management. In *International conference on disaster risk mitigation* (pp. 1–5).
47. Rathore, M. M., Ahmad, A., Paul, A., & Rho, S. (2016). Urban planning and building smart cities based on the internet of things using big data analytics. *Computer Networks, 101*, 63–80.
48. Rjab, A. B., & Mellouli, S. (2021). Smart cities in the era of artificial intelligence and internet of things: promises and challenges. In: *Smart cities and smart governance: Towards the 22nd century sustainable city* (pp. 259–288).

49. Saha, H. N., Auddy, S., Pal, S., Kumar, S., Pandey, S., Singh, R., Saha, S., et al. (2017). Waste management using internet of things (IoT). In *2017 8th Annual industrial automation and electromechanical engineering conference (IEMECON)* (pp. 359–363). IEEE.
50. Salam, A., & Salam, A. (2020). Internet of things in water management and treatment. In: *Internet of things for sustainable community development: Wireless communications, sensing, and systems* (pp. 273–298).
51. Samaei, S. R. (2024). Using artificial intelligence to increase urban resilience: a case study of Tehran. In *13th International conference on advanced research in science, engineering and technology*, Brussels, Belgium.
52. Sanchez-Miralles, A., Calvillo, C., Martín, F., & Villar, J. (2014). Use of renewable energy systems in smart cities. In: *Use, operation and maintenance of renewable energy systems: Experiences and future approaches* (pp. 341–370).
53. Schaffers, H., Komninos, N., & Pallot, M. (2012). Smart cities as innovation ecosystems sustained by the future internet.
54. Selvaraj, R., Kuthadi, V. M., & Baskar, S. (2023). Smart building energy management and monitoring system based on artificial intelligence in smart city. *Sustainable Energy Technologies and Assessments, 56*, Article 103090.
55. Seyedan, M., & Mafakheri, F. (2020). Predictive big data analytics for supply chain demand forecasting: Methods, applications, and research opportunities. *Journal of Big Data, 7*(1), 53.
56. Shin, S. Y., Kim, D., & Chun, S. A. (2021). Digital divide in advanced smart city innovations. *Sustainability, 13*(7), 4076.
57. Simonofski, A., Asensio, E. S., De Smedt, J., & Snoeck, M. (2017). Citizen participation in smart cities: Evaluation framework proposal. In *2017 IEEE 19th conference on business informatics (CBI)* (Vol. 1, pp. 227–236). IEEE.
58. Soomro, S., Miraz, M. H., Prasanth, A., & Abdullah, M. (2018). Artificial intelligence enabled IoT: Traffic congestion reduction in smart cities.
59. Szpilko, D., Naharro, F. J., Lăzăroiu, G., Nica, E., & de la Torre Gallegos, A. (2023). Artificial intelligence in the smart city—a literature review. *Engineering Management in Production and Services, 15*(4), 53–75.
60. Thornbush, M. J., & Thornbush, M. J. (2015). Reduced traffic congestion and air pollution. In: *Vehicular air pollution and urban sustainability: An assessment from Central Oxford, UK* (pp. 11–23).
61. Trivellato, B. (2017). How can 'smart'also be socially sustainable? Insights from the case of Milan. *European Urban and Regional Studies, 24*(4), 337–351.
62. Ullah, R., Faheem, Y., & Kim, B. S. (2017). Energy and congestion-aware routing metric for smart grid AMI networks in smart city. *IEEE Access, 5*, 13799–13810.
63. Ullah, Z., Al-Turjman, F., Mostarda, L., & Gagliardi, R. (2020). Applications of artificial intelligence and machine learning in smart cities. *Computer Communications, 154*, 313–323.
64. Vishnu, S., Ramson, S. J., Senith, S., Anagnostopoulos, T., Abu-Mahfouz, A. M., Fan, X., Kirubaraj, A. A., et al. (2021). IoT-enabled solid waste management in smart cities. *Smart Cities, 4*(3), 1004–1017.
65. Zhang, X. Q. (2016). The trends, promises and challenges of urbanisation in the world. *Habitat International, 54*, 241–252.
66. Zheng, Y., Capra, L., Wolfson, O., & Yang, H. (2014). Urban computing: Concepts, methodologies, and applications. *ACM Transactions on Intelligent Systems and Technology (TIST), 5*(3), 1–55.
67. Zhu, L., Yu, F. R., Wang, Y., Ning, B., & Tang, T. (2018). Big data analytics in intelligent transportation systems: A survey. *IEEE Transactions on Intelligent Transportation Systems, 20*(1), 383–398.

Application of Digital Technologies for Quality Control Practices in Smart Construction: A Scoping Review

Abdulkabir Opeyemi Bello

Abstract The integration of digital technologies is revolutionising quality control (QC) practice in smart construction, offering innovative solutions to longstanding inefficiencies while presenting notable challenges. Through a comprehensive desktop review, this study critically examines the application of digital technologies in QC, focusing on technological advancements, impacts on project outcomes, barriers to adoption, and research gaps. The findings reveal that technologies such as Internet-of-Things (IoT)-enabled sensors, Artificial Intelligence (AI)-driven predictive analytics, blockchains, and digital twins significantly enhance precision, efficiency, and collaboration in construction workflows. These tools improve defect detection, resource optimisation, and transparency and establish new benchmarks for quality assurance. However, its widespread adoption has been hindered by high implementation costs, interoperability challenges, technical expertise shortages, and resistance to change. Additionally, gaps in scalability and the lack of empirical studies on long-term implications highlight the need for additional investigation and the establishment of standardised frameworks. This review provides actionable insights for researchers, practitioners, and policymakers, emphasising the importance of a collaborative and standardised approach to overcoming these challenges. By addressing these critical issues, this study advances the understanding of digital QC practices and contributes to the broader goals of improved performance, sustainability, and innovation in smart infrastructure development.

Keywords Construction automation · Digital technologies in construction · Infrastructure development · Quality management practices · Smart construction quality control · Smart quality monitoring

---

A. O. Bello (✉)
Smart and Sustainable Built Environment Research Lab, Bellium R&D, Ilorin, Nigeria
e-mail: abdulkabiropeyemi@gmail.com

P. Danish et al. (eds.), *Innovative Technologies for Smart and Sustainable Infrastructure*,
https://doi.org/10.1007/978-3-031-92801-7_7

1 Introduction

The construction industry is transforming with the emergence of smart construction practices, which leverage advanced digital technologies to optimise efficiency, sustainability, and collaboration across project lifecycles [1, 2]. Central to these practices is quality control (QC), a systematic process that monitors, inspects, and evaluates construction activities to ensure compliance with predefined quality standards, regulatory requirements, and project specifications. QC is critical for reducing defects, minimising rework, and ensuring the structural integrity and functionality of construction projects [3]. However, in traditional construction settings, QC is mainly manual, relying on labour-intensive inspections, paper-based documentation, and fragmented communication systems [4]. While foundational, these traditional methods are increasingly inadequate for addressing the demands of modern construction projects, leading to inefficiencies, delays, and inaccuracies [5–7].

The growing complexity and scale of construction projects have exacerbated the limitations of the traditional QC methods. Projects now demand greater precision, real-time decision-making, and alignment with global sustainability goals. Digital technologies such as Building Information Modeling (BIM), the Internet of Things (IoT), Artificial Intelligence (AI), blockchain, and digital twins offer innovative solutions to these issues [8–10]. These tools enable real-time monitoring, automated defect detection, and predictive analytics, thereby transforming QC practices and setting new benchmarks for quality assurance in construction [11, 12]. Furthermore, advancements in drones, 3D laser scanners, and augmented reality (AR) have enhanced inspection accuracy and efficiency, allowing proactive defect identification and resolution [13, 14]. These technologies promote resource optimisation, stakeholder collaboration, and improved decision-making, all of which are essential for the success of smart construction projects.

Despite their transformative potential, the integration of digital technologies into QC frameworks remains inconsistent and faces significant barriers. According to [2, 15], high implementation costs, technical complexity, and interoperability challenges hinder widespread adoption, this could be more particular for small and medium-sized enterprises (SMEs). The lack of standardised frameworks and protocols further complicates the seamless integration of these tools, resulting in fragmented workflows and inefficiency. Moreover, there is limited empirical evidence on the scalability of these technologies for large-scale and complex projects and their long-term impacts on sustainability, workforce dynamics, and overall project outcomes [16]. These challenges underscore the need for comprehensive research to bridge the existing knowledge gaps and develop actionable strategies for implementing digital technologies in QC.

This study addresses these critical gaps by conducting a scoping review of the applications of digital technologies for QC in smart construction. It examines their technological advantages, impact on project outcomes, barriers to adoption, and research gaps. This research synthesises extant knowledge to yield valuable insights for academic researchers, industry professionals, and policymakers. The findings contribute to the development of standardised, scalable, and sustainable QC practices that align with the principles of smart construction, ensuring enhanced project performance, sustainability, and long-term viability.

2 Quality Control Practices in Construction Projects

2.1 Traditional Methods of Quality Control Practices in Construction

Traditional construction quality control (QC) practices are predominantly manual and rely on observational inspections, paper-based records, and fragmented communication systems. Although these practices were foundational to the early construction processes, they are increasingly viewed as inadequate for managing the complexities of modern projects [6, 7]. Inspections often involve labour-intensive tasks such as manual dimensional checks and defect identification, which are prone to human error. For instance, precast concrete inspections depend heavily on on-site evaluations that lack integration with automated systems, thereby leading to delays and inaccuracies [6]. Modular integrated construction (MiC) workflows are similarly constrained by disconnected systems across production and assembly phases, making it challenging to ensure consistent quality [11, 12].

In road construction, manual monitoring of parameters such as asphalt temperature and paving conditions during transportation and installation fails to provide real-time feedback, diminishing the effectiveness of quality management [15]. Mega construction projects (MCPs), such as the Hong Kong-Zhuhai-Macao Bridge, have revealed the inefficiency of paper-based inspections and delayed stakeholder communication, which hinders the timely resolution of quality issues [17].

Traditional QC methods also lack collaboration and data integration mechanisms, often resulting in rework and increased project costs. The absence of standardised frameworks exacerbates inefficiencies, particularly in large-scale projects that require dynamic solutions [7, 18]. Additionally, inadequate training and resistance to modern systems further constrain the effectiveness of these methods [19].

These limitations emphasise the need for technology-driven solutions. Advanced tools, such as BIM, IoT, and AI, offer improved accuracy, real-time monitoring, and enhanced collaboration, thereby addressing the inefficiencies of traditional QC practices [11, 12].

2.2 *Digital Method of Quality Control Practices in Construction*

Digital technologies have transformed quality control practices in construction by facilitating real-time monitoring, automated inspection, and improved stakeholder collaboration. These innovations address the inefficiencies of traditional QC methods by introducing a dynamic data-driven approach to quality management. BIM is a foundational technology that integrates the design, construction, and QC processes into a centralised platform. Authors in [11, 12] demonstrated that BIM enhances defect detection, design validation, and stakeholder collaboration, thereby ensuring adherence to the project specifications. Authors in [15, 20] further highlighted that when combined with Internet of Things (IoT) technologies, BIM enables real-time data collection and monitoring through embedded sensors, allowing precise tracking of structural and environmental conditions.

The blockchain technology adds a layer of security and transparency by maintaining immutable transaction records and modifications. Authors in [12, 21] note that smart contracts embedded within blockchain systems automate compliance checks and streamline approval processes, enhancing efficiency. Similarly, authors in [2, 8] emphasised that AI and machine learning facilitate predictive analytics and automated defect detection by leveraging data from IoT and BIM platforms to anticipate quality risks and optimise decision-making. Digital twins replicate physical construction sites within virtual environments, enabling real-time quality assessment and predictive maintenance. Authors in [2, 22] identified digital twins as transformative tools for continuous monitoring and quality control. Drones and 3D laser scanners, according to [14, 20], automate inspections and provide high-resolution images for defect tracking. Augmented reality (AR), as noted by authors in [6], overlays digital models onto physical structures, facilitating interactive inspections. Radio Frequency Identification (RFID) tags and IoT devices also play crucial roles in the lifecycle monitoring of prefabricated components. Authors in [11, 16] highlighted how these technologies ensure consistent quality across manufacturing, transportation, and assembly stages. Authors in [17, 23] described Cyber-Physical Systems (CPS) as integrating physical and digital processes to enhance communication and decision-making through real-time data synchronisation. The challenges persist despite the progress, wherein authors in [12, 15] noted that high costs, integration complexity, and interoperability issues hinder adoption. However, the ability of these technologies to automate QC, reduce errors, and promote sustainability underscores their critical role in advancing construction practices.

3 Methodology

This study adopts a scoping review methodology to evaluate the application of digital technologies in construction QC. A scoping review enables the exploration of broad research questions, mapping of key concepts, identifying gaps in the literature, and synthesising findings across diverse sources [24]. This review focused on scholarly articles published between 2018 and 2024. Academic databases, such as Scopus and Google Scholar, were searched using following keywords; "quality control", "smart construction", "quality control in smart construction", "BIM for quality management", "IoT-enabled quality monitoring", and "blockchain in quality control". Relevant literature was selected based on empirical evidence, theoretical contributions, and alignment with themes. A thematic technique was used to categorise the findings into five areas: technologies/tools for QC, technological advantages, impact on project outcomes, barriers to adoption, and research gaps.

4 Analysis of Key Themes

4.1 Technological Advantages

Integrating digital technologies into construction quality control (QC) has been identified as a transformative development that addresses inefficiencies inherent in traditional methods. Authors in [13, 25] emphasised that these technologies facilitate enhanced precision, real-time monitoring, and improved stakeholder collaboration, aligning construction practices with the evolving demands of modern projects. Authors in [20, 22] argued that such advancements would enable dynamic and adaptive workflows in complex construction environments. However, as [26] noted, these technologies' transformative impacts and challenges warrant a critical examination to understand their implications fully.

One of the most significant advantages of digital technologies is the automation of quality assurance processes. Authors in [2, 27, 28] highlighted tools such as BIM and AI, to streamline defect detection, design validation, and compliance checks. Authors in [12] noted that BIM reduces manual oversight and accelerates decision-making by centralising data integration. Similarly, authors in [8, 29] have proven that AI algorithms enable predictive analytics and automated defect identification, thereby significantly minimising human errors.

While these technologies enhance operational efficiency, authors in [19] pointed out that their implementation requires substantial investment in infrastructure and training. This financial burden may limit their accessibility, particularly for small and medium-sized enterprises (SMEs), further highlighting disparities in technological adoption within the construction industry.

Real-time monitoring capabilities facilitated by Internet of Things (IoT) devices and digital twins were identified as transformative for quality control practices, offering unparalleled transparency. Authors in [20] highlighted that IoT-enabled sensors monitor critical parameters such as structural stability and material quality, enabling the early detection of deviations and the implementation of timely corrective measures. Similarly, authors in [12, 15] emphasised that these sensors provide real-time data, streamline QC processes, and reduce the risk of project delays. The integration of IoT with building information Modeling platforms, as discussed by authors in [9, 30], facilitates real-time progress tracking, enabling enhanced monitoring during on-site assembly phases.

Authors in [31] noted that advanced analytics and AI integration within digital skins allows stakeholders to base decisions on real-time and historical data. Technologies such as drones further improve QC efficiency by automating data collection and analysis, which, according to [14], results in faster decision-making and improved project timelines. As authors in [2] explained, digital twins replicate physical environments in virtual spaces, thereby supporting continuous monitoring and predictive maintenance. However, authors in [23] cautioned that integrating these systems with legacy technologies would often present significant interoperability challenges.

Predictive analytics also play a pivotal role in QC. Authors in [2] demonstrated that historical and real-time data enable the anticipation of potential quality risks, whereas AI-driven models integrated with IoT and BIM platforms provide actionable insights into mitigating defects and reducing rework. Authors in [32, 33] further highlighted that AI-powered systems enhance QC accuracy by delivering high-resolution imagery and insights using tools such as 3D laser scanners and drones.

Authors in [34] argued that the convergence of AI and the IoT enables the creation of smart construction objects capable of delivering safer, more sustainable, and more efficient structures than traditional methods. Smart devices automate inspection and data collection, as noted by authors in [33], thereby significantly reducing the time required for QC processes. However, authors in [16, 23] point out that the reliance of these technologies on sophisticated hardware and software necessitates technical expertise and robust data governance frameworks, complicating their adoption.

Blockchain technology has been identified as a critical tool for addressing trust and accountability in construction quality control (QC). Authors in [10, 35] highlighted that blockchain creates immutable records for quality inspections, ensuring data integrity and reducing the likelihood of disputes. Furthermore, authors in [21, 36] emphasise the role of smart contracts within blockchain systems, which automate compliance processes and foster collaboration and transparency among project stakeholders. However, as authors in [31] argued, the scalability of blockchain applications in construction remains constrained by high deployment costs and a lack of standardised protocols.

Despite the significant benefits of digital technologies in enhancing the efficiency, accuracy, and sustainability of QC processes, their adoption is fraught with several challenges. The authors in [16, 18] noted that barriers, such as high implementation costs, technical complexity, and the need for specialised expertise, inhibit widespread adoption. Authors in [2, 31] observed that these limitations are particularly acute in projects involving multiple stakeholders where integration challenges often arise. To

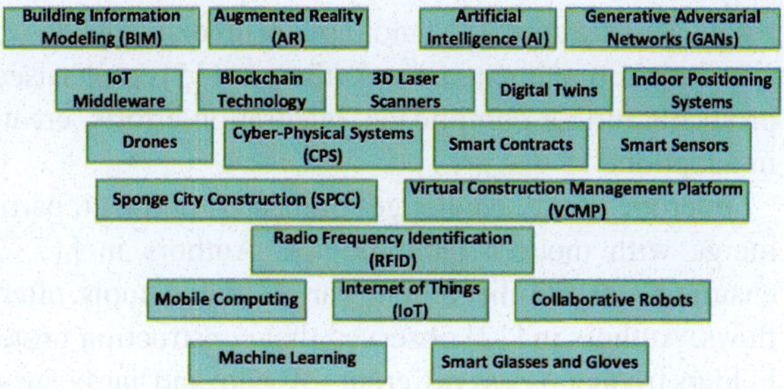

Fig. 1 Cluster of technologies used in smart construction

unlock the full potential of these technologies, the construction industry must prioritise investments in developing standardised frameworks and scalable solutions. Such measures would ensure seamless adoption across diverse projects and stakeholders, ultimately advancing the effectiveness and accessibility of the digital QC processes. Figure 1 shows the common technologies applied to QC in smart constructions.

4.2 Barriers to the Adoption of Digital Technologies for QC in Smart Construction

The adoption of digital technologies in construction quality control has been transformative; however, numerous barriers impede their widespread implementation. Authors in [12, 33] identify high costs, technical complexity, integration challenges, and regulatory gaps as key obstacles to an industry's digital transformation. These challenges collectively hinder the seamless adoption and scalability of advanced technology.

Economic constraints are among the most frequently cited challenges in the adoption of digital technology. Authors in [9, 23, 33] emphasised that implementing systems such as BIM, IoT, and blockchain require substantial financial investment in infrastructure, hardware, and workforce training. Small and medium-sized enterprises (SMEs), in particular, face significant disadvantages because of the prohibitive costs of acquiring and maintaining these technologies, as noted by authors in [10, 29]. Furthermore, authors in [35, 36] argue that unclear returns on investment, particularly in regions with limited digital literacy or low awareness of the long-term benefits of digital technologies, exacerbate these financial constraints.

Another significant barrier is the technical complexity of integrating digital technologies into existing construction workflows. Authors in [19, 20] highlighted that technologies such as blockchain, IoT, and digital twins demand advanced technical expertise, which is often lacking in traditional construction teams. Authors in [9, 37] noted that combining BIM platforms with IoT devices and blockchain systems requires high interoperability and customisation, which is a challenge for many organisations. Authors in [22] further observed that an inadequate infrastructure for integrating digital technologies with existing QC workflows results in inefficiencies during implementation. Authors in [15] emphasised that the lack of standardised protocols further complicates integration efforts, creating delays and inefficiencies in adoption.

Interoperability challenges are also prominent, particularly when legacy systems merge with modern technologies. Authors in [17, 23] identified difficulties in ensuring compatibility among various digital tools, often leading to fragmented workflows. Authors in [12] observed that construction projects involving multiple stakeholders frequently use different software and hardware systems, resulting in miscommunication, inconsistent data management, and increased costs. Authors in [11] concluded that these inefficiencies undermine the potential of digital technologies to fully optimise QC processes.

Regulatory and standardisation gaps were identified as critical impediments to the effective adoption of digital technologies in construction. Authors in [2, 10, 14] highlighted that the absence of universally accepted standards for implementing technologies such as blockchain, IoT, and drones results in variability in application and performance. Authors in [15, 28] further argued that inconsistent regulations across regions exacerbate these challenges, particularly in global projects that must comply with multiple legal frameworks. Authors in [16, 36] observed that these gaps deter investment as firms are reluctant to adopt systems that may require costly modifications to meet evolving regulatory standards.

Human factors are another significant barrier to digital technology adoption. Authors in [19] emphasised that resistance to change is pervasive within the construction industry, where traditional workflows are deeply entrenched. Authors in [23, 32, 37] noted that many construction professionals lack the necessary digital skills to effectively utilise advanced technologies, creating a substantial training burden for organisations. Additionally, authors in [22, 33] highlighted that fear of job displacement due to automation intensifies this resistance, particularly among workers accustomed to manual methods. Authors in [8] argued that these fears further hinder the acceptance of emerging technologies.

Concerns regarding the reliability of digital systems persist. Authors in [12, 15, 26] observed that IoT devices are often affected by environmental conditions, which can compromise their accuracy and functionality. Similarly, [21] noted that blockchain systems face significant scalability issues, particularly in large-scale construction projects with extensive data volumes. These technical limitations raise questions regarding the dependability of digital technologies to consistently deliver optimal performance.

Despite these barriers, the transformative potential of digital technologies in revolutionising QC processes in construction remains evident. As suggested by authors in [16, 36], addressing these challenges requires the development of standardised protocols, targeted investments in workforce training, and the exploration of cost-effective solutions tailored to SMEs. Collaborative efforts among governments, industry leaders, and technology providers are essential for bridging regulatory gaps and fostering a culture of innovation in the construction sector.

4.3 Impact of Digital Technologies for QC in Smart Construction

The adoption of digital technologies in construction quality control significantly influences project outcomes by improving efficiency, sustainability, and collaboration across various stages of construction. Authors in [1, 38] argued that these advancements address the key challenges in traditional construction practices by redefining quality, safety, and resource management standards.

One of the most notable impacts of digital technology is the enhancement of quality assurance. As authors in [8] highlighted, technologies such as BIM, IoT, and AI enable real-time monitoring and automated inspections, ensuring adherence to design specifications while reducing the likelihood of defects. Authors in [2, 23, 27] further emphasised that these technologies streamline QC processes, while authors in [17] observed that their application improves consistency in quality standards. According to [21, 37], blockchain technology enhances accountability by maintaining immutable inspection records, reducing disputes, and fostering trust between stakeholders.

Automation tools such as drones and 3D laser scanners, as noted by [14, 20], have been instrumental in reducing the time required for inspections and defect detection by automating data collection and analysis. As demonstrated by [12, 15], IoT-enabled systems provide real-time updates on construction parameters, allowing proactive adjustments that minimise delays and optimise workflows. Furthermore, [16] highlighted that RFID-BIM integration ensures adherence to quality standards and reduces the need for rework, while [27] noted that BIM-AI integration enhances quality assurance through rigorous compliance checks. Authors in [2] concluded that these technologies accelerate project timelines and reduce costs associated with rework and inefficiency.

Centralised platforms such as BIM-IoT systems facilitate seamless communication among stakeholders. Authors in [11] demonstrated that these systems ensure all parties can access accurate, real-time project data, reduce miscommunication, and improve decision making. As authors in [12] noted, blockchain technology supports collaborative decision-making by providing secure, transparent records of transactions and quality assessments, fostering trust across the supply chain.

The adoption of digital technologies has contributed significantly to advancing the sustainability of construction practices. Authors in [16] argued that these technologies align construction workflows with global environmental goals by optimising resource utilisation and minimising waste. As highlighted by [15], IoT systems play a pivotal role in monitoring material usage and environmental conditions, ensuring efficient resource deployment while reducing ecological impact. Furthermore, authors in [16] emphasised that technologies such as RFID and BIM facilitate the reuse of construction components by providing detailed lifecycle information, thereby promoting circular economy practices within the sector.

Despite these substantial benefits, integrating digital technologies into sustainable construction processes remains a challenge. Authors in [12, 23] noted that high implementation costs, interoperability issues, and the need for extensive workforce training present significant challenges to its adoption. Overcoming these challenges is crucial for achieving optimal results and seamlessly embedding digital technologies into construction workflows. This necessitates strategic investments in cost-effective solutions, standardised protocols, and targeted training initiatives to ensure that the sustainability benefits of digital technologies are realised across diverse construction environments.

4.4 Gap-Spotting in Smart Construction Quality Control

Despite the transformative potential of digital technologies in construction quality control (QC), significant gaps persist in research and implementation, hindering their widespread adoption and scalability. Authors in [2, 12] emphasised that these gaps limit the ability of digital technologies to achieve their full potential and underscore the need for further investigation into their long-term impacts. However, a critical gap lies in the scalability of these technologies for complex and large-scale construction projects. Authors in [15, 21] noted that although IoT, BIM, and blockchain have demonstrated efficacy in modular and small-scale applications, their integration into megaprojects remains underexplored. Authors in [2] further highlighted that the computational requirements and energy consumption of blockchain systems pose significant challenges for scalability in large-scale implementations.

The integration of emerging technologies is another unresolved research area. Authors in [8] observed that current studies primarily focus on the isolated application of technologies, such as AI, machine learning, and predictive analytics, with limited attention given to their combined potential to enhance QC processes across the construction lifecycle. Authors in [20] argued that the interoperability between digital tools, such as BIM, IoT, and blockchain, requires the development of standardised frameworks that are currently lacking and impede seamless integration.

Furthermore, there is a paucity of empirical research on the long-term impacts of digital technologies on project outcomes. Authors in [16] noted that although existing studies often emphasise short-term benefits, critical questions regarding the durability, adaptability, and social implications of these technologies remain unanswered. These gaps extend to sustainability, cost efficiency, and workforce dynamics, highlighting the need for comprehensive investigation. Addressing these gaps is crucial to unlock the full potential of digital technologies in construction QC. As authors in [2, 12] concluded, future research should prioritise scalability, integration, and the long-term impact of these innovations to ensure their consistent and equitable adoption across diverse construction environments.

5 Key Findings from the Study

This study critically examines the application of digital technologies in construction QC, focusing on five interrelated themes: technologies/tools for QC, technological advantages, impacts on project outcomes, barriers to adoption, and research gaps. The findings highlight the transformative potential of digital tools, while emphasising the challenges that must be addressed to unlock their full potential. Various technologies have emerged as key enablers of QC in smart constructions. BIM provides centralised platforms for defect detection, design validation, and data integration. IoT devices with sensors enable the real-time monitoring of construction parameters, whereas blockchain technology enhances traceability and automation through immutable records and smart contracts. AI drives predictive analytics and defect detection, improves efficiency, and reduces manual oversight. Tools such as 3D laser scanners, drones, AR, and RFID enhance precision and automate inspections, create an integrated ecosystem that streamlines QC processes, and ensures quality assurance.

These technological advancements have revolutionised QC by enabling real-time monitoring, predictive capabilities, and automation. IoT systems analyse data for timely adjustments, AI proactively identifies defects, and digital twins provide virtual replicas of sites for continuous quality assessments. Blockchain ensures accountability with transparent records, whereas tools, such as drones and AR, improve efficiency and accuracy. Together, these technologies optimise resource utilisation, reduce human errors, and set new benchmarks for construction precision. The impacts of these advantages are profound, real-time monitoring minimises defects and rework, reducing delays and costs, and enhanced collaboration through platforms like BIM and blockchain fosters transparent communication and effective decision-making. Digital technologies align construction practices with sustainability goals by optimising resource use; minimising waste; enabling material reuse; and improving quality, efficiency, and environmental responsibility.

However, there are significant barriers to adoption, such as high implementation costs and limited accessibility, particularly for SMEs. The complexity of integrating tools, such as blockchain, IoT, and AI, requires advanced expertise, which is often lacking in traditional teams. Interoperability challenges and the absence of standardised frameworks hinder scalability and seamless integration. Resistance to change and fear of job displacement further exacerbate these challenges. Regulatory and standardisation gaps add uncertainty, deterring firms from adopting digital transformation. Critical research gaps also persist, with a limited number of studies exploring the scalability of these technologies for large and complex projects. The integration of tools such as machine learning with existing systems remains underexplored. Most studies have prioritised short-term benefits over long-term impacts on sustainability, workforce dynamics, and project outcomes. Addressing these gaps requires empirical studies, standardised frameworks, and cost-effective solutions tailored to diverse project needs.

6 Conceptual Framework

Building on the findings of this study, we proposed a conceptual framework for implementing digital technologies in QC for smart construction, as shown in Fig. 2. The framework emphasises leveraging technological enablers, such as BIM, IoT, blockchain, AI, and digital twins, to create an integrated ecosystem for quality management. Implementation begins with planning and assessment in which project-specific needs and readiness for technology adoption are evaluated.

The integration of digital tools must ensure seamless interoperability and scalability within the existing workflows. Training and change management are critical in equipping teams with technical skills while addressing resistance to change and fostering a culture of innovation. Data management and analytics, driven by IoT-enabled monitoring and AI-based predictive insights, form the backbone of this framework, enabling real-time decision-making and quality assurance. Validation

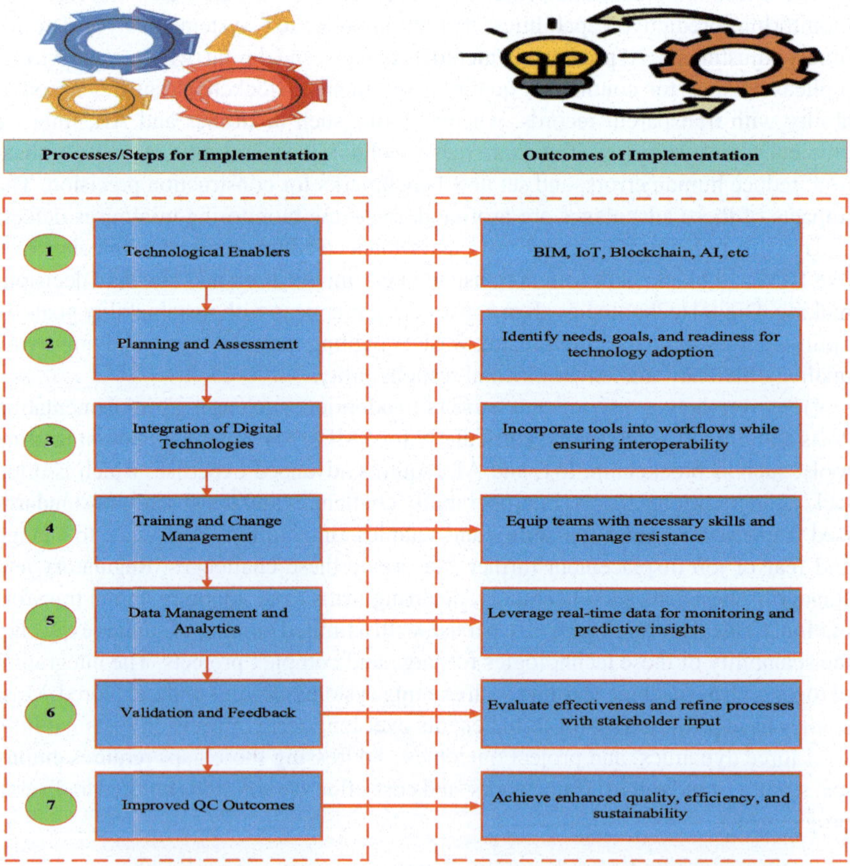

Fig. 2 Conceptual framework for integrating digital tools in QC for smart construction

and feedback mechanisms, facilitated by tools such as digital twins and blockchain, ensure iterative refinement of QC processes. The framework aims to enhance the quality, operational efficiency, and sustainability of smart construction projects. It also incorporates feedback loops to address barriers, such as high costs, technical complexity, and regulatory gaps, ensuring continuous improvement and adaptability.

7 Conclusion and Recommendations

This study underscores the pivotal role of digital technologies in shaping the future of QC in smart constructions. Tools such as BIM, IoT, blockchain, AI, and digital twins have propelled the construction sector toward precision-driven, efficient, and collaborative workflows. These technologies address longstanding inefficiencies in traditional QC methods while aligning construction practices with the modern demands for sustainability and transparency. However, this study also highlights the dichotomy between the potential of these technologies and the challenges hindering their widespread adoption. High costs, technical complexity, interoperability issues, and a lack of regulatory clarity continue to pose significant barriers. Additionally, research gaps in scalability and the long-term impacts of these tools emphasise the need for deeper investigations into their application across diverse construction environments. A multipronged approach is essential for achieving a digitally empowered construction ecosystem. Industry leaders must champion technological innovation, policymakers should establish enabling regulatory frameworks, and academia must expand empirical research to address the existing knowledge gaps. Collaboration among these stakeholders is critical for translating technological advancements into universally accessible, scalable, and sustainable solutions, ensuring that the transformative potential of digital technologies is fully realised.

References

1. Osunsanmi, T., Aigbavboa, C., & Thwala, W. (2018). Smart technologies in construction management: A literature synthesis. *Building Research Journal, 45*(1), 1–16.
2. Pan, Y., & Zhang, L. (2023). Integrating BIM and AI for smart construction management: Current status and future directions. *Archives of Computational Methods in Engineering, 30*(2), 1081–1110.
3. Wang, J., Sun, W., Shou, W., Wang, X., Wu, C., Chong, H. Y., & Sun, C. (2015). Integrating BIM and LiDAR for real-time construction quality control. *Journal of Intelligent & Robotic Systems, 79*, 417–432.
4. Puri, N. (2019). *Progress monitoring and quality assessment/quality control of construction projects using Lidar and BIM*. Oregon State University.
5. Abdulraheem, A. A., Abdulazeez, A. S., Bello, A. O., Faruq, M. Y., Musa, A. K., & Shabi, M. O. (2025). Exploring the factors affecting quality of project delivery in the Nigerian construction industry. *The TQM Journal*.

6. Hajdukiewicz, M., Goggins, J., de la Torre, O., Holleran, D., & Keane, M. M. (2019). An automated standard-based life cycle quality inspection methodology for smart precast concrete solutions in buildings. *Journal of Structural Integrity and Maintenance, 4*(3), 123–134.
7. Ma, Z., Cai, S., Mao, N., Yang, Q., Feng, J., & Wang, P. (2018). Construction quality management based on a collaborative system using BIM and indoor positioning. *Automation in Construction, 92*, 35–45.
8. Baduge, S. K., Thilakarathna, S., Perera, J. S., Arashpour, M., Sharafi, P., Teodosio, B., & Mendis, P. (2022). Artificial intelligence and smart vision for building and construction 4.0: Machine and deep learning methods and applications. *Automation in Construction, 141*, 104440.
9. Ghosh, A., Edwards, D. J., & Hosseini, M. R. (2021). Patterns and trends in internet of things (IoT) research: Future applications in the construction industry. *Engineering, Construction and Architectural Management, 28*(2), 457–481.
10. Li, J., & Kassem, M. (2021). Applications of distributed ledger technology (DLT) and Blockchain-enabled smart contracts in construction. *Automation in Construction, 132*, Article 103955.
11. Li, X., Lu, W., Xue, F., Wu, L., Zhao, R., Lou, J., & Xu, J. (2022). Blockchain-enabled IoT-BIM platform for supply chain management in modular construction. *Journal of Construction Engineering and Management, 148*(2), 04021195.
12. Wu, L., Lu, W., Xue, F., Li, X., Zhao, R., & Tang, M. (2022). Linking permissioned blockchain to internet of things (IoT)-BIM platform for off-site production management in modular construction. *Computers in Industry, 135*, Article 103573.
13. Hunde, B. R., & Woldeyohannes, A. D. (2022). Future prospects of computer-aided design (CAD)—a review from the perspective of artificial intelligence (AI), extended reality, and 3D printing. *Results in Engineering, 14*, Article 100478.
14. Li, Y., & Liu, C. (2018). Applications of multirotor drone technologies in construction management. *International Journal of Construction Management, 19*(5), 401–412.
15. Sharif, M., Mercelis, S., Van Den Bergh, W., & Hellinckx, P. (2017). Towards real-time smart road construction: Efficient process management through the implementation of internet of things. In: *Proceedings of the international conference on big data and internet of thing* (pp. 174–180).
16. Iacovidou, E., Purnell, P., & Lim, M. K. (2018). The use of smart technologies in enabling construction components reuse: A viable method or a problem creating solution. *Journal of Environmental Management, 216*, 214–223.
17. Zhou, H., Wang, H., & Zeng, W. (2018). Smart construction site in mega construction projects: A case study on island tunneling project of Hong Kong-Zhuhai-Macao Bridge. *Frontiers of Engineering Management, 5*(1), 78–87.
18. Built2Spec Project. (2017). Methodology for smart material self-inspection and quality checks. Horizon 2020 project deliverable.
19. Závadská, Z., & Závadský, J. (2018). Quality managers and their future technological expectations related to industry 4.0. *Total Quality Management & Business Excellence, 31*(7–8), 717–741.
20. Xu, J., & Lu, W. (2018). Smart construction from head to toe: A closed-loop lifecycle management system based on IoT. In: *Construction research congress* (pp. 157–168).
21. Lu, W., Wu, L., Zhao, R., Li, X., & Xue, F. (2021). Blockchain technology for governmental supervision of construction work: Learning form digital currency electronic payment systems. *Journal of Construction Engineering and Management, 147*(10), 04021122.
22. Yevu, S. K., Owusu, E. K., Chan, A. P., Sepasgozar, S. M., & Kamat, V. R. (2023). Digital twin-enabled prefabrication supply chain for smart construction and carbon emissions evaluation in building projects. *Journal of Building Engineering, 78*, Article 107598.
23. Tambare, P., Meshram, C., Lee, C. C., Ramteke, R. J., & Imoize, A. L. (2022). Performance measurement system and quality management in data-driven industry 4.0: A review. *Sensors, 22*(1), 224.

24. Munn, Z., Peters, M. D., Stern, C., Tufanaru, C., McArthur, A., & Aromataris, E. (2018). Systematic review or scoping review? Guidance for authors when choosing between a systematic or scoping review approach. *BMC Medical Research Methodology, 18*, 1–7.
25. O'Grady, T. M., Brajkovich, N., Minunno, R., Chong, H. Y., & Morrison, G. M. (2021). Circular economy and virtual reality in advanced BIM-based prefabricated construction. *Energies, 14*(13), 4065.
26. Tian, Y., & Spatari, S. (2022). Environmental life cycle evaluation of prefabricated residential construction in China. *Journal of Building Engineering, 57*, Article 104776.
27. Rane, N. (2023). Integrating building information modelling (BIM) and artificial intelligence (AI) for smart construction schedule, cost, quality, and safety management: Challenges and opportunities.
28. Wang, M., Wang, C. C., Sepasgozar, S., & Zlatanova, S. (2020). A systematic review of digital technology adoption in off-site construction: Current status and future direction towards industry 4.0. *Buildings, 10*(11), 204.
29. Yuan, S., Wang, Y., Kang, L., Yu, Z., & Feng, Y. (2018). Discussion on quality management and control in construction engineering. *Smart Construction Research, 2*(4), 1–6.
30. Li, C. Z., Zhong, R. Y., Xue, F., Xu, G., Chen, K., Huang, G. G., & Shen, G. Q. (2017). Integrating RFID and BIM technologies for mitigating risks and improving schedule performance of prefabricated house construction. *Journal of Cleaner Production, 165*, 1048–1062.
31. Edirisinghe, R. (2019). Digital skin of the construction site: Smart sensor technologies towards the future smart construction site. *Engineering, Construction and Architectural Management, 26*(2), 184–223.
32. Darko, A., Chan, A. P., Yang, Y., & Tetteh, M. O. (2020). Building information modeling (BIM)-based modular integrated construction risk management–Critical survey and future needs. *Computers in Industry, 123*, Article 103327.
33. Silverio-Fernández, M., Renukappa, S., & Suresh, S. (2021). Strategic framework for implementing smart devices in the construction industry. *Construction Innovation: Information, Process, Management, 21*(2), 218–243.
34. Turner, C.J., Oyekan, J., Stergioulas, L., & Griffin, D. (2020). Utilizing industry 4.0 on the construction site: Challenges and opportunities. *IEEE Transactions on Industrial Informatics, 17*(2), 746–756.
35. Loo, B. P., & Wong, R. W. (2023). Towards a conceptual framework of using technology to support smart construction: The case of modular integrated construction (MiC). *Buildings, 13*(2), 372.
36. Zhao, S., Zhang, P., & Li, W. (2021). A study on evaluation of influencing factors for sustainable development of smart construction enterprises: Case study from china. *Buildings, 11*(6), 221.
37. Zhong, B., Wu, H., Ding, L., Luo, H., Luo, Y., & Pan, X. (2020). Hyperledger fabric-based consortium blockchain for construction quality information management. *Frontiers of Engineering Management, 7*(4), 512–527.
38. Okpala, I., Parajuli, A., Nnaji, C., & Awolusi, I. (2020). Assessing the feasibility of integrating the internet of things into safety management systems: A focus on wearable sensing devices. In: *Construction research congress* (pp. 236–245). American Society of Civil Engineers.

Bacteria-Induced Self-Healing Concrete: A Critical Review for the Construction Industries

Peerzada Danish, Jasir Mushtaq, Imen Ben Salem, and Towseef Ahmed Gilani

Abstract Self-healing concrete is an innovative material that autonomously repairs cracks and damage, increasing the durability and longevity of structures. This innovative construction material employs various mechanisms to initiate and facilitate healing, reducing maintenance costs and increasing infrastructure sustainability. One key approach involves incorporating encapsulated healing agents within the concrete matrix. Microcapsules contain materials such as bacteria, minerals, or polymers. These capsules rupture when cracks form in the concrete, releasing the healing agents into the damaged areas. The development and integration of self-healing concrete represent a significant stride towards sustainable and resilient infrastructure. As the construction industry continues to embrace innovative solutions, self-healing concrete stands out as a promising advancement, offering a proactive and efficient means of maintaining the integrity of our built environment. This review paper explores the intriguing domain of Self-Healing concrete, and examines how it can transform infrastructure resilience. The study delves into the methods and materials utilized in self-healing concrete technology, examining its capability to heal the cracks autonomously, and other damages brought on by environmental stresses. It explores various methods, including bacteria-based systems and encapsulated healing agents, highlighting their effectiveness in enhancing the durability and lifespan of concrete structures. This review highlights the role of self-healing concrete in enhancing the mechanical and durability properties of structures with the help of MICP (Mechanically Induced Calcite Precipitation) technology using ureolytic and non-ureolytic bacteria.

P. Danish
Department of Civil Engineering, Indian Institute of Science, Bangalore, India

J. Mushtaq (✉) · T. A. Gilani
Department of Civil Engineering, Institute of Technology, University of Kashmir, Jammu and Kashmir, Srinagar, India
e-mail: drjasirnitsri@gmail.com

I. B. Salem
College of Natural and Health Sciences, Zayed University, Abu Dhabi, United Arab Emirates

P. Danish et al. (eds.), *Innovative Technologies for Smart and Sustainable Infrastructure*,
https://doi.org/10.1007/978-3-031-92801-7_8

Keywords Bacillus megaterium · Bacillus cohnii · Bacillus subtilis · MICP (Microbially Induced Calcite Precipitation)

1 Introduction

The self-healing capability of concrete refers to its ability to autonomously repair cracks without external intervention. This process not only seals the cracks but also restores the mechanical properties of the structural elements, either partially or completely. Due to its relatively low tensile strength compared to other building materials, concrete is particularly susceptible to surface cracks. These cracks compromise the concrete's durability by allowing harmful gases and liquids to penetrate. Self-healing is a well-known phenomenon, primarily attributed to concrete's natural autogenous healing ability. The three main types of autonomous self-healing include microbiological, vascular, and capsule-based methods. Autogenous healing in cementitious materials facilitates crack closure, enhancing the physical and mechanical performance of composites while extending their lifespan. This property has contributed to the longevity of historic structures.

During cracking, calcium ions from fracture surfaces support "surface-controlled crystal growth," forming calcite. As an initial calcite layer develops, calcium ion availability decreases, shifting the process to slower, diffusion-controlled crystal growth. In composite cement containing pozzolanic additives, a portion of calcium hydroxide, a key source of Ca^{2+} ions, undergoes pozzolanic reactions to form calcium silicate hydrate (C–S–H), which slows down the precipitation of calcium carbonate and diminishes its intensity.

Microencapsulation (with a diameter of less than 1 mm) continues to be a widely used method for creating self-healing components in cementitious systems. Microcapsules are directly integrated into the matrix and release their core material into cracks as they form, reacting with a dispersed catalyst to repair the damage. Rising economic concerns over concrete deterioration have driven the adoption of innovative maintenance methods. While conventional techniques like epoxy fillers, latex binders, and organic/inorganic coatings enhance performance, their irreversible nature and limited long-term effectiveness have sparked debate [1]. The efficiency of concrete reduces with time due to the penetration of water through pores [2]. Pores permit the entry of harmful substances into the concrete matrix, potentially disrupting the passive layer between the concrete and steel, thereby initiating steel corrosion[3]. Concrete's pore structure has a major impact on how long it lasts [4].

Microbially induced calcium carbonate precipitation (MICP) is a sustainable approach to enhancing the durability of building materials [5]. Researchers worldwide support using microorganisms as self-healing agents for crack repair [6, 7]. Studies show bacteria can fix atmospheric CO_2 into carbonate minerals like calcite and dolomite [8, 9]. MICP involves bacterial-induced calcium carbonate crystal formation under suitable conditions [10]. Studies have highlighted the successful application of MICP in fields like soil stabilization, dust mitigation, and concrete rehabilitation [11–15]. MICP relies on biological processes like urea hydrolysis, photosynthesis, sulfate reduction, and nitrate reduction [16]. Ureolytic MICP, a common method, utilizes urease-producing bacteria to hydrolyze urea, producing CO_2 and ammonium ions that facilitate calcium carbonate crystal formation [16, 17].

Bacteria act as effective nucleation sites for mineral precipitation, thanks to their negatively charged functional groups, such as carboxyl, phosphate, and amine, which readily bind with carbonate ions [18]. Self-healing technology offers a sustainable solution to enhance structural durability, complementing concrete's limited natural self-healing abilities [19, 20]. Microbial self-healing using bacteria, primarily from the Bacillus genus, promotes calcium carbonate deposition through MICP [21, 22]. Bacillus is preferred for its spore-forming ability, allowing survival in alkaline conditions for up to 50 years [23, 24]. Studies show that 84% of research on bacterial self-healing in cement-based materials focuses on Bacillus [23].

The objective of this study is to review various literature on concrete incorporating bacteria to assess its ability to synthesize calcite and improve the properties of ordinary concrete using bio-based agents. This study highlights the global research conducted for the efficiency of bacteria-induced concrete using various species of "Bacillus" individually and in combination. Concrete is responsible for up to 8% of total anthropogenic carbon emissions due to use of cement. During the production of cement clinker, the calcium carbonate gets converted into calcium oxide and carbon dioxide leading to an increase in carbon dioxide emission.

Concrete is one of the most widely used materials, its increased use leads to increased production of cement which in-turn increases the concentration of carbon dioxide in the atmosphere. Also, the development of cracks in concrete structures limits their life span and necessitates the need for repair or replacement of structure which further increases the concrete demand. The use of bacteria belonging to the genus "Bacillus" has proven to enhance the mechanical properties of concrete in addition to repairment of cracks due to calcite precipitation which leads to increased life of concrete structures. Therefore, the utilization of bacteria in concrete ends the need for the repair of structures for a certain period and aids the reduction of carbon emissions as well.

2 Literature Review

2.1 Mechanism of Calcite Formation

Urease catalyzes the hydrolysis of urea into carbonate and ammonia. This reaction breaks down one mole of urea into one mole each of ammonia (NH_3) and carbamic acid (NH_2COOH) [Eq. 1] Carbamic acid is further hydrolyzed into 1 mol of ammonia and carbonic acid (H_2CO_3) [Eq. 2] [25, 26]

$$CO(NH_2)_2 + H_2O \rightarrow NH_2COOH + NH_3 \quad (1)$$

$$NH_2COOH + H_2O \rightarrow NH_3 + H_2CO_3 \tag{2}$$

The equilibration of the end products (NH_3 and H_2CO_3) in water results in the formation of bicarbonate [Eq. 3], along with two moles of ammonium and two moles of hydroxide ions [Eq. 4] The increase in pH due to hydroxide ions shifts the bicarbonate equilibrium, leading to the formation of carbonate ions. [Eq. 5] [27]. This shifting results in metal ion precipitation. The presence of NH_4^+ leads to increased pH in the medium enabling voluntary formation of calcium carbonate [28, 29].

$$H_2CO_3 \leftrightarrow HCO_3^- + H^+ \tag{3}$$

$$2NH_3 + 2H_2O \leftrightarrow 2NH_4^- + 2OH^- \tag{4}$$

$$HCO_3^- + H^+ + 2OH^- \leftrightarrow CO_3^{2-} + 2H_2O \tag{5}$$

$CaCO_3$ precipitation will occur at the bacterial cell membrane if the required amount of Ca^{2+} and $CO_3{}^{2-}$ are present in the solution [Eq. 6,7] [30].

$$Ca^{2-} + \text{Bacterial cell} \rightarrow \left(Ca^{2+} - \text{Cell}\right) \tag{6}$$

$$\left(Ca^{2+} - \text{Cell}\right) + CO_3^{2-} \rightarrow (\text{Cell} - CaCO_3) \tag{7}$$

2.2 *MICP Mechanism*

MICP is an environmentally friendly technology recognized for its crack-healing properties [30]. It enables the synthesis of calcium carbonate from a highly saturated solution through microbial activity and biochemical processes. During MICP, microorganisms release metabolic byproducts that interact with surrounding ions, resulting in mineral precipitation [31]. The process is straightforward, efficient, and capable of producing significant calcium carbonate within a short time [32]. Urease activity affects biomineralization through four key factors: pH, dissolved inorganic carbon, calcium concentration, and the availability of nucleation sites. While the first three factors influence the concentration of carbonate ions ($CO_2{}^{2-}$), nucleation sites are essential for the stable formation of $CaCO_2$ [33] (Fig. 1).

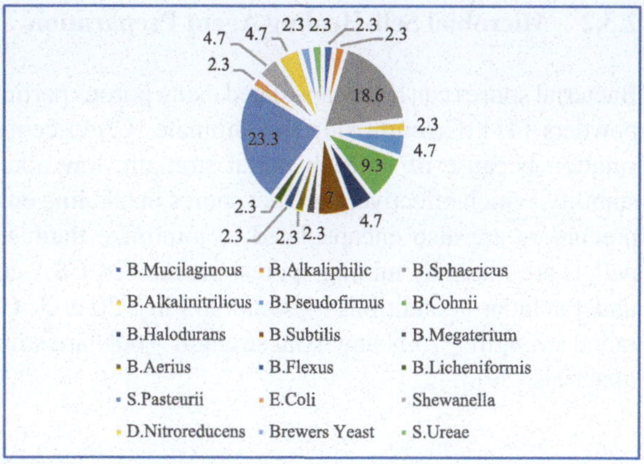

Fig. 1 Statistics of different bacteria application in cementitious material [34]

2.3 *Bacillus Megaterium as Enhancer*

Bacillus megaterium is an aerobic, gram-positive, spore-forming bacterium that grows as pairs or chains of cells linked by polysaccharides on their cell walls. It exhibits a broad temperature tolerance, thriving between 3 and 45 °C [35, 36]. Studies show that incorporating *B. megaterium* into concrete enhances its mechanical properties, demonstrating it's effectiveness in such applications [37, 38]. Common cultivation media for *B. megaterium* include various combinations of beef extract, peptone, yeast extract, sodium chloride, and wheat bran, typically at pH 8 [39, 40]. For calcium carbonate ($CaCO_3$) precipitation, a culture medium containing 2% (w/v) urea, 25 mM $CaCl_2$, and bacterial cells (10^7 cells/ml) is incubated at 37 °C in a rotating shaker (120 rpm) for 24 h to stimulate urease production [41].

2.3.1 Optimum Concentration of *B. Megaterium*

The required quantity of culture medium is determined through an optical density test using spectrophotometry [42]. Studies involving bacterial concentrations ranging from 10^3 to 10^7 cfu/ml in concrete have shown that a dilution factor of 10^5 cfu/ml notably enhanced compressive strength at various curing ages. SEM and XRD analyses showed that concrete with 10^5 cfu/ml bacteria had reduced porosity due to calcite filling, compared to other concentrations. Additionally, study tests demonstrated that bacterial concentrations between 10×10^5 and 50×10^5 cfu/ml revealed that 30×10^5 cfu/ml optimized both compressive and flexural strengths. Higher concentrations, however, led to reduced strength due to nutrient competition among the bacteria [37].

2.3.2 Microbial Self-Healing Agent Preparation

Bacterial spores can be encapsulated using porous particles, polymers, or compressed powders [43]. Calcium sulpho-aluminate (CSA) cement is well-suited for encapsulation because of its high initial strength, low alkalinity, water resistance, and stability, which effectively protect spores in alkaline environments [39, 44]. Nutrient precursors are also encapsulated to minimize their effect on concrete. Microbial pellets are made by mixing spores, precursors, CSA cement, and water, then cured under standard conditions ($\geq$90% humidity, 20 $\pm$ 3 °C) for three days. Pellets with radial strength $\geq$75 N and axial strength $\geq$60N are suitable for use in cement-based materials [39].

2.3.3 Efficiency of Encapsulated Spores

The viability of spores, both with and without encapsulation, is evaluated using OD600 values. Germination tests are performed after 28 and 56 days of immersion in a simulated pore solution consisting of $Ca(OH)_2$ (0.002 mol/L), NaOH (0.25 mol/L), KOH (0.54 mol/L), and K_2SO_4 (0.003 mol/L) [45]. After 28 days, non-encapsulated spores show an OD600 of 0.5, significantly lower than normal, and barely germinate after 56 days, confirming that high alkalinity reduces germination over time [46]. Encapsulated spores, however, maintain consistent germination rates comparable to unsoaked spores, indicating effective protection from encapsulation [39].

2.3.4 Preparation of Cracks

Cracks are prepared by breaking the specimen and embedding it [45, 47]. Microcracks, 0.4–0.5 mm wide, are created using 0.4 mm copper wire. A dry–wet cycle, involving 12 h of water immersion followed by 12 h of air exposure at 25$\pm$1 °C, facilitates oxygen penetration [39]. For beam specimens, a 0.3 mm thick copper plate is inserted 10 mm deep into fresh concrete and removed before the final setting [48].

2.3.5 Effect on Mechanical Properties

Compressive Strength

B. megaterium species show the ability to improve compressive strength by 24% [21]. The compressive strength of the concrete specimen with *Bacillus Megaterium* MTCC 1684 was obtained to be 16.1% higher than the control specimen at 28 days of age. Also, for a concrete specimen with B. megaterium BSKAU the 28-day compressive strength was found to be 12.1% greater than the control specimen [48].

Split Tensile Strength

The concrete specimen containing *B. megaterium* exhibits an 18.29% increase in split tensile strength compared to conventional concrete [38].

Permeability

Total porosity of the bacteria-treated specimen is estimated to be 17.4% while for control specimens it's about 25.3% which indicates a decrease of 31% in porosity due to calcification in the bacterial specimen [41].

2.3.6 Characterization of Crack Healing Efficiency

The self-healing properties of concrete can be assessed through several methods, such as the diffusion coefficient of chloride ions, the capillary suction water coefficient before and after repair, ultrasonic pulse velocity, and the corrosion rate of reinforcement under accelerated electromigration [31, 49].

Repairing Efficiency and Permeability

Stereomicroscope is employed to capture images of cracks from the same height, magnification, and light source both before and after the cracks are repaired. These images can then be binarized using software, and by analyzing the grey threshold and pixel statistics, the number of pixels in the crack area can be determined. The area repair ratio, which indicates the efficiency of the healing process, can then be calculated.

$$\text{Area repairing ratio } \frac{A - At}{A} \times 100\% \text{ [39, 47]}$$

A number of initial crack area pixel dots
A_t number of crack area pixel dots after healing.

Cracks create pathways for water to infiltrate concrete, but self-healing reduces these cracks, thereby lowering seepage velocity, which serves as an indicator of self-healing effectiveness [39]. Water permeability testing follows Luo's method [47], where a cylindrical specimen (100 × 50 mm) is sealed with epoxy resin and connected to a PVC pipe (100 × 1200 mm). The water column height is kept constant, and water flow through cracks is measured over 5 min.

Rapid Chlorine Permeability Test (RCPT) and Healing Depth Testing Method

RCPT is widely used to assess self-healing by evaluating chloride transmission through cracks [45, 50]. After 28 days of repair, a specimen is placed in a sealed chamber, and a 60 V voltage is applied. The electric flux is measured over 6 h, with changes in flux indicating the extent of repair, as conductivity correlates with chloride permeability [39, 51]. The area repairing ratio reflects surface crack healing but not internal repair. Similarly, permeability tests and RCPT assess durability improvements after healing. X-CT evaluates full crack repair but struggles to distinguish repair products from the matrix. Alternatively, splitting the specimen and observing healing depth under a stereomicroscope offers better clarity [39, 52].

SEM Analysis and Crack Healing Property

SEM analysis, equipped with an Energy Dispersive Spectrometer (EDS), is used to visualize cementation, analyze pores, and detect $CaCO_3$ precipitates in samples with and without bacteria [53]. EDS provides spot or field chemical analysis, measuring calcium content [37]. SEM delivers high-resolution, high-depth images, enabling the study of morphology and material composition in a non-destructive, cost-effective manner. It uses a low-energy electron beam to scan specimen surfaces, with emitted photons and electrons detected to form images. Typical SEM setups include an electron gun (e.g., tungsten) for high-resolution imaging, electromagnetic lenses for focused beams, and high-vacuum environments to ensure free electron movement [54]. Broken mortar samples are powdered, oven-dried, and gold/platinum-coated before analysis. SEM imaging reveals micropores in expanded perlite, making it suitable as a carrier for bacterial spores [55, 56].

However, concrete heals through autogenous and autonomous healing. Autogenous healing occurs through the hydration of desiccated cement, while autonomous healing requires an activation trigger [57, 58]. Bacteria in concrete form endospores that withstand stresses during mixing [59]. When cracks develop, water and air ingress trigger the germination of endospores into vegetative cells, initiating calcite precipitation [48]. For cracks of 0.4–0.5 mm, 95% of healing occurred in *B. megaterium*-induced concrete after 28 days, compared to 24% for control specimens undergoing autogenous healing. The healing depth was 100 μm for the reference mix, 1850 μm for *B. megaterium* with glucose, and 4000 μm with calcium lactate after 28 days [39].

2.3.7 Use of Bacillus Megaterium with Other Bacteria

Compressive Strength and Crack Healing

The concrete specimen with co-cultured bacteria, *B. megaterium* and *B. cohnii*, showed an increase in compressive strength of 41.04%, while specimens with *B.*

megaterium and *B. sphaericus* exhibited a 27.20% increase. In contrast, the individual application of each bacterium resulted in a compressive strength increase of 28.03% for *B. megaterium*, 39.63% for *B. cohnii*, and 19.89% for *B. sphaericus* [60].

After 28 days of curing, the best crack healing was observed in the co-cultured mix of *B. megaterium* and *B. cohnii*. In contrast, the mix of the two ureolytic bacteria (*B. megaterium* and *B. sphaericus*) showed incomplete crack filling with less precipitate. SEM analysis revealed that the highest calcite formation occurred in the mix combining the ureolytic bacterium *B. megaterium* and the non-ureolytic bacterium *B. cohnii* [60].

2.3.8 Use of Metakaolin in Bacillus Megaterium (MTCC 3353) Induced Concrete

Supplementary cementitious materials (SCMs) are increasingly used as cement replacements due to their pozzolanic and filler effects [61], which enhance concrete properties [62]. Metakaolin, a reactive nanocrystalline aluminosilicate produced by calcining kaolin clay, reacts with cement hydration byproducts, accelerating calcium hydroxide utilization and improving concrete's mechanical properties [63]. Studies show that bacterial specimens with 10%, 12.5%, and 15% metakaolin replacement exhibit compressive strength increases of 22%, 13.8%, and 9.7%, respectively, while flexural strength improves by 19.04% with 10% replacement. Although acidic conditions typically reduce concrete strength, 28-day acid immersion results indicate strength improvements of 43%, 28.6%, and 14.3% for the respective replacements, attributed to bacterial interactions with metakaolin's secondary cementitious products [64].

2.4 Bacillus Cohnii

Bacillus Cohnii, a non-ureolytic alkaliphilic bacterium that survives up to 70 °C [65, 66], produces $CaCO_3$ through bio-mineralization [67]. Its spores can withstand high mechanical forces and remain viable for up to 200 years under unfavourable conditions. Calcium compounds such as calcium lactate, calcium glutamate, and calcium chloride support bacterial growth. To culture *Bacillus Cohnii*, a nutrient medium was prepared with 10 g polypeptone, 2 g yeast extract, 0.5 g $MgSO_4 \cdot 7H_2O$, and 10 g Na_2CO_3 per liter of distilled water. The medium was sterilized by autoclaving at 120 °C for 15 min and then distributed into six flasks. The pH of each flask was adjusted to 9, 10, 11, 12, and 12.5 using HCl. *Bacillus Cohnii* was inoculated ($A^{600} = 0.1$) and incubated at 37 °C with shaking at 100 rpm for 48 h. Maximum bacterial growth was indicated by the highest absorbance, with optimal growth observed at pH 10 [55].

2.4.1 Culturing Process and Bacteria Encapsulation

Kumari et al. [55] cultured *Bacillus Cohnii* in liquid medium for 48 h at various pH levels, using Colony Forming Units (CFUs) and the spread plate method to assess growth. After serial dilution and plating, CFUs were counted at 37 °C. Besides the growth was 0.15×10^7 CFU/ml at pH 9, 6.05×10^7 CFU/ml at pH 10, 2.79×10^7 CFU/ml at pH 11, 2.95×10^7 CFU/ml at pH 12, and 0.21×10^7 CFU/ml at pH 12.5. The optimal pH for *Bacillus Cohnii* growth was 10 ($6.05 \times 10_7$ CFU/ml).

Additionally, the Encapsulating bacterial spores with nutrients enhance their performance and longevity, making bacterial concrete viable in harsh environments. The outer capsule protects the bacteria from damage during mixing and setting. Encapsulation has proven more effective than direct addition, enabling efficient self-healing through calcium carbonate precipitation [68]. Encapsulation methods include: Diatomaceous Earth (DE) have been used to protect bacteria from high pH [69], with optimal immobilization at 60%. Further, the coated expanded clay with geopolymer heals the cracks up to 0.79 mm [70–72]. Materials like metakaolin and activating fluids to encapsulate bacteria ensured proper interaction with concrete [73]. Graphite Nano Particles (GNP) heal cracks up to 0.38 mm and increase compressive strength by 9.8% [74].

2.4.2 Bacteria in Concrete in Spore Form

Bacterial spores remain dormant in unfavourable conditions but activate upon crack formation, interacting with water and nutrients to precipitate calcium carbonate, filling the cracks and returning to a dormant state once healing is complete [75]. X-ray diffraction (XRD) characterizes the material's structure and composition, revealing that calcium carbonate is the precipitated mineral, with the highest peak at $2\theta = 29.42\%$. Microstructure analysis shows that expanded perlite (EP) particles effectively immobilize bacterial spores [54, 56, 76].

Pal et al. (2022) incorporated *Bacillus cohnii* into M30 grade concrete, resulting in a 39.63% increase in compressive strength. One researcher reported a 21% increase in compressive strength at 7 and 28 days in specimens containing *Bacillus cohnii* without calcium lactate, and a 25% increase with the addition of calcium lactate. Other research has reported a 59.74% increase at 7 days and 49.18% at 28 days with *Bacillus Cohnii*. Additionally, observations showed similar strength gains with *Bacillus cohnii*, with a 39.63% increase in compressive strength at 7 days and a 41.04% increase at 28 days, attributed to calcium carbonate precipitation.

2.4.3 Self-Healing Ability

Researchers have evaluated the performance of MICP using *Bacillus cohnii* in bio-concrete [56]. Crack widths (0.27–1.24 mm) healed almost completely after 28 days, with healing indicated by white deposits from bacterial activity. Specimens without bacteria showed no healing, while those with bacteria started healing at 7 days, improving significantly by 28 days. *B. cohnii* and mixed cultures (*B. cohnii* + *B. megaterium*) demonstrated superior results, particularly in fly ash-blended concrete [60, 60]. Immobilized bacteria converted organic compounds into calcium carbonate, sealing cracks. EP-B specimens healed cracks up to 0.79 mm, EC-B up to 0.45 mm, and control specimens showed no healing [72]. A 16-h wet and 8-h dry cycle enhanced healing was observed using a Leica M205C microscope over 7 to 84 days [77, 78].

Cracks (0.02–2 mm) achieved up to 60% sealing with *B. cohnii* [79]. Complete healing occurred for cracks $\leq$ 1.2 mm within 28 days [78, 80]. Encapsulated bacterial capsules (PNC) improved healing consistency and water transport recovery, achieving 100% permeability reduction and full crack closure (<350 microns) after 40–50 cycles [75, 81–83]. Bacterial concrete showed a self-healing efficiency of 74.5%, outperforming reference concrete (63.8%) [9, 84–87].

2.4.4 Split Tensile Strength and Drying Shrinkage

Studies have reported tensile strength increases of 13% and 25% in concrete with *B. cohnii* bio-agents, without, and with calcium lactate, respectively [88]. Also, some of the observations have significantly reduced drying shrinkage (0.8%–0.05%) in bacterial specimens compared to control samples (0.38%) when tested per ASTM C531:2005.

2.4.5 Decrease in Water Absorption and Permeability

Jiang et al. (2020) [56] have utilized *Bacillus Cohnii* (1.1×10^9 cells/ml) with a mixed nutrient solution of yeast extract, inosin, calcium lactate of 1.5 g/L, 2.68 g/L, and 20.8 g/L respectively. This has depicted about 94% of the decrement of water absorption. In the same experiment, permeability coefficients were found to decrease gradually. As healing time reached 14 days, specimens could withstand water pressure up to 1.1 cm head height, and at 28 days specimens could withstand up to 1.7 cm head height [74, 89].

2.4.6 Chloride Penetration and Flexural Strength

In chloride migration tests, the penetration depth of chloride ions was shallower in Bio-AAFC (17 mm) compared to AAFC (26 mm), showing that MICP effectively drops the pore networks, enhancing durability [85, 90], and percentage increment in flexure strength after 3, 7, and 28 days respectively [91, 92].

BC10^5	−2%, −3%, 4%
BC10^7	−19%, 16%, 22%
BC10^9	−12%, 22%, 14%

DBC10^5 −18%, 16%, 19%
DBC10^7 −24%, 24%, 6%
DBC10^9 −21%, 9%, −4%.

2.4.7 B. Cohnii with Ureolytic Bacteria

Pal et al. [60] investigated the use of Bacillus sphaericus, Bacillus megaterium, and non-ureolytic *Bacillus cohnii*, both individually and in combination, in concrete specimens. Concrete mixes (M1–M7) were prepared, including combinations like *B. cohnii* + *B. megaterium* (M6) and *B. sphaericus* + *B. cohnii* (M5). M6 showed the highest compressive strength increase (47.29 MPa, 41.04%), while *B. sphaericus* alone had the least improvement. Fly ash-blended mixes showed reduced strength due to void filling and inadequate hydration. Chaurasia et al. (2019) tested *B. megaterium*, *B. pasteurii*, and *B. cohnii* (MTCC strains) in M30 concrete. The mix with *B. megaterium* achieved the highest compressive strength [88]. Mixed bacterial cultures consistently outperformed single strains in enhancing compressive strength.

2.5 *Bacillus Subtilis*

Bacillus subtilis, a Gram-positive, rod-shaped, spore-forming bacterium, is highly adaptable to harsh environments and is commonly found in soil as well as in the digestive tracts of humans and mammals. It thrives at 25–35 °C, taking 3–4 days to grow. When concrete cracks are exposed to moisture, *B. subtilis* metabolizes calcium lactate into calcium carbonate, aiding self-healing. Researchers introduced bacteria into concrete via direct and encapsulation methods, incorporating *B. subtilis* with nutrient agar, LB media, or soybean media into conventional concrete materials.

2.5.1 Compressive Strength

Compressive strength of bacterial concrete depicts a significant improvement over conventional concrete. After 28 days, the strength improvement is 14.36%, 12.58%, and 15.86% for *B. subtilis*, *B. magaterium*, and their combination, respectively. *B. magaterium* initially showed lower strength but surpassed *B. subtilis* after 28 days due to slower calcium carbonate precipitation. Adding bacteria, nutrients, and oxygen reduced porosity and increased strength, with calcite filling inner cracks [38, 74, 82].

B. subtilis improved compressive strength regardless of incorporation methods, with bacterial concrete showing a 20% strength increase over control specimens. Fly ash and coir fibres also enhanced strength, with bacterial and coir fibre mixes achieving a 42% improvement. Optimal bacterial cell concentrations (10^3 to 10^5 cells/ml) yielded a 15–27% increase after 28 days. Higher concentrations (10^7) reduced strength due to non-reactive materials limiting hydration [80, 83, 86].

Silica fume addition improved compressive strength by 18%. Specimens with calcium lactate and *B. subtilis* showed a 12% strength increase due to calcium carbonate filling. Dead bacterial cells did not enhance strength, while live cells decreased it at higher concentrations. Advanced-grade bacterial concrete achieved up to 45 MPa with these modifications [78, 87, 90, 92–99]

2.5.2 Split Tensile and Flexure Strength

Split tensile strength: It was observed indirectly employing compressive strength. The result showed that there was an increase in tensile strength concerning conventional concrete by 25.3%, 18.29% and 19.5% for the *B. subtilis*, *B. magaterium* and the addition of two bacteria respectively for 28 days [38]. The enhancement in split tensile strength is a result of the EPS layer formation facilitated by the bacterial strain [82]. Split tensile strength of bacteria concrete value was higher as compared to control specimens at 3, 7 and 28 days. But BNMPS attained lower 28 days split tensile strength [92].

The enrichment of flexural strength is attributed to the urease enzyme secreted by the biomaterial [82]. It can regain flexure strength up to 14%.[93] Maximum Flexural strength was shown up to 15% replacement of coarse aggregate after 28 days of curing at 2% of *B. subtilis*. Flexural strength noted was 10.2 N/mm^2 [87].

2.5.3 Crack Healing

Concrete specimens with bacteria added directly to the mix showed no crack healing. However, using GNP as a carrier improved bacterial protection and distribution, achieving maximum crack healing efficiency in specimens pre-cracked at 3 and 7 days. In later days, healing efficiency decreased. Specimens with lightweight aggregates as carriers were less effective for early-age cracks but showed consistent healing for later-age cracks [74]. *B. subtilis* healed cracks up to 0.3 mm but struggled with deeper cracks due to high alkalinity and low pH in inner layers [93]. Iron Oxide Micro Particles (INMPS) sealed cracks up to 1.2 mm, regaining 85% compressive strength. In contrast, BNMPS reduced healing efficiency but served as the immobilization material [92].

Fibre-reinforced bacterial concrete exhibited 75–85% surface healing at 7 days and 60–65% at 28 days, with *B. sphaericus* showing superior calcite formation compared to *B. subtilis* and *B. cohnii* [78]. Pre-cracked specimens healed cracks up to 0.8 mm in 7 days and 0.5 mm in 28 days, aided by agglomerated and lamellar calcium carbonate crystals. Microcracks (~0.2 mm) underwent autogenous healing through CO_2 reacting with calcium hydroxide to form calcium carbonate, facilitating self-healing [98–100].

2.5.4 Water Absorption and Chloride Penetration

Bacterial concrete exhibits reduced water absorption compared to normal concrete due to pore and crack filling by bacterial calcium carbonate precipitates [82]. With *B. subtilis* and barchip fibres, water penetration depth decreased by up to 50% when a calcium source was present. Bacterial surface treatment with calcium lactate solution reduced water penetration depth by 60% [94]. Specimens with S. pasteurii exhibited lower water absorption than those with B. subtilis, with reductions of up to 37% [90]. Bacterial concrete reduced pore continuity, leading to lower water absorption [86, 95, 101]. Lightweight concrete with bacteria and steel fibres, cured in urea-calcium lactate, with further decrement of water absorption. After 28 days of curing, bacterial treatments enhanced water resistance in shotcrete [102].

Chloride penetrations were decreased in bacteria-induced specimens than in control specimens[86]. It was noted that specimens containing S. pasteurii had lower steel corrosion, and chloride ion penetration value than in specimens of *B. subtilis* [90].

2.5.5 Water Penetration

Penetration depth was reduced by 44.3% in bs-p-c mix design as compared to the control specimen [101]. The addition of bacteria, fiber reinforcement (barchip fibers) and calcium source in concrete mixture results in 50% reduction in water penetration depth [103, 104].

3 Conclusion

The literature analysis indicated that self-healing concrete effectively extends the lifespan of structures while minimizing maintenance and repair needs through the autonomous repair of cracks. This advanced material enhances both mechanical properties, such as compressive and tensile strength, and durability by utilizing Mechanically Induced Calcite Precipitation (MICP) technology with ureolytic and non-ureolytic bacteria. Additionally, it reduces water absorption by sealing voids, thereby improving overall structural stability. Consequently, crack expansion is prevented, ensuring structural integrity and enhanced performance. Furthermore, self-healing concrete improves resistance to environmental factors such as freezing, water ingress, and reinforcement corrosion, contributing to the long-term sustainability of concrete structures. Henceforth, different researchers have demonstrated that the combined utilization of ureolytic and non-ureolytic bacteria results in more effective crack healing compared to their application.

References

1. De Muynck, W., De Belie, N., & Verstraete, W. (2010). Microbial carbonate precipitation in construction materials: A review. *Ecological Engineering, 36*, 118–136. https://doi.org/10.1016/j.ecoleng.2009.02.006
2. Muhammad, N. Z., Keyvanfar, A., Muhd, M. Z., Shafaghat, A., & Mirza, J. (2015). Waterproof performance of concrete: A critical review on implemented approaches. *Construction and Building Materials, 101*, 80–90. https://doi.org/10.1016/j.conbuildmat.2015.10.048
3. Justo-Reinoso, I., Arena, N., Reeksting, B. J., Gebhard, S., & Paine, K. (2023). Bacteria-based self-healing concrete—a life cycle assessment perspective. *Developments in the Built Environment, 16*. https://doi.org/10.1016/j.dibe.2023.100244.
4. Khan, M. I. (2003). Isoresponses for strength, permeability and porosity of high-performance mortar. www.sciencedirect.comwww.elsevier.com/locate/buildenv
5. Dhami, N. K., Reddy, M. S., & Mukherjee, M. S. (2013). Biomineralization of calcium carbonates and their engineered applications: A review. *Frontiers in Microbiology, 4*. https://doi.org/10.3389/fmicb.2013.00314
6. Van Tittelboom, K., & De Belie, N. (2013). Self-healing in cementitious materials—a review. *Materials, 6*, 2182–2217. https://doi.org/10.3390/ma6062182
7. Liu, C., Xing, L., Liu, H., Huang, W., Nong, X., & Xu, X. (2021). Experimental on repair performance and complete stress-strain curve of self-healing recycled concrete under uniaxial loading. *Construction and Building Materials, 285*. https://doi.org/10.1016/j.conbuildmat.2021.122900
8. Dhami, N. K., Reddy, M. S., & Mukherjee, A. (2014). Synergistic role of bacterial urease and carbonic anhydrase in carbonate mineralization. *Applied Biochemistry and Biotechnology, 172*, 2552–2561. https://doi.org/10.1007/s12010-013-0694-0
9. Bharti, R. K., Srivastava, S., & Thakur, I. S. (2014). Isolation, purification, characterization and mass spectroscopic analysis of carbonic anhydrase from Serratia sp. for sequestration of carbon dioxide and formation of calcite. *Journal of Environmental Chemical Engineering, 2*, 31–39. https://doi.org/10.1016/j.jece.2013.11.018
10. Jiang, N. J., Wang, Y. J., Chu, J., Kawasaki, S., Tang, C. S., Cheng, L., Du, Y. J., Shashank, B. S., Singh, D. N., Le Han, X., & Wang, Y. Z. (2022). Bio-mediated soil improvement: An introspection into processes, materials, characterization and applications. *Soil Use and Management, 38*, 68–93. https://doi.org/10.1111/sum.12736
11. Zhu, T., & Dittrich, M. (2016). Carbonate precipitation through microbial activities in natural environment, and their potential in biotechnology: A review. *Frontiers in Bioengineering and Biotechnology, 4*. https://doi.org/10.3389/fbioe.2016.00004
12. Whiffin, V. S., van Paassen, L. A., & Harkes, M. P. (2007). Microbial carbonate precipitation as a soil improvement technique. *Geomicrobiology Journal, 24*, 417–423. https://doi.org/10.1080/01490450701436505
13. Martin, K., Tirkolaei, H. K., & Kavazanjian, E. (2021). Enhancing the strength of granular material with a modified enzyme-induced carbonate precipitation (EICP) treatment solution. *Construction and Building Materials, 271*. https://doi.org/10.1016/j.conbuildmat.2020.121529
14. Wu, M., Hu, X., Zhang, Q., Zhao, Y., Sun, J., Cheng, W., Fan, Y., Zhu, S., Lu, W., & Song, C. (2020). Preparation and performance evaluation of environment-friendly biological dust suppressant. *Journal of Cleaner Production, 273*. https://doi.org/10.1016/j.jclepro.2020.123162
15. Jagannathan, P., Satya Narayanan, K. S., Arunachalam, K. D., & Annamalai, S. K. (2018). Studies on the mechanical properties of bacterial concrete with two bacterial species. *Materials Today Proceedings,* Elsevier Ltd, pp 8875–8879. https://doi.org/10.1016/j.matpr.2017.12.320
16. Castro-Alonso, M. J., Montañez-Hernandez, L. E., Sanchez-Muñoz, M. A., Macias Franco, M. R., Narayanasamy, R., & Balagurusamy, N. (2019). Microbially induced calcium carbonate precipitation (MICP) and its potential in bioconcrete: Microbiological and molecular concepts. *Frontiers in Materials, 6*. https://doi.org/10.3389/fmats.2019.00126

17. Mwandira, W., Nakashima, K., & Kawasaki, S. (2017). Bioremediation of lead-contaminated mine waste by Pararhodobacter sp. based on the microbially induced calcium carbonate precipitation technique and its effects on strength of coarse and fine grained sand. *Ecological Engineering, 109*, 57–64. https://doi.org/10.1016/j.ecoleng.2017.09.011
18. Wang, Z., Su, J., Ali, A., Zhang, R., Yang, W., Xu, L., & Zhao, T. (2021). Microbially induced calcium precipitation based simultaneous removal of fluoride, nitrate, and calcium by Pseudomonas sp. WZ39: Mechanisms and nucleation pathways. *Journal of Hazardous Materials, 416*. https://doi.org/10.1016/j.jhazmat.2021.125914
19. Reinhardt, H. W., & Jooss, M. (2003). Permeability and self-healing of cracked concrete as a function of temperature and crack width. *Cement and Concrete Research, 33*, 981–985. https://doi.org/10.1016/S0008-8846(02)01099-2
20. Huang, H., & Ye, G. (2012). Simulation of self-healing by further hydration in cementitious materials. *Cement and Concrete Composites, 34*, 460–467. https://doi.org/10.1016/j.cemconcomp.2012.01.003
21. Algaifi, H. A., Bakar, S. A, Sam, A. R. M., Ismail, M., Abidin, A. R. Z., Shahir, S., & Altowayti, W. A. H. (2020). Insight into the role of microbial calcium carbonate and the factors involved in self-healing concrete. *Construction and Building Materials, 254*. https://doi.org/10.1016/j.conbuildmat.2020.119258
22. Ruan, S., Qiu, J., Weng, Y., Yang, Y., Yang, E. H., Chu, J., & Unluer, C. (2019). The use of microbial induced carbonate precipitation in healing cracks within reactive magnesia cement-based blends. *Cement and Concrete Research, 115*, 176–188. https://doi.org/10.1016/j.cemconres.2018.10.018
23. Nguyen, T. H., Ghorbel, E., Fares, H., & Cousture, A. (2019). Bacterial self-healing of concrete and durability assessment. *Cement and Concrete Composites, 104*, Article 103340. https://doi.org/10.1016/j.cemconcomp.2019.103340
24. Wiktor, V., & Jonkers, H. M. (2016). Bacteria-based concrete: From concept to market. *Smart Materials and Structures, 25*. https://doi.org/10.1088/0964-1726/25/8/084006
25. Burne, R. A., & Chen, Y. Y. M. (2000). Bacterial ureases in infectious diseases, *2*(5), 533–542. https://doi.org/10.1016/s1286-4579(00)00312-9
26. Hammes, F., Boon, N., de Villiers, J., Verstraete, W., & Siciliano, S. D. (2003). Strain-specific ureolytic microbial calcium carbonate precipitation. *Applied and Environment Microbiology, 69*, 4901–4909. https://doi.org/10.1128/AEM.69.8.4901-4909.2003
27. Fujita, Y., Taylor, J. L., Gresham, T. L. T., Delwiche, M. E., Colwell, F. S., Mcling, T. L., Petzke, L. M., & Smith, R. W. (2008). Stimulation of microbial urea hydrolysis in groundwater to enhance calcite precipitation. *Environmental Science and Technology, 42*, 3025–3032. https://doi.org/10.1021/es702643g
28. Fujita, Y., Ferris, F. G., Lawson, R. D., Colwell, F. S., & Smith, R. W. (2000). Subscribed content calcium carbonate precipitation by ureolytic subsurface bacteria. *Geomicrobiology Journal, 17*(4), 305–318
29. Mitchell, A. C., & Ferris, F. G. (2005). The coprecipitation of Sr into calcite precipitates induced by bacterial ureolysis in artificial groundwater: Temperature and kinetic dependence. *Geochimica et Cosmochimica Acta, 69*, 4199–4210. https://doi.org/10.1016/j.gca.2005.03.014
30. Anbu, P., Kang, C. H., Shin, Y. J., & So, J. S. (2016). Formations of calcium carbonate minerals by bacteria and its multiple applications. *Springerplus, 5*, 250. https://doi.org/10.1186/s40064-016-1869-2
31. Qian, C. X., Zheng, T. W., & Rui, Y. F. (2021). Living concrete with self-healing function on cracks attributed to inclusion of microorganisms: Theory, technology and engineering applications—a review. *Science China Technological Science, 64*, 2067–2083. https://doi.org/10.1007/s11431-021-1879-6
32. Dhami, N. K., Reddy, M. S., & Mukherjee, A. (2013). Biomineralization of calcium carbonates and their engineered applications: A review. *Frontiers in Microbiology, 4*. https://doi.org/10.3389/fmicb.2013.00314

33. Hammes, F., & Verstraete, W. (2002). Key roles of pH and calcium metabolism in microbial carbonate precipitation. http://welcome.to/labmet
34. Nguyen, T. H., Ghorbel, E., Fares, H., & Cousture, A. (2019). Bacterial self-healing of concrete and durability assessment. *Cement and Concrete Composites 104,* 103340.
35. Yatish Reddy, P. V., Ramesh, B., & Prem Kumar, L. (2020). Influence of bacteria in self-healing of concrete—a review. *Materials Today Proceedings*. Elsevier Ltd, pp 4212–4218. https://doi.org/10.1016/j.matpr.2020.07.233
36. Pu, L., Zeng, Y. J., Xu, P., Li, F. Z., Zong, M. H., Yang, J. G., & Lou, W. Y. (2020). Using a novel polysaccharide BM2 produced by Bacillus megaterium strain PL8 as an efficient bio-flocculant for wastewater treatment. *International Journal of Biological Macromolecules, 162*, 374–384. https://doi.org/10.1016/j.ijbiomac.2020.06.167
37. Andalib, R., Abd Majid, M. Z., Hussin, M.W., Ponraj, M., Keyvanfar, A., Mirza, J., & Lee, H. S. (2016). Optimum concentration of Bacillus megaterium for strengthening structural concrete. *Construction and Building Materials, 118*, 180–193. https://doi.org/10.1016/j.conbuildmat.2016.04.142
38. Nain, N., Surabhi, R., Yathish, N. V., Krishnamurthy, V., Deepa, T., & Tharannum, S. (2019). Enhancement in strength parameters of concrete by application of Bacillus bacteria. *Construction and Building Materials, 202*, 904–908. https://doi.org/10.1016/j.conbuildmat.2019.01.059
39. Su, Y., Zheng, T., & Qian, C. (2021). Application potential of Bacillus megaterium encapsulated by low alkaline sulphoaluminate cement in self-healing concrete. *Construction and Building Materials, 273*. https://doi.org/10.1016/j.conbuildmat.2020.121740
40. Vashisht, R., & Shukla, A. (2020). Potential application of bacteria to improve the self-healing and strength of concrete. *Journal of Building Pathology and Rehabilitation, 5*. https://doi.org/10.1007/s41024-020-0073-5
41. Dhami, N. K., Reddy, M. S., & Mukherjee, A. (2013). Bacillus megaterium mediated mineralization of calcium carbonate as biogenic surface treatment of green building materials. *World Journal of Microbiology & Biotechnology, 29*, 2397–2406. https://doi.org/10.1007/s11274-013-1408-z
42. Vijay, K., & Murmu, M. (2019). Effect of calcium lactate on compressive strength and self-healing of cracks in microbial concrete. *Frontiers of Structural and Civil Engineering, 13*, 515–525. https://doi.org/10.1007/s11709-018-0494-2
43. Justo-Reinoso, I., Heath, A., Gebhard, S., & Paine, K. (2021). Aerobic non-ureolytic bacteria-based self-healing cementitious composites: A comprehensive review. *Journal of Building Engineering, 42*. https://doi.org/10.1016/j.jobe.2021.102834
44. Xu, J., & Wang, X. (2018). Self-healing of concrete cracks by use of bacteria-containing low alkali cementitious material. *Construction and Building Materials, 167*, 1–14. https://doi.org/10.1016/j.conbuildmat.2018.02.020
45. Zheng, T., Su, Y., Qian, C., & Zhou, H. (2020). Low alkali sulpho-aluminate cement encapsulated microbial spores for self-healing cement-based materials. *Biochemical Engineering Journal, 163*, Article 107756. https://doi.org/10.1016/j.bej.2020.107756
46. Jonkers, H. M., Thijssen, A., Muyzer, G., Copuroglu, O., & Schlangen, E. (2010). Application of bacteria as self-healing agent for the development of sustainable concrete. *Ecological Engineering, 36*, 230–235. https://doi.org/10.1016/j.ecoleng.2008.12.036
47. Luo, M., Qian, C., & Li, R. (2015). Factors affecting crack repairing capacity of bacteria-based self-healing concrete. *Construction and Building Materials, 87*, 1–7. https://doi.org/10.1016/j.conbuildmat.2015.03.117
48. Krishnapriya, S., & Babu, D. L. V. (2015). Isolation and identification of bacteria to improve the strength of concrete. *Microbiological Research 174*, 48–55. https://doi.org/10.1016/j.micres.2015.03.009
49. Ling, H., & Qian, C. (2017). Effects of self-healing cracks in bacterial concrete on the transmission of chloride during electromigration. *Construction and Building Materials, 144*, 406–411. https://doi.org/10.1016/j.conbuildmat.2017.02.160

50. Siddique, R., Nanda, V., Kunal, E. H., Kadri, M. I., Singh, M., & Rajor, A. (2016). Influence of bacteria on compressive strength and permeation properties of concrete made with cement baghouse filter dust. *Construction and Building Materials, 106*, 461–469. https://doi.org/10.1016/j.conbuildmat.2015.12.112
51. Achal, V., Mukerjee, A., & Sudhakara Reddy, M. (2013). Biogenic treatment improves the durability and remediates the cracks of concrete structures. *Construction and Building Materials, 48*, 1–5. https://doi.org/10.1016/j.conbuildmat.2013.06.061
52. Wang, J., Dewanckele, J., Cnudde, V., Van Vlierberghe, S., Verstraete, W., & De Belie, N. (2014). X-ray computed tomography proof of bacterial-based self-healing in concrete. *Cement and Concrete Composites, 53*, 289–304. https://doi.org/10.1016/j.cemconcomp.2014.07.014
53. Park, S. S., Choi, S. G., & Nam, I. H. (2014). Effect of plant-induced calcite precipitation on the strength of sand. *Journal of Materials in Civil Engineering, 26*. https://doi.org/10.1061/(asce)mt.1943-5533.0001029
54. Sampath Kumar, T. S. (2013). Physical and chemical characterization of biomaterials. In: *Characterization of biomaterials* (pp. 11–47). Elsevier Inc. https://doi.org/10.1016/B978-0-12-415800-9.00002-4
55. Kumari, C., Das, B., Jayabalan, R., Davis, R., & Sarkar, P. (2016). Effect of nonureolytic bacteria on engineering properties of cement mortar. https://doi.org/10.1061/(ASCE)
56. Jiang, L., Jia, G., Jiang, C., & Li, Z. (2020). Sugar-coated expanded perlite as a bacterial carrier for crack-healing concrete applications. *Constructions and Building Materials, 232*. https://doi.org/10.1016/j.conbuildmat.2019.117222
57. Tang, W., Kardani, O., & Cui, H. (2015). Robust evaluation of self-healing efficiency in cementitious materials—a review. *Construction and Building Materials, 81*, 233–247. https://doi.org/10.1016/j.conbuildmat.2015.02.054
58. Zhang, W., Zheng, Q., Ashour, A., & Han, B. (2020). Self-healing cement concrete composites for resilient infrastructures: A review. *Composites B Engineering, 189*. https://doi.org/10.1016/j.compositesb.2020.107892
59. Sagripanti, J. L., & Bonifacino, A. (1996). Mailing address: Molecular biology branch (HFZ-113), Center for Devices and Radiological Health, Food and Drug Administration, 5600 Fishers Ln. https://journals.asm.org/journal/aem
60. Pal, S., Tiwari, S., Puria, R., Nain, V., & Prasad Pathak, R. (2022). Impact of ureolytic and nonureolytic bacteria on self healing of artificial cracks in biofortified concrete. *Advances in Materials and Processing Technologies, 8*, 841–860. https://doi.org/10.1080/2374068X.2021.1953919
61. Yildirim, G., Sahmaran, M., & Ahmed, H. U. (2015). Influence of hydrated lime addition on the self-healing capability of high-volume fly ash incorporated cementitious composites. *Journal of Materials in Civil Engineering, 27*. https://doi.org/10.1061/(asce)mt.1943-5533.0001145
62. Ahmed, H. U., Abdalla, A. A., Mohammed, A. S., & Mohammed, A. A. (2022). Mathematical modeling techniques to predict the compressive strength of high-strength concrete incorporated metakaolin with multiple mix proportions. *Cleaner Materials, 5*. https://doi.org/10.1016/j.clema.2022.100132
63. Poon, C. S., Lam, L., Kou, S. C., Wong, Y. L., & Wong, R. (2001). Rate of pozzolanic reaction of metakaolin in high-performance cement pastes. *Cement and Concrete Research, 31*(9), 1301–1306.
64. Rameshkumar, V., Prabhath Ranjan Kumar, S., Poornima, V., Venkatasubramani, R., & Sreevidya, V. (2022). Improvements in mechanical and durability parameters of bio-engineered concrete with metakaolin as a partial substitute for cement. *European Journal of Environmental and Civil Engineering, 26*, 2753–2766. https://doi.org/10.1080/19648189.2020.1767696
65. Spanka, R., & Fritze, D. (1993). *Bacillus cohnii* sp. nov., a new, obligately alkaliphilic, oval-spore-forming bacillus species with ornithine and aspartic acid instead of diaminopimelic acid in the cell wall.

66. Ghorbel, R. E., Maktouf, S., Ben Massoud, E., Bejar, S., & Chaabouni, S. E. (2009). New thermostable amylase from *Bacillus cohnii* US147 with a broad pH applicability. *Applied Biochemisty and Biotechnology, 157*, 50–60. https://doi.org/10.1007/s12010-008-8278-0
67. Xu, J., Yao, W., & Jiang, Z. (2014). Non-ureolytic bacterial carbonate precipitation as a surface treatment strategy on cementitious materials. https://doi.org/10.1061/(ASCE)MT
68. Shahid, S., Aslam, M. A., Ali, S., Zameer, M., & Faisal, M. (2020). Self-healing of cracks in concrete using bacillus strains encapsulated in sodium alginate beads. *ChemistrySelect, 5*, 312–323. https://doi.org/10.1002/slct.201902206
69. Wang, J. Y., De Belie, N., & Verstraete, W. (2012). Diatomaceous earth as a protective vehicle for bacteria applied for self-healing concrete. *Journal of Industrial Microbiology and Biotechnology, 39*, 567–577. https://doi.org/10.1007/s10295-011-1037-1
70. Wiktor, V., & Jonkers, H. M. (2011). Quantification of crack-healing in novel bacteria-based self-healing concrete. *Cement and Concrete Composites, 33*, 763–770. https://doi.org/10.1016/j.cemconcomp.2011.03.012
71. Alazhari, M., Sharma, T., Heath, A., Cooper, R., & Paine, K. (2018). Application of expanded perlite encapsulated bacteria and growth media for self-healing concrete. *Construction and Building Materials, 160*, 610–619. https://doi.org/10.1016/j.conbuildmat.2017.11.086
72. Zhang, J., Liu, Y., Feng, T., Zhou, M., Zhao, L., Zhou, A., & Li, Z. (2017). Immobilizing bacteria in expanded perlite for the crack self-healing in concrete. *Construction and Building Materials, 148*, 610–617. https://doi.org/10.1016/j.conbuildmat.2017.05.021
73. De Koster, S. A. L., Mors, R. M., Nugteren, H. W., Jonkers, H. M., Meesters, G. M. H., & Van Ommen, J. R. (2015). Geopolymer coating of bacteria-containing granules for use in self-healing concrete. *Procedia Engineering*, Elsevier Ltd, pp 475–484. https://doi.org/10.1016/j.proeng.2015.01.193
74. Khaliq, W., & Ehsan, M. B. (2016). Crack healing in concrete using various bio influenced self-healing techniques. *Construction and Building Materials, 102*, 349–357. https://doi.org/10.1016/j.conbuildmat.2015.11.006
75. Jiang, L., Jia, G., Wang, Y., & Li, Z. (2020). Optimization of sporulation and germination conditions of functional bacteria for concrete crack-healing and evaluation of their repair capacity. *ACS Applied Materials & Interfaces, 12*, 10938–10948. https://doi.org/10.1021/acsami.9b21465
76. Omidi, M., Fatehinya, A., Farahani, M., Akbari, Z., Shahmoradi, S., Yazdian, F., Tahriri, M., Moharamzadeh, K., Tayebi, L., & Vashaee, D. (2017). Characterization of biomaterials. In: *Biomaterials for oral and dental tissue engineering* (pp. 97–115). Elsevier. https://doi.org/10.1016/B978-0-08-100961-1.00007-4
77. Tan, L., Reeksting, B., Ferrandiz-Mas, V., Heath, A., Gebhard, S., & Paine, K. (2020). Effect of carbonation on bacteria-based self-healing of cementitious composites. *Constructions and Building Materials, 257*. https://doi.org/10.1016/j.conbuildmat.2020.119501
78. Rauf, M., Khaliq, W., Khushnood, R. A., & Ahmed, I. (2020). Comparative performance of different bacteria immobilized in natural fibers for self-healing in concrete. *Constructions and Building Materials, 258*. https://doi.org/10.1016/j.conbuildmat.2020.119578
79. Sarkar, M., Maiti, M., Xu, S.,& Mandal, S. (2023). Bio-concrete: Unveiling self-healing properties beyond crack-sealing. *Journal of Building Engineering, 74*. https://doi.org/10.1016/j.jobe.2023.106888
80. Abdulkareem, M., Ayeronfe, F., Majid, M. Z. A., Abdul, A. R., & Kim, J. H. J. (2019). Evaluation of effects of multi-varied atmospheric curing conditions on compressive strength of bacterial (*Bacillus subtilis*) cement mortar. *Construction and Building Materials, 218*, 1–7. https://doi.org/10.1016/j.conbuildmat.2019.05.119
81. Cappellesso, V. G., Van Mullem, T., Gruyaert, E., Van Tittelboom, K., & De Belie, N. (2023). Bacteria-based self-healing concrete exposed to frost salt scaling. *Cement Concrete Composites, 139*. https://doi.org/10.1016/j.cemconcomp.2023.105016
82. Durga, C. S. S., Ruben, N., Chand, M. S. R., & Venkatesh, C. (2020). Performance studies on rate of self healing in bio concrete. *Materials Today Proceedings*, Elsevier Ltd, pp 158–162. https://doi.org/10.1016/j.matpr.2019.09.151

83. Pei, R., Liu, J., Wang, S., & Yang, M. (2013). Use of bacterial cell walls to improve the mechanical performance of concrete. *Cement and Concrete Composites, 39*, 122–130. https://doi.org/10.1016/j.cemconcomp.2013.03.024
84. Xiao, X., Unluer, C., Chu, S., & Yang, E. H. (2023). Single bacteria spore encapsulation through layer-by-layer self-assembly of poly(dimethyldiallyl ammonium chloride) and silica nanoparticles for self-healing concrete. *Cement and Concrete Composites, 140*. https://doi.org/10.1016/j.cemconcomp.2023.105105
85. Sarkar, M., Maiti, M., Malik, M. A., & Xu, S. (2022). Evaluation of the crack-healing performance and durability of bacteria integrated alkali-activated fly ash composites. *Journal of Building Engineering, 54*. https://doi.org/10.1016/j.jobe.2022.104642
86. Nosouhian, F., Mostofinejad, D., & Hasheminejad, H. (2016). Concrete durability improvement in a sulfate environment using bacteria. *Journal of Materials in Civil Engineering, 28*. https://doi.org/10.1061/(asce)mt.1943-5533.0001337
87. Rohini, I., & Padmapriya, R. (2020). Effect of *Bacteria subtilis* on e-waste concrete. *Materials Today Proceedings*, Elsevier Ltd, pp 465–474. https://doi.org/10.1016/j.matpr.2020.10.192
88. Chaurasia, L., Bisht, V., Singh, L. P., & Gupta, S. (2019). A novel approach of biomineralization for improving micro and macro-properties of concrete. *Construction and Building Materials, 195*, 340–351. https://doi.org/10.1016/j.conbuildmat.2018.11.031
89. Dipendraditya, K., & Singh, B. K. (2023). Performance of bacillus cohnii added microbial concrete. *Civil Engineering and Architecture, 11*, 525–530. https://doi.org/10.13189/cea.2023.110141
90. Tayebani, B., & Mostofinejad, D. (2019). Penetrability, corrosion potential, and electrical resistivity of bacterial concrete. *Journal of Materials in Civil Engineering, 31*.
91. Skevi, L., Reeksting, B. J., Hoffmann, T. D., Gebhard, S., & Paine, K. (2021). Incorporation of bacteria in concrete: The case against MICP as a means for strength improvement. *Cement and Concrete Composites, 120*. https://doi.org/10.1016/j.cemconcomp.2021.104056
92. Shaheen, N., Khushnood, R. A., Khaliq, W., Murtaza, H., Iqbal, R., & Khan, M. H. (2019). Synthesis and characterization of bio-immobilized nano/micro inert and reactive additives for feasibility investigation in self-healing concrete. *Construction and Building Materials, 226*, 492–506. https://doi.org/10.1016/j.conbuildmat.2019.07.202
93. Feng, J., Chen, B., Sun, W., & Wang, Y. (2021). Microbial induced calcium carbonate precipitation study using *Bacillus subtilis* with application to self-healing concrete preparation and characterization. *Construction and Building Materials, 280*, 122460.
94. Akindahunsi, A. A., Adeyemo, S. M., & Adeoye, A. (2021). The use of bacteria (*Bacillus subtilis*) in improving the mechanical properties of concrete. *Journal of Building Pathology and Rehabilitation, 6*(1). https://doi.org/10.1007/s41024-021-00112-7
95. Rao, M. V. S., Reddy, V. S., & Sasikala, C. (2017). Performance of microbial concrete developed using *Bacillus subtilus* JC3. *Journal of the Institution of Engineers (India) Series A, 98*(4), 501–510. https://doi.org/10.1007/s40030-017-0227-x
96. Vijay, K., & Murmu, M. (2019). Effect of calcium lactate on compressive strength and self-healing of cracks in microbial concrete. *Frontiers of Structural and Civil Engineering, 13*(3), 515–525. https://doi.org/10.1007/s11709-018-0494-2
97. Khaliq, W., & Ehsan, M. B. (2016). Crack healing in concrete using various bio-influenced self-healing techniques. *Construction and Building Materials, 102*, 349–357. https://doi.org/10.1016/j.conbuildmat.2015.11.006
98. Mondal, S., Ghosh, A. D. (2018). Investigation into the optimal bacterial concentration for compressive strength enhancement of microbial concrete. *Construction and Building Materials, 183*, 202–214. https://doi.org/10.1016/j.conbuildmat.2018.06.176
99. Rauf, M., Khaliq, W., Khushnood, R. A., & Ahmed, I. (2020). Comparative performance of different bacteria immobilized in natural fibers for self-healing in concrete. *Construction and Building Materials, 258*(119578), Article 119578. https://doi.org/10.1016/j.conbuildmat.2020.119578
100. Pachaivannan, P., Hariharasudhan, C., Mohanasundram, M., & Anitha Bhavani, M. (2020). Experimental analysis of self-healing properties of bacterial concrete. *Materials Today: Proceedings, 33*, 3148–3154. https://doi.org/10.1016/j.matpr.2020.03.782

101. Salmasi, F., & Mostofinejad, D. (2020). Investigating the effects of bacterial activity on compressive strength and durability of natural lightweight aggregate concrete reinforced with steel fibers. *Construction and Building Materials, 251*(119032), Article 119032. https://doi.org/10.1016/j.conbuildmat.2020.119032
102. Kalhori, H., & Bagherpour, R. (2017). Application of carbonate precipitating bacteria for improving properties and repairing cracks of shotcrete. *Construction and Building Materials, 148*, 249–260. https://doi.org/10.1016/j.conbuildmat.2017.05.074
103. Karimi, N., & Mostofinejad, D. (2020). Bacillus subtilis bacteria used in fiber-reinforced concrete and their effects on concrete penetrability. *Construction and Building Materials, 230*(117051), Article 117051. https://doi.org/10.1016/j.conbuildmat.2019.117051
104. Almohana, A. I., Abdulwahid, M. Y., Galobardes, I., Mushtaq, J., & Almojil, S. F. (2022). Producing sustainable concrete with plastic waste: A review. *Environmental Challenges, 9*, 100626. https://doi.org/10.1016/j.envc.2022.100626

Evaluation on the Impact of Gap Graded Fine Aggregate on the Strength and Porosity Behaviour of Novel Pervious Mortar

P. Anima, R. T. Arjun Siva Rathan, and K. Jayakesh

Abstract Pervious concrete is an emerging material primarily used in waterlogged areas, in walkways and car parks due to its unique property such as efficient runoff drainage and ground water recharge. The main constituents of conventional pervious concrete are water, cement, coarse aggregates and small amount or no fine aggregates. This study aims to develop a pervious mortar by excluding coarse aggregates and using a gap-graded fine aggregate approach. The fine aggregate gradation for the study includes sizes of 4.75 mm, 2.36 mm, 1.18 mm and 600micron. The research explores the preparation of pervious mortar by excluding one specific gradation at a time. For example, a gradation with fine aggregates of 2.36 mm, 1.18 mm, and 600 microns was prepared by omitting the 4.75 mm grade. The aim of excluding different gradations is to develop meso- and macro-pores in the mortar, which could be used to create paving flags for walkways and parking areas. The study evaluates the mechanical, permeable and functional properties of different grades of pervious mortar. The optimum aggregate gradation of pervious mortar is determined based on the strength and porosity properties with the aid of statistical analysis. The result shows that the sample excluding the 2.36 mm gradation exhibits better strength and infiltration properties.

Keywords Pervious mortar · Fine aggregate · Gradation · Compressive strength · Porosity

P. Anima · R. T. A. S. Rathan (✉)
Department of Civil Engineering, Amrita School of Engineering, Amrita Vishwa Vidyapeetham, Coimbatore, India
e-mail: rt_arjunsivarathan@cb.amrita.edu

K. Jayakesh
Department of Civil Engineering, National Institute of Technology Arunachal Pradesh, Jote, India

P. Danish et al. (eds.), *Innovative Technologies for Smart and Sustainable Infrastructure*,
https://doi.org/10.1007/978-3-031-92801-7_9

1 Introduction

Construction technology has seen a vigorous growth over time. Many complex structures can be constructed within a short span using advanced construction methods. The drainage characteristic for a pavement is vital to drain the surface runoff which aids in ground water recharge [1]. It is noted that the flexible and rigid pavements are found to be low permeable which results in accumulation of water on pavement surface when not properly designed [2]. Lack of impermeable surfaces is also one of the major issues faced by the urbanization. Building impermeable surfaces serves as a deterrent to ground water recharge. The impervious surfaces of typical concrete have a number of drawbacks, including preventing precipitation of water into the soil, destroying the natural ecosystem's equilibrium by lowering the water table, and causing more water pollution and erosion [4].

Porous concrete (PC) is considered as an emerging material made of cement and coarse aggregate with less amount or no fines with a porous macro structure. PC is a type of permeable and porous concrete with voids greater than 50 nm in diameter [8]. Because of its capacity to regulate storm water, porous concrete, which was first employed as a construction material, later attracted interest as a pavement medium[10]. This causes the pavement to have an enhanced skid resistance, noise absorption characteristics, and lessen the heat island effect [4]. Usually pervious concrete contains voids in between 15 and 35% [3, 5]. The porosity of pervious concrete is typically in the range of 15–35% [2, 5, 13, 15, 16] with infiltration rate between 0.001 and 0.012 m/sec and a water permeability of about 2–6 mm/sec [13, 15]. Akkay Ahmet, and İsmail Hakkı Çağatay [2] states that the three types of voids can be found in the previous concrete namely, voids between the aggregates, voids between the sample and the interior of the mould, and voids within the aggregates. Vaishnavi et al. [6] explain that the amount of fibres added to the mortar mix is a factor that affects the strength and workability of mortar. Peyman Mehrabi et al. [12] described that the aggregate gradation types, water to cement (W/C), cement to aggregate, and compaction procedures with differences in the attempts, all involve in a significant part in achieving the ideal consistency in the previous concrete mixtures.

The benefits of pervious concrete include enhancing water quality through percolation, absorbing noise produced from tyre-pavement interaction, and minimising road puddles. When compared with conventional concrete, the pervious concrete has less or no fine aggregates in the gradation. This results in porosity ranges from 15 to 35%, thereby limiting the strength parameters of pervious concrete in comparison with normal concrete. As a result, only low volume highways, parking spaces, cycling paths, pedestrian pathways, and residential alleys can use pervious concrete pavements, which together make up the majority of the pavement network in urban areas. Thus, the usage of pervious concrete pavements are restricted to less density roads, parking spaces, motorcycle lanes, pedestrian bridges, residential areas, where these types of classes of roadways account for the majority of the pavement system in urban areas [4, 6]. Kia et al. [2] mentioned that clogging due to surface blockage and fine particle infiltration, is a serious issue with permeable concrete which results in

Table 1 Test results of cement

Characteristics	Values	Range/limits	IS specification
Normal consistency (%)	32	28–33	IS 4031-1968 (IV)
Initial setting time (Mins)	60	>30	IS 4031-1968 (V)
Specific gravity	3.13	3.13–3.15	IS 2720 (III)
Strength of cement(N/mm^2)	54.5	53	IS 4031-1987

permeability loss and performance degradation. The study also states that utilization of fine aggregates can reduce the clogging effect in pervious concrete.

Based on the literature findings, it is inferred that most of the studies in the previous concrete are carried out by eliminating the fine aggregates. Limited studies were focus on developing pervious mortar by eliminating coarse aggregates. Therefore, the current study aims to develop a pervious mortar without coarse aggregate. The purpose of the recent research is to develop the pervious mortar which is prepared with water, cement and gap graded fine aggregate. The gradation of the fine aggregate used for this study includes 4.75 mm, 2.36 mm, 1.18 mm and 600micron. Pervious mortar was made by eliminating single sized fine aggregate. The compressive strength, permeability test, infiltration test and the Cantabro test were carried out for all the mix prepared.

2 Material and Mix Design

2.1 Material

In the present study cement, fine aggregate and water were used to prepare the specimens.

2.1.1 Cement

Portland cement with 53 grade adhering to IS 4031-1987 was used. The cement used for experiments was obtained from a single consignment and of same grade and from same source. The characteristics of the cement are listed in Table 1.

2.1.2 Fine Aggregates

Manufacturing sand (M Sand) available locally was used as fine aggregate. The M sand was then sieved and segregated with a maximum nominal size of 4.75 mm,

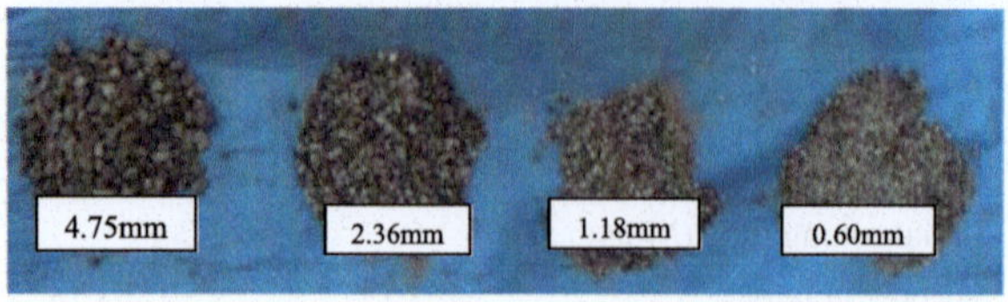

Fig. 1 Grades of M sand

Fig. 2 Specimens for compression, Cantabro, infiltration and permeability test

2.36 mm, 1.18 mm and 600 microns. Different grades of M sand used for the study are shown in Fig. 1.

2.2 *Mix Design & Sample Preparation*

Pervious mortar samples were prepared at the laboratory as per the IS standards. Four gradations were selected and the mortar is prepared using cement sand ration 1:3.Samples was termed as PM4.75, PM2.36, PM1.18 and PM0.6. PM4.75 refers to pervious mortar with gradation eliminating 4.75 mm size. The sample is prepared in the laboratory using hand missing and hand compaction. The samples were demoulded and kept for curing in a tank. Cube of size 70.7 mm × 70.7 mm × 70.7 mm was used to find the compressive strength. Cylinder of diameter 100 mm and height 120 mm was used for the permeability and abrasion characteristics. Specimen of size 450mmX 350mm × 60 mm was used to evaluate the infiltration characteristics (Fig. 2).

3 Experimental Assessment

3.1 *Compressive Strength*

The compressive testing machine (CTM) of capacity 2000 kN is used to determine the compressive strength of the pervious mortar. To determine 7th day and 28th day compressive strength of the pervious mortar, the testing is carried out on a sample of 70.7 mm × 70.7 mm × 70.7 mm size cube. A set of three specimens was tested for

Fig. 3 Test setup for compressive strength

every design mix to find the 7th and 28th day compressive strength. The strength of the specimen is computed as the ratio of failure load to the loaded area. Test set up is shown in Fig. 3.

3.2 Density and Porosity

ASTM C1754 is used to determine the density and porosity of the pervious mortar samples. Density is determined by dividing the weight of the specimen by their corresponding volume. The porosity is obtained by the volumetric method. Experimental setup is shown in Fig. 4. The volume of the sample was measured. Keep the sample in water for 24 h and measure the mass of the fully immersed sample (M1). The samples are taken out of the water and left for 5 min and then measure the surface-dried samples (M2). The porosity is measured using the below equation,

$$PC = \left(1 - \frac{(M2 - M1)/\rho w}{V}\right) \times 100\% \quad (1)$$

3.3 Permeability Test

Permeability test is done is by Constant Head and Falling Head methods. A cylinder of dimension 100 mm diameter and 120 mm height is used for testing. The constant-head permeability determines the amount of water passing through the sample in a specified time t (sec). The falling-head permeability test estimates the time t (sec) taken by the water from height h_1 (cm) to pass through a sample of pervious concrete. The initial time and the final time is noted when water reaches a preferred height. The experimental set up for constant and falling head are shown in Fig. 5.

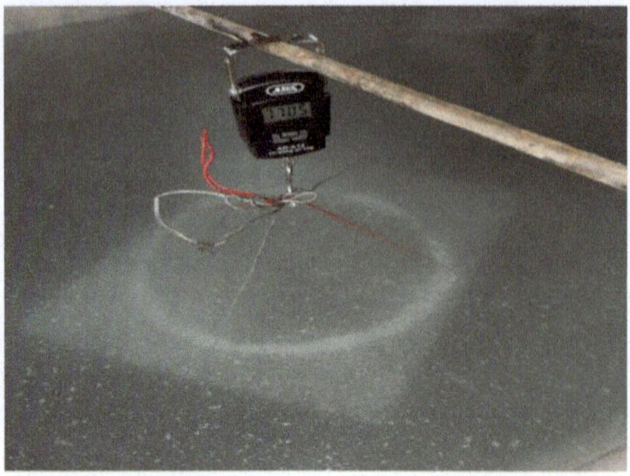

Fig. 4 Test setup for porosity

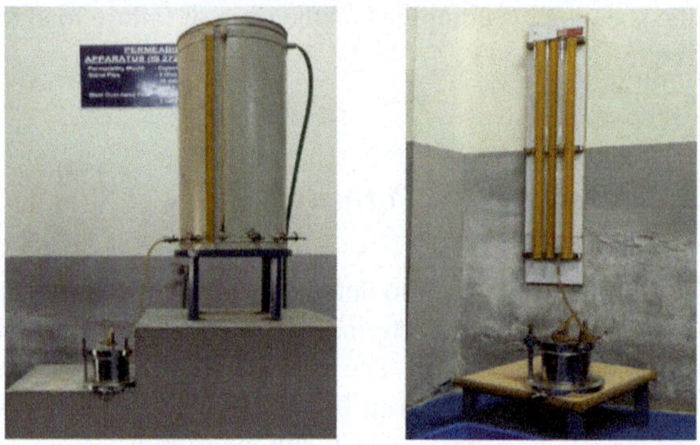

Fig. 5 Experimental setup for permeability test

The coefficient of permeability by constant head is calculated by

$$K = QL/Aht \tag{2}$$

Q Amount of water, in cumec
L Length of the sample in cm
h Hydraulic head, in cm
A Cross-sectional areas of the sample in cm^2
a Cross-sectional areas of the tube in cm^2.

The permeability coefficient by falling head method is calculated by

$$K = 2.303\frac{aL}{At}\log_{10}\left(\frac{h_2}{h_1}\right) \tag{3}$$

h_1 Initial head of water in cm
h_2 Final head of water in cm.

Fig. 6 Test setup for Infiltration test

3.4 Infiltration Characteristics

Infiltration is the vertical flow of water to the specimen on the top surface and its subsequent downward movement. The infiltration test is carried out to ascertain the rate of infiltration of a pervious mortar. The test involves the principle to observe the connectivity into the sample surface. Two points are marked at 10 mm and 15 mm from bottom on the interior of the ring and the time interval taken by water to reach from the top level to bottom level is noted. The amount of water needed for the actual test is shown by this elapsed time.. The same procedure was repeated for every sample. The quantity of the actual water used in this phase and the time elapsed between the two levels was then measured (Fig. 6).

Rate of infiltration is determined using the given formula,

$$I = KM/D^2t \tag{4}$$

I rate of Infiltration in mm/sec,
M Weight of infiltrated water, kg
D The inside diameter of infiltration ring in mm
T Time interval taken by water to reach from the top level to bottom level
k Constant.

3.5 Abrasion Resistance

Skidding, sliding, and breaking control test are critical elements in determining the functional performance of the pavement, which is evaluated using the coefficient of friction determined from Cantabro test. The test was adopted to find aggregate breakdown on the surface of the pavement and reduction in weight due to abrasion. Los Angles abrasion machine without steel balls was used for the test. Cylinder of height 120 mm and diameter 100 mm with 28 days of curing was prepared for the

Fig. 7 Test setup for Cantabro test

test. The cylindrical specimen was rotated in the machine, loss of weight of each sample was measured after every 50 revolutions up to 300revolution and the weight loss is measured (Fig. 7).

Cantabro loss in weight is calculated from the equation

$$\text{Cantabro weight reduction} = \frac{(W_1 - W_2)}{W_1} \times 100 \tag{5}$$

where,

W_1 Initial weight (kg)
W_2 Weight after each revolution (kg)

4 Results and Discussion

4.1 Effect of Gradation on the Density, Porosity and Compressive Strength

Aggregate gradation has significant influence on the structural properties of porous mortar. Effect of aggregate gradation on the compressive strength is shown in Table 2 and Fig. 8. The sample PM1.18 has more compressive strength than the other samples and sample PM4.75 exhibited the least compressive strength. PM1.18 compiles aggregates of 4.75 mm, 2.36 mm and 600 microns and higher portion of 4.75 mm exhibits a higher compressive strength of 11.13 MPa. The reason for the rise in the compressive strength is due the dense packing of aggregate which results in lesser voids and high density, thus enhancing the compressive strength. Samples

Table 2 Properties of pervious mortar

Sl no	Constant head (cm/s)	Falling head (cm/s)
PM4.75	0.0032	0.096
PM2.36	0.0059	0.111
PM1.18	0.0095	0.013
PM0.60	0.003	0.086

PM4.75,PM2.36, PM1.18 and PM0.60 have density 1317.91kg/m^3,1509.90kg/m^3, 1941.85 kg/m^3 and 1418.78 kg/m^{3}respectively. It is evident that PM1.18 shows higher density because of densely packed and less void results in higher compressive strength. Sample PM4.75 exhibits low density also exibit lower compressive strength. Sample PM0.06 showed high porosity of 10.31% and PM4.75 shows lower porosity of 8.94%. It is evident that the porosity sustainably reduces the voids results in increase in compressive strength. This is due to the fact that as decrease in porosity declines, the pores get reduced, and thus increasing the bonding points or areas, which results in an increase in compressive strength.

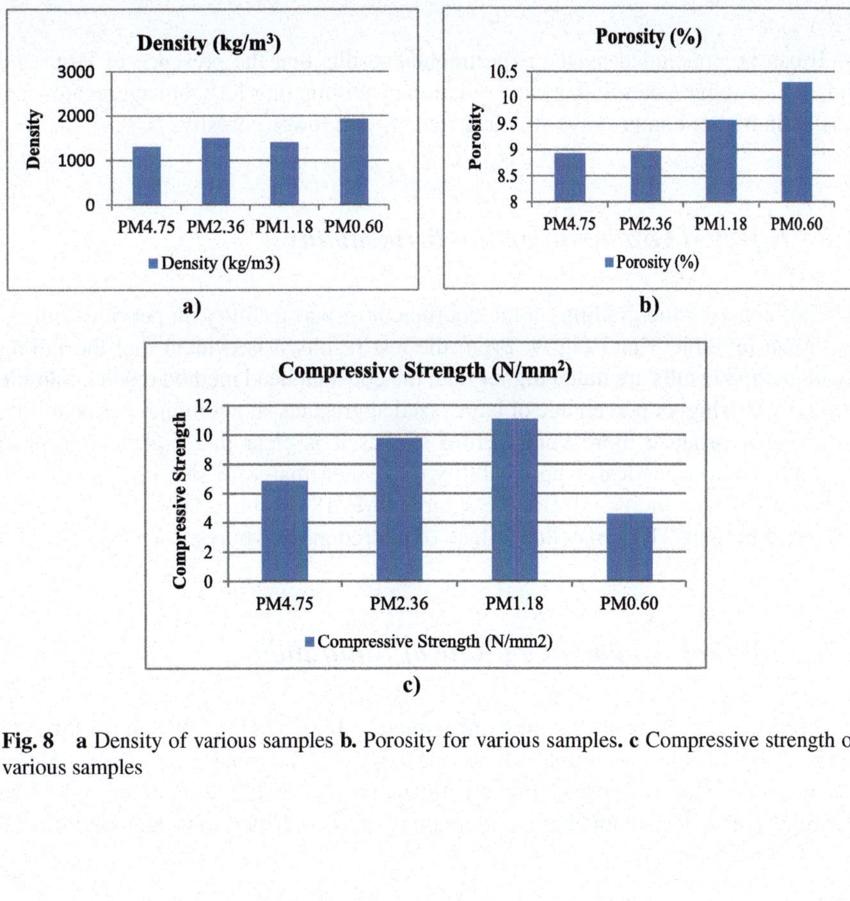

Fig. 8 **a** Density of various samples **b.** Porosity for various samples**. c** Compressive strength of various samples

Table 3 Coefficient of permeability by both methods

Samples	Porosity (%)	Compressive Strength (N/mm^2)	Density (kg/m^3)
PM4.75	8.94	6.93	1317.91
PM2.36	8.99	9.83	1509.90
PM1.18	9.57	11.13	1941.85
PM0.60	10.31	4.66	1418.78

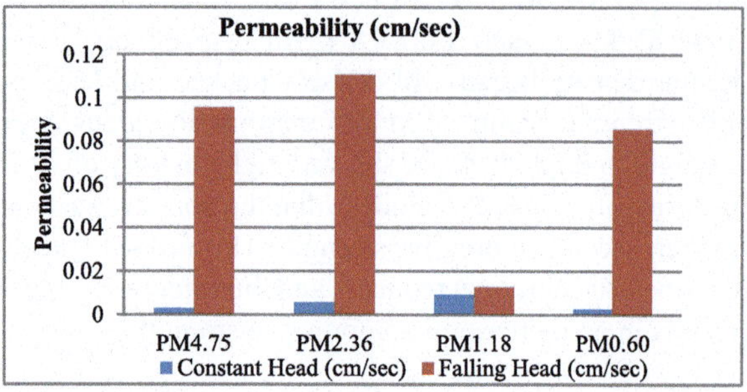

Fig. 9 Coefficient of permeability for different samples

It can be concluded by the experimental results that the presence of large and smaller size aggregates with a gap gradation by eliminating 1.18 mm aggregates size results in higher compressive strength, density and lower porosity.

4.2 *Effect of Gradation on the Permeability*

The influence of the grading on the coefficient of permeability on pervious mortar is shown in Table 3 and Fig. 9. From the test results it is evident that the falling head method results are much higher than the constant head method results. Sample PM2.36 with higher percentage of large sized aggregates shows higher permeability, as larger it produce more voids. From studies it is clear that high compressive strength experiences lesser permeability. It is seen that with rise in porosity, the permeability also increased. But the sample PM1.18 exhibited lesser value despite of higher porosity. This may due to lack of interconnected pores.

4.3 *Effect of Gradation on Rate of Infiltration*

Infiltration rates of various samples are given in Table 4 and Fig. 10. Sample PM2.36 showed higher infiltration rate than all the samples. The trend was same as that of permeability. PM 4.75 shows higher infiltration rate as the absence of large size aggregate made the sample high void, making the sample pervious. Sample PM1.18

Table 4 Infiltration test results of samples

Sample	Infiltration Rate (m/sec)
PM4.75	0.012
PM2.36	0.015
PM1.18	0.002
PM0.60	0.007

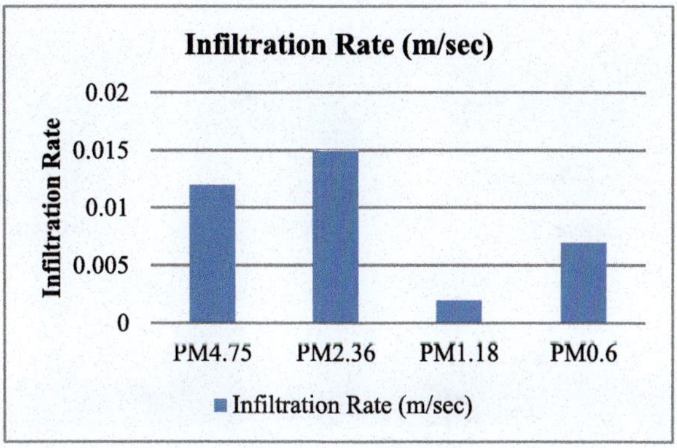

Fig. 10 Infiltration rate for different samples

has a lower infiltration rate though it has a higher value of porosity. The reason may be due to reduced size of void in the sample. Presence of fewer voids in PM0.6 made the sample less porous.

4.4 *Effect of Gradation on Abrasion*

Test results are shown in Table 5 and Fig. 12. During the test trial it is inferred that the specimen PM4.75 split in to pieces after 200 revolutions. The test results states that PM4.75 experienced maximum weight loss of 41.55% after 300 revolutions. This is due to the absence of the aggregate size 4.75 mm which reduces the abrasion resistance results in poor bonding between finer particles. PM2.36 and PM1.18 had shown a better abrasion resistance compared to PM4.75. This may be due to the presence of large sized aggregates, which enhances dense packing. From figure it is evident that compressive strength and the resistance to abrasion are directly proportional. The samples with higher compressive strength showed higher resistance to abrasion resistance because of the less void and dense packing. Change in shape of sample after each revolution is given in Fig. 11.

Table 5 Weight loss after each revolution

Samples	50 rev	100 rev	150 rev	200 rev	250 rev	300 rev
PM4.75	7.40	13.93	18.84	28.58	35.51	41.55
PM2.36	4.85	8.64	12.58	16.44	19.33	22.21
PM1.18	7.76	12.35	16.56	19.78	23.33	24.37
PM0.60	6.43	12.64	18.32	22.36	26.70	30.82

Fig. 11 Change in shape after each revolution

Fig. 12 Percentage weight loss of each sample after 300 revolutions

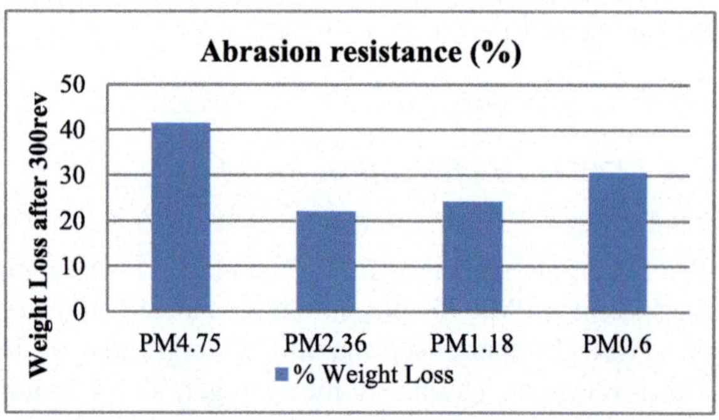

4.5 Optimum Gradation

Test results showed that the sample PM2.36 is having a better performance in permeability, infiltration and abrasion characteristics among the fore samples. But in compressive strength sample PM1.18 shows a higher strength than sample PM2.36. In order to ensure the strength compatibility statistical analysis t-test was carried out between the compressive strength of both the samples. The null hypothesis is stated as difference in the compressive strength has no significance and the alternative hypothesis statement is the difference in the compressive strength has significance.

The test was performed for a total of 12 samples. From Table 6, the t-critic value is higher than the t-static value, which confirmed the acceptance of null hypothesis with confidence level of 95%. Thus the statistical results conclude that the differences in the compressive strength of PM2.36 and PM1.18 have no significance (Figs. 13 and 14).

Table 6 Comparison of compressive strength of PM2.36 and PM1.18

Sample	7th day compressive strength	28th day compressive strength	No of samples	t-static	t-critic	Remarks
PM2.36	7.4	9.94	6	0.767	1.833	Accept the null hypothesis
	7.1	9.67				
	7.2	9.81				
PM1.18	7.48	11.29	6			
	7.39	11.09				
	7.51	10.98				

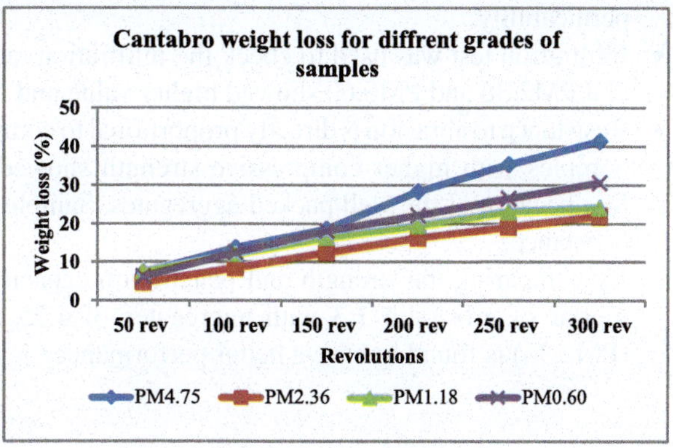

Fig. 13 Cantabro weight loss after each revolution

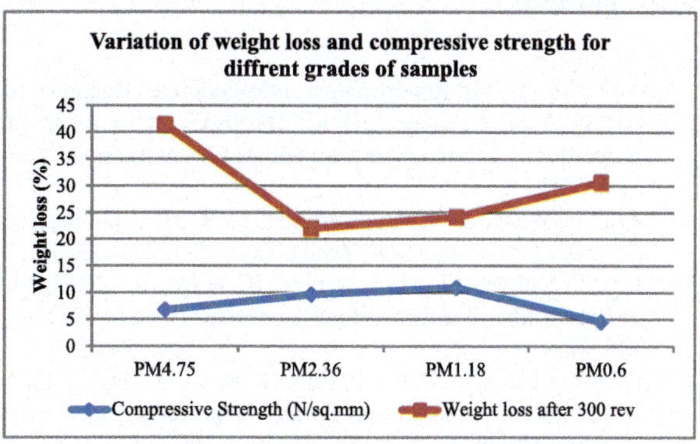

Fig. 14 Variation of weight loss and compressive strength

5 Conclusions

The current study evaluates the effect of various aggregate gradations on mechanical, permeability and surface characteristics. The gradation of the fine aggregate used for the study is 4.75 mm, 2.36 mm, 1.18 mm and 600micron. A different method is adopted in the preparation of pervious mortar. The method involves the preparation of pervious mortar with fine aggregate by eliminating one particular gradation. The followings are the conclusions based on the experimental study conducted.

- It is found that the sample PM1.18 possess higher compressive strength. Results of samples showed that compressive strength is in a range of 4N/mm^2 to 12N/mm^2. As per ACI 522–2011 the compressive strength of pervious concrete pavements is in the range 2.8 N/mm^2 to 28 N/mm^2. Sample with high density showed high compressive strength and lesser porosity. Dense packing plays a very important role in porosity. PM0.60 with lesser density experienced higher porosity and lesser compressive strength.
- Permeability was found out using two different methods, constant head and falling head. Presence of large voids in the sample PM2.36 showed higher permeability, studies revealed that samples having high compressive strength experience lesser permeability.
- Infiltration test was used to check the infiltration rate of the samples. It is found that PM2.36 and PM0.60 showed higher value and lower value of infiltrations.
- Resistance to abrasion is directly proportional to compressive strength and density. Samples with higher compressive strength showed better resistance to weight loss because of the well packed aggregates. Sample PM2.36 got higher abrasion resistance.
- By comparing the strength and penetration capacity of each samples, pervious mortar of mix ratio 1:3 with aggregates of 4.75 mm, 1.18 mm and 0.6 mm (PM2.36) is found to have a better performance.

References

1. ACI PRC-522-10 Report on pervious concrete (Reapproved 2011)
2. Akkay, A., & Çağatay, İ. H. (2021). Investigation of the density, porosity, and permeability properties of pervious concrete with different methods. *Construction and Building Materials, 294*, 123539.
3. Kia, A., Wong, H. S., & Cheeseman, C. R. (2017). Clogging in permeable concrete: A review. *Journal of Environmental Management, 193*, 221–233.
4. Raj, B., Sathyan, D., Madhavan, M. K., & Raj, A. (2020). Mechanical and durability properties of hybrid fiber reinforced foam concrete. *Construction and Building Materials, 245*, Article 118373.
5. Bilal, H., Chen, T., Ren, M., Gao, X., & Su, A. (2021). Influence of silica fume, metakaolin & SBR latex on strength and durability performance of pervious concrete. *Construction and Building Materials, 275*, 122–124.

6. Huang, J., Zhang, Y., Sun, Y., Ren, J., Zhao, Z., & Zhang, J. (2021). Evaluation of pore size distribution and permeability reduction behavior in pervious concrete. *Construction and Building Materials, 290*, 123228.
7. Yang, J., & Jiang, G. (2003). Experimental study on properties of pervious concrete pavement materials. *Cement and Concrete Research, 33*, 381–386.
8. Mehta, P. K., & Monteiro, P. J. M. (2006). *Concrete microstructure, properties, and materials* (3rd ed.). McGraw-Hill.
9. Vaishnavi, M., Aswathi, A., Saarani, S. S., Varghese, A., Sathyan, D., & Mini, K. M. (2021). Strength and workability characteristics of coir and nylon fiber reinforced self-compacting mortar. *Materials Today: Proceedings, 46*, 4696–4701.
10. Ghafoori, N., & Shivaji, D. (1995). Development of no-fines concrete pavement applications. *Journal of Transportation Engineering, 121*(3), 283–288.
11. Saboo, N., Shivhare, S., Kori, K. K., & Chandrappa, A. K. (2019). Effect of fly ash and metakaolin on pervious concrete properties. *Construction and Building Materials, 223,* 322–328.
12. Mehrabi, P., Shariati, M., Kabirifar, K., Jarrah, M., Rasekh, H., Trung, N. T., Shariati, A., & Jahandari, S. (2021). Effect of pumice powder and nano-clay on the strength and permeability of fiber-reinforced pervious concrete incorporating recycled concrete aggregate. *Construction and Building Materials, 287,* 122652.
13. Tennis, P. D., Leming, M. L., & Akers, D. J. (2004). Pervious concrete pavements. EB302.02, Portland Cement Association, Skokie, Illinois, and National Ready Mixed Concrete Association, Silver Spring, Maryland, USA.
14. Sahdeo, S. K., Ransinchung, G. D., Rahul, K. L., & Debbarma, S. (2020). Effect of mix proportion on the structural and functional properties of pervious concrete paving mixtures. *Construction and Building Materials, 255*, 119260.
15. Hari, R., Zhuge, Y., & Mini, K. M. (2024). Performance assessment of pervious concrete incorporated with calcium silicate hydrates (C–S–H) cultivated on rice husk ash substrates—a trend surface analysis interpretation. *Construction and Building Materials, 446,* 138050.
16. Hari, R., & Mini K. M. (2024). Mechanical and clogging characteristics of SBR modified pervious concrete reinforced with geogrids—a functional data analysis approach. *Construction and Building Materials, 412,* 134780.
17. Schaefer, V. R., Wang, K., Suleiman, M. T., & Kevern, J. T. (2006). Mix design development for pervious concrete in cold weather climates, Final Report. National Concrete Pavement Technology Center, Iowa State University, Ames, IA.
18. Zhang, Y., Li, H., Abdelhady, A., & Yang, J. (2020). Comparative laboratory measurement of pervious concrete permeability using constant-head and falling-head permeameter methods. *Construction and Building Materials, 263*, 120614.

Sustainable Construction

Recycle and Reuse: Demolished Concrete Coarse Aggregate as Partial Replacement in the Conventional Concrete

Peerzada Danish, S. Ganesh, Akshat Mahajan, and Sarvdaman Sharma

Abstract The idea of sustainable development encourages the prudent use of rapidly depleting natural resources. Reusing damaged concrete as recycled aggregates (RA) also aims to decrease the need of additional depletable resources, such as aggregates which are available in natural form commonly said as natural aggregates (NA). Recycled aggregates attained by devastating the concrete are used as substitute materials in the present study for aggregate substitutions. The recycled coarse aggregates are attained from 30 year-old demolished structure. Natural aggregates have been substituted by 0%, 25%, and 50% with recycled aggregates. Physical and mechanical properties including the water absorption and stress–strain characteristics of concrete made by consuming recycled aggregates have been premeditated. The testing of concrete cylinders for stress–strain characteristics and prisms for flexural strength have been carried out in a 2500 kN servo controlled UTM. From the outcomes it is experiential that with the addition of natural aggregates up to 25% there is no substantial result on the mechanical properties of concrete casted by consuming RA. However, with 50% replacement of NA the firmness of demolished RA concrete is declined. Water absorption of RA concrete is also high, when associated with conventional concrete. The longitudinal breaking strain of RA concrete rises with the amount of field demolished recycled concrete coarse particles. In addition, the stress–strain curve appears to be similar to standard concrete.

Graphical Abstract Detailed Overview

In the present work, laboratory experiments are accomplished to provide a comprehensive experimental evaluation of RAC with different replacement percentage along with nominal concrete that are prepared. The recovered coarse aggregates were composed from a 30-year-old demolished structure. Two different size portions of

P. Danish
Department of Civil Engineering, Indian Institute of Science, Bangalore, India

S. Ganesh (✉) · A. Mahajan
School of Civil Engineering, Lovely Professional University, Phagwara, India
e-mail: ganeshrajan08@gmail.com

S. Sharma
School of Civil Engineering, Model Institute of Engineering and Technology, Jammu, J&K, India

P. Danish et al. (eds.), *Innovative Technologies for Smart and Sustainable Infrastructure*,
https://doi.org/10.1007/978-3-031-92801-7_10

crushed natural coarse aggregate (NA) were used: one by a minimal size of 10 mm and another with a minimal proportion of 20 mm. The compacted bulk density was determined to be 1712.53 kg/m^3 and 1679.91 kg/m^3. Similarly, two different size portions of Recycled Concrete Aggregate (RCA) were used one with a minimal proportion of 10 mm another with a minimal proportions of 20 mm. The compacted bulk density was initiated to be 1390.06 and 1528.66 kg/m^3. Nominal size of 10 and 20 mm aggregates respectively are mixed in the ratio of 40:60 in the concrete mix. Natural aggregates have been substituted by (0, 25, 50 and 100) percentage with RA. The stress–strain analysis was performed to know the ultimate stress value and the ultimate strain in longitudinal and lateral direction of cylindrical specimens.

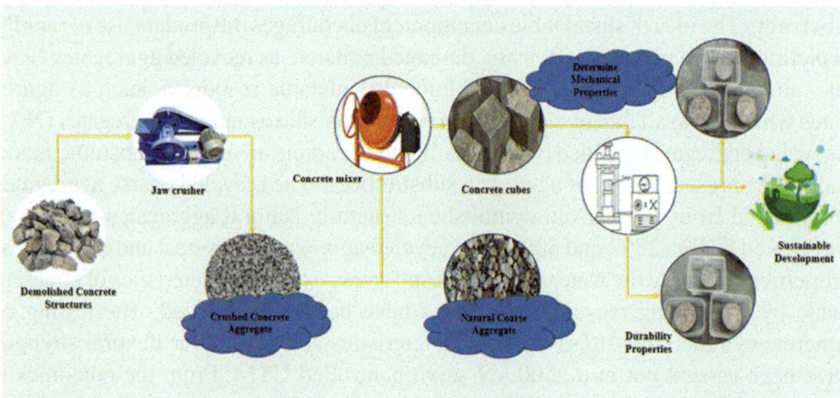

Keywords Recycled aggregates · Compressive strength · Flexural strength · Water absorption · Stress–strain curves

1 Introduction

Due to urbanization, population growth, and natural calamities, a significant amount of waste is produced during construction and demolition. These wastes include things like concrete, masonry, iron, ceramics, and wood. As a result, managing and disposing of such building and demolition debris is a challenge for the majority of industrialized and developing nations. It is difficult to dispose of these wastes safely due to a lack of dumping sites and rising transportation costs. In light of this, it is advantageous to reuse the waste produced in order to save the environment and ensure efficient utilization [1]. Environmental attentions are progressively affecting the source of aggregate, extending from oppositions to quarrying to glitches with the dumping of construction demolition waste and with increasing the shortage of lands for the dumping. RCA is a more naturally responsive alternatives to virgin aggregate meanwhile it lowers the need to quarry fresh resources. It can also assist to save spaces in landfills by distracting concrete debris from them. Due to its extra pointed shape and rougher exterior texture than NA, recycled aggregate can advance the binding and firmness of RA concrete [2]. They asserted that oven-drying recycled aggregates will increase their compressive strength by creating an interfacial relationship between the fragments of aggregate with cement paste. In comparison to the parent concrete found a 30% decrease in modulus of elasticity reported a 20% decrease, and found a 15% decrease [3–5]. It was determined that when the quantity of RA rises, the RAC Modulus of Elasticity will always decrease. Using a water-cement ratio of 0.43 as a constant, experimented with RA concrete using five different mix proportions [6]. The percentages of RCA replacements that make up the mix proportion are 0%, 30%, 50%, 70%, and 100%, respectively. The experimental research indicates that the shear mode is the primary mechanism of failure for RAC and that the failure method is quite quick. All five mixes achieved target strength, by the possible exception of nearly about 50 Mpa. RAC, in someplace the measurable value was a little less than the target strength. For RAC and NAC, at 56th day concrete strength was characteristically 5% and 3% greater than the 28th day strength, respectively. A 10% reduction in the strength under compression of coarse aggregate RAC was found, whilst 20%, 30%, and 45% declines in the same were seen in structures that used (30, 50, and 100) % of the coarse aggregate substitution, respectively [7, 8]. Three sets of concrete specimens through compressive force levels close to 18, 37, and 48 MPa were the subject of experiments [9]. A normal granular crushed stone and two different kinds of used aggregates, denoted as A and B by grinding old concrete materials that correspond to more strength and ordinary strength concrete, respectively, are included in each series. The study on load–deflection curves which were used to govern the fracture energy under maximum load is determined. It was shown that both high strength and normal strength concrete behaved similarly when crushed to produce the RA. According to the results, using recycled aggregate instead of natural coarse aggregate increases the mortar's elastic affinity for the coarse aggregate, which alters the fracture process by strengthening the connection and reducing stiffness.

Using durability indices like water sorptivity, oxygen permeability, and chloride conductivity, authors in [10] examined the act of concrete made with RA with three distinct concrete mixes that contained (0, 50, and 100)% RA after the curing on 3rd, 7th, 28th and 56th days. The concrete mix's goal compressive strength of 30 MPa was part of its design focusing for the 28th day. The RA mixes were created by simply substituting 50% and 100% of the NA percentage in the conventional mix with RA using OPC, so partially replacing it (by mass). For a specific cure time of concrete mixes, chloride conductivity rose as replacement amounts of RA increased. After three days of curing, the concrete mix comprising 100% RA had a 41.4% higher chloride conductivity value than the mix containing 0% RA. As the percentage of RA in concrete mixes grew, so did its water sorptivity. The water sorptivity rose by 47.3% at the 3-day curing age, from 21.21 mm/h1/2 for the 0% RA concrete to 31.24 mm/h1/2 for the 100% RA concrete. For mixes that were cured for 7, 28, and 56 days, the corresponding increases were 43.6%, 38.5%, and 28.8%. As the curing time increased, it should be observed that these percentage changes declined, suggesting that the sorptivity values would not have changed much if the mixes were left to cure for an extended period of time [11]. We investigated the workability, durability, fire resistance, strength, on the structural effectiveness of recycled aggregate-based concrete. For regular concrete, the water absorption was between 4.3 and 4.7% after two months of air curing. As the replacement ratio rose, recycled concrete's water content increased as well, reaching 6.9–7.1% at 100% replacement ratio. In order to validate the structural integrity of the recycled reinforced concrete (RC) component, an evaluation was conducted to control the material characteristics impacted by the bond property. The water sorptivity rose by 47.3% at the 3-day curing age, from 21.21 mm/h1/2 for the 0% RA concrete to 31.24 mm/h1/2 for the 100% RA concrete. For mixes that were cured for 7, 28, and 56 days, the corresponding increases were 43.6%, 38.5%, and 28.8%. Investigated the carbonation, total pore volume, and water absorption of recycled aggregate concrete [12]. RA from demolished concrete structures were used to make concrete mixtures. It was experiential that when more RA are used in place of NA, concrete's ability to absorb water rises. The carbonation depth is still less than that of reference natural aggregate-based concrete, even when all recycled ingredients are used.

2 Experimental Investigation

The primary way that recycled coarse aggregate varies from natural aggregates is that it is made up of two distinct components associated with cement mortar and NA. The quality of the novel concrete and how it was crushed determine how much cement mortar is applied. The ACI method is used to strategy the amounts of the concrete mix [13]. In accordance with IS: 269-987 [14], ordinary Portland cement of grade 53 was utilized. River sand that is readily available locally is utilized for the fine aggregate. A minimal size of 10 mm were employed in one size fraction of crushed NA, while a minimal size of 20 mm was added in the other. Additionally, it was discovered that the compacted bulk density was 1712.53 and 1679.91 kg/m^3. The thirty-year-old building that was dismantled provided the RCA. In the similar way the dual size fractions of RCA which were used were nominally 10 mm and nominally 20 mm. Additionally, it was discovered that the compacted bulk density was 1390.06 and 1528.66 kg/m^3. In a 40:60 ratio, the concrete mix includes aggregates with nominal diameters of 10 and 20 mm. Recycled aggregates have been used to replace natural aggregates at percentages of (0, 25, 50, and 100) %. A pile of irregularly shaped gray and black rocks, with a single small yellow leaf visible among them. The rocks vary in size and texture, creating a rough and uneven surface. The image captures the natural arrangement and color contrast between the rocks and the leaf is shown in the Fig. 1.

Between 0.4 and 0.8% of cement was mixed with a super plasticizer to increase the mix's workability. Natural coarse aggregate absorbs water at a rate of 0.7%, whereas recycled aggregate absorbs water at a rate of 4.2% of aggregate weight. The specimens cast for the study are displayed in Fig. 2.

Fig. 1 Recycled coarse aggregate (RCA)

Fig. 2 Specimens casted for the studies

2.1 Compressive Strength

After the curing period of 7, 28 and 56 days the concrete crushing compressive strengths were determined using standard a laboratory cube specimen size of (100 × 100 × 100) mm. Figure 3 displays the compression configuration. When associated with the outcomes it was shown that the addition of RA considerably reduced the concrete's compressive strength when related to the conventional concrete mix. When 25% of the coarse aggregate was substituted with a comparable volume of recycled aggregate, a 10.67% decrease in compressive strength was noted. The strength dropped by 16.81% when recycled material was used in place of 50% of the coarse aggregate. For RA concrete with 100% by quantity of recycled aggregate substitution, a 23.11% reduction in strength was observed.

The type of crack formation was another finding made during the compressive strength test. Because the relationship among RA and cement paste is poor when compared to the concrete made by natural aggregates, fracture formation in RA

Fig. 3 Compression test on cubes

concrete differs from that of plain concrete. By way of a outcome, the RA initially developed fractures surrounding them. Although the addition of RA meaningfully condensed the compressive strength values, considering that a 10.67% decline in compressive strength was seen in this category, these results are still within a reasonable range for a 25% replacement.

2.2 *Split Tensile Strength*

A cylindrical concrete sample is being compressed between two metal plates in a hydraulic press. The sample shows visible cracks, indicating stress. Figure 4 depicts the cylinder breaking and the setup for the split tensile strength test which is calculated according to the guidance of IS 5816:1999 [15]. According to the results for RA concrete, the tensile strength decreased as the proportion of recycled aggregate rose, which is consistent with the results of the compressive strength test. There is a minimum decrease of around 9 for 25%, a maximum reduction of approximately 28 for 100% replacement, and a reduction of 18 for 50% replacement of natural coarse aggregate.

2.3 *Flexural Strength*

A flexural assessment was made on a (500 × 100 × 100) mm beam specimen at 28 days of life. A universal testing equipment with a 1000 kN capacity was used for

Fig. 4 Split tensile strength

the test as made known in the Fig. 5. The samples were placed under the three points loading, and the flexural load was noted at the site of failure. When compared with concrete samples with less recycled aggregate, the results indicate that the specimens of concrete with the maximum substitution of recycled material had the least flexural strength.

Fig. 5 Testing of specimen for flexural strength

3 Stress—Strain Characteristic of Recycled Aggregate Concrete

Cast cylinders of 150 mm in diameter and 300 mm in height were left to cure in water for 28 days. The cylinders' top and bottom surfaces were smoothed using a grinding machine before being covered in sulfur to make them perfectly horizontal. The capped specimens are displayed in Fig. 6.

A UTM controlled by a 2500 kN servo was used for the tests. To measure the displacement, two LVDTs were positioned diametrically opposite each other. The distance between the two platens was measured. Two 2 mm electrical strain gauges were adhered; one was positioned horizontally to determine the lateral strain and the other vertically to amount the longitudinal strain. An online data collection system was linked to the strain gauges and LVDTs. In the testing matching, the cylinder was positioned concentrically. A typical perspective of ongoing testing is depicted in Fig. 7. Fifteen cylinders in all were examined. Three control specimens and three specimens for 25, 50%, and 100% RA were tested. The usual stress–strain charts for the compressed cylinders are shown in Fig. 8. The results show that replacing 25%, 50%, and 100% of the aggregates with recycled aggregates reduced the strength by 16%, 26%, and 28%, respectively. As the proportion of recycled coarse particles in the concrete surges, so does the longitudinal strain. Additionally, it is noted that the stress–strain profile resembles that of regular concrete in appearance. The breakage patterns of the several cylinders are displayed in Fig. 9.

Fig. 6 Capped cylinderical specimens

Fig. 7 Typical views of testing in progress

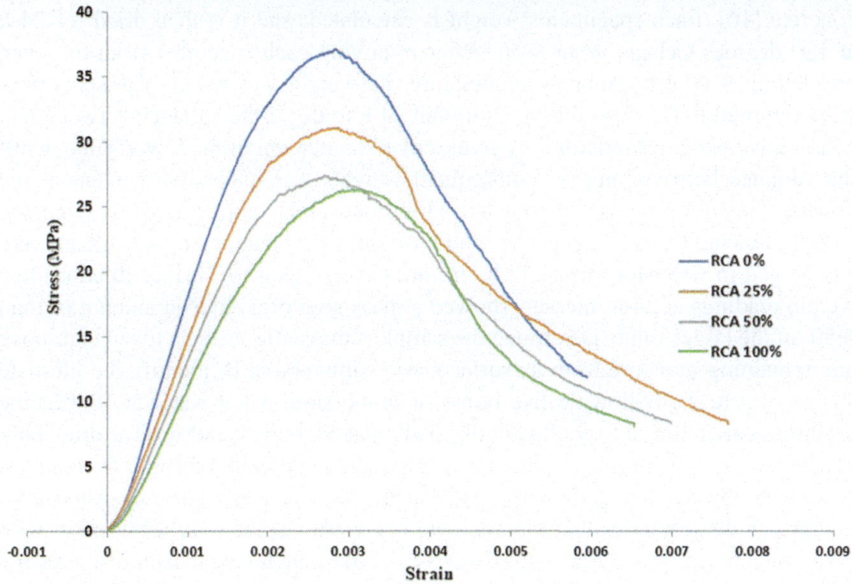

Fig. 8 Plots of typical stress and strain for the compressed cylinders

Fig. 9 Failure patterns of Concrete cylinders with 100% recycled aggregates

4 Density and Water Absorption Test

ASTM C 642 is used to calculate the density and percentage absorption of cured concrete [16]. Each specimen's weight is calculated, and it is then dried for 24 h at 100 degrees Celsius in an oven. After removing each specimen from the oven and letting it cool to ambient temperature, its weight was noted. The specimens were returned to the oven for an additional 24 h to dry if the variance between two consecutive weight standards was greater than 0.5 percent of the lower value. Until the variance between any two subsequent weights was under five percent of the lowest value, this process was repeated. This final weight value should be designated as A. Following final drying, cooling, and weight determination, the specimens were submerged in water for atleast 48 h, or until two consecutive surface-dried sample weight readings at 24-h intervals showed a mass gain of a reduced amount of than 0.5% of the larger value. Determine the sample's mass after using a towel to remove any remaining moisture from its surface. After immersion B, identify the ultimate surface-dry mass. Following five hours of immersion in hot water at 1000C, the specimens are allowed to cool naturally. This soaked, boiled, and surface-dried bulk is denoted as C. After the item has been submerged and boiled, hang it from a wire bucket and measure its apparent mass in water. This mass should be designated as D.

Table 1 demonstrates that the concrete results in rise of water absorption with addition of recycled material in the concrete. Unlike concrete with natural aggregate, which had aggregate water absorption of 3.78%, RA concrete with 100% recycled aggregate showed 6.14% water absorption. As the number of recycled aggregates in concrete surges, its density falls. Concrete with 100% recycled aggregates had a density of 2228 kg/m^3, compared to 2386 kg/m^3 for control concrete.

Table 1 Water absorption and bulk density of RCA

Specimen details	Water absorption (%)	Bulk density (kg/m^3)
Control mix	3.780	2386
RCA 25%	4.972	2275
RCA 50%	5.541	2235
RCA 100%	6.149	2228

5 Conclusion

Around the world, rapid manufacturing expansion leads to major issues including the depletion of natural aggregates and the production of massive amounts of garbage from demolition and building. The use of RCA in concrete production is one method to lessen this issue. To demonstrate that reused concrete particles might be a dependable substitute for aggregate in the manufacturing of concrete, numerous important studies have been conducted. A tried-and-true method for replacing some of the natural material in new concrete construction is the devastating and reusing of building and destruction debris as aggregate.

The following conclusions can be made in light of the studies:

- Strength loss of 23.11% was noted for RCA with 100% RA substitution.
- The split tensile force of RCA is lesser than that of control regular concrete; for a 25% replacement, the strength drop is around 9.16%, and for a 100% replacement, it is approximately 28.68%.
- Concrete's bulk density decreases as the amount of recycled material increases over time.
- The recycled aggregate concrete's stress–strain characteristic curve resembles that of regular concrete in form.
- 100% RCA was found to have a water absorption rate of 6.14%, 62% greater than that of the control concrete.

Acknowledgements I want to express my gratitude to all authors for their help and advice during this research. For their invaluable support and encouragement, I am also grateful to Lovely Professional University and CSIR-SERC, Chennai.

References

1. Rao, M. C., Bhattacharyya, S. K., & Barai, S. V. (2011). Recycled coarse aggregate and its use in concrete. *Indian Concrete Institute Journal, 11*(4).
2. Poon, C. S., Lam, L., Fok, H., & Kou, S. C. (2004). Influence of moisture states of natural and recycled aggregates on the slump and compressive strength of concrete. *Cement and Concrete Research, 34*(1), 31–36.
3. Ravindrarajah, R., & Tam, C. T. (1985). Properties of concrete made with crushed concrete as coarse aggregate. *Magazine of Concrete Research, 37*, 29–38.

4. Topcu, I. B., & Guncan, N. F. (1995). Using waste concrete as aggregate. *Cement and Concrete Research, 25*(7), 1385–1390.
5. Domingo-Cabo, A. (2009). Creep and shrinkage of recycled aggregate concrete. *Construction and Building Materials, 23*, 2545–2553.
6. Xiao, J., Li, J., & Zhang, Ch. (2005). Mechanical properties of RAC under uniaxial loading. *Cement and Concrete Research, 35*, 1187–1194.
7. Rahal, K. (2007). Mechanical properties of concrete with recycled coarse aggregate. *Building and Environment, 42*, 407–415.
8. Yamato, T. (1998). Mechanical properties, drying shrinkage and resistance to freezing and thawing of concrete using recycled aggregate. ACI Special Publication, SP 179-7, pp. 105–121.
9. Casuccio, M., Torrijos, M. C., Giaccio, G., & Zerbino, R. (2008). Failure mechanism of recycled aggregate concrete. *Construction and Building Materials, 22*, 1500–1506.
10. Olorunsogo, F. T., & Padayachee, N. (2002). Performance of recycled aggregate concrete monitored by durability indexes. *Cement and Concrete Research, 32*, 179–185.
11. Eguchi, K., Teranishi, K., Nakagome, A., Kishimoto, H., Shinozaki, K., & Narikawa, M. (2007). Application of recycled coarse aggregate by mixture to concrete construction. *Construction and Building Materials, 21*, 1542–1551.
12. Levy, S. M., & Helene, P. (2004). Durability of recycled aggregates concrete a safe way to sustainable development. *Cement and Concrete Research, 34*, 1975–1980.
13. ACI Committee 211 1-91. (1994). Standard practice for selecting proportions for normal, heavyweight, and mass concrete, part 1. ACI Manual of Concrete Practice.
14. IS: 269-1987. Specification for ordinary portland 53 grade cement.
15. Bureau of Indian Standards (BIS). (1999). Splitting tensile strength of concrete: Method of test, IS: 5816. BIS, New Delhi, India.
16. ASTM C 642-06. Standard test method for density, absorption and voids in hardened concrete.

Optimization of Lime, Cement and Bagasse Content for the Stabilization of Subgrade Soil

S. Sruthi, Attili Venkatasatya Rajapranav, Bandaru Himaja, Pathuri Raghu, Pasupula Bhavana, and R. T. Arjun Siva Rathan

Abstract Soil stabilisation is mainly used for enhancing soil strength, durability, and load-bearing capacity by blending it with stabilizing agents to improve its engineering properties for construction and infrastructure projects. This process enhances the bearing capacity, shear strength, and durability of the soil. The current investigation focuses on optimizing the dosage of lime, cement, and bagasse content to be used for the stabilization of the subgrade. The study examined material combinations including bagasse at 8%, 10%, and 12%, lime at 1%, 3%, and 5%, and cement at 3%, 5%, and 7%. This research also seeks to explore the pozzolanic behaviour of the three-material combination in terms of heat of hydration at various curing stages. The hydrated products at different curing periods are also determined using SEM analysis. The tests conducted for the present study are California Bearing Ratio (CBR) and Unconfined Compressive Strength (UCS). Additionally, the investigation examines crack behaviour in the samples with different proportions using image analysis. The research determines the optimal percentages of lime, cement, and bagasse based on strength and hydration rate. The study concluded that a cement, lime, and bagasse ratio of 3:7:8 is found to be the most effective proportion.

Keywords Stabilization · Lime · Bagasse · Heat of hydration · Image processing · SEM analysis

1 Introduction

Flexible pavements are pavement types widely used across the world. The subgrade is one of the critical components of flexible pavement. The process of improving soil strength, load-bearing capacity and durability by blending and mixing it with stabilizing agents or binder materials is called soil stabilization. Stabilizing the subgrade remains a significant challenge, particularly in India. Expansive clay as a subgrade

S. Sruthi · A. V. Rajapranav · B. Himaja · P. Raghu · P. Bhavana · R. T. Arjun Siva Rathan (✉)
Department of Civil Engineering, Amrita School of Engineering, Amrita Vishwa Vidyapeetham, Coimbatore, India
e-mail: rt_arjunsivarathan@cb.amrita.edu

P. Danish et al. (eds.), *Innovative Technologies for Smart and Sustainable Infrastructure*,
https://doi.org/10.1007/978-3-031-92801-7_11

material poses significant challenges due to its swelling upon wetting and shrinking upon drying, which can result in pavement cracking, non-uniform settlement, and structural instability [1]. To mitigate these issues, various treatment and stabilization techniques are employed to enhance the performance of expansive clay subgrades in road and foundation construction. This process enhances the soil's shear strength, bearing capacity, and durability while reducing permeability, compressibility, and settlement, making unsuitable soils viable for construction. The most common soil stabilization methods consist of mechanical and chemical methods [2]. Mechanical stabilization involves physically altering soil properties through compaction or the addition of materials, while chemical stabilization uses additives like lime, cement, or other binders to improve soil strength and durability. Lime and cement stabilization is particularly effective, as it improves soil properties through a pozzolanic reaction.

Sugarcane Bagasse Ash (SBA), a byproduct of sugarcane processing, contributes significantly to subgrade stabilization by utilizing its pozzolanic properties to enhance soil strength and durability while minimizing plasticity and swelling, thereby improving stability and resistance to deformation under load [3]. The present study aims to optimise lime, cement, and bagasse content to stabilise subgrade soil [4–7]. Lime, a traditional soil stabilizer, improves strength, durability, compressibility, and workability by facilitating cation exchange and pozzolanic reactions in high-pH environments. Hydrated lime is more suitable and efficient than quick lime, with an optimal content range of 7–10% for stabilization [8–10]. Cement stabilization is ideal for gravel and coarse-grained soils, enhancing their strength, and durability by binding soil particles through cement hydration [11, 12]. The process involves hydration, cation exchange, flocculation, and soil aggregation, which modify in-situ soil characteristics. The factors which influence the effectiveness of this method depend such as soil composition, cement content, compaction degree, and curing conditions, with strength improving as cement content, density, and curing time increase. However, the cement pricing is higher necessitating the use of a pozzolanic material that can decrease the cement content during the stabilization process. SBA is one of the alternative pozzolanic materials which is produced through the controlled combustion of sugarcane bagasse at temperatures between 700 °C and 900 °C, SBA is rich in amorphous silica, has a low carbon content, and possesses a large specific surface area, making it highly reactive and efficient in enhancing soil properties. The incorporation of bagasse ash highlights the viability of using agro-industrial waste as a sustainable and cost-effective supplement to traditional stabilizers, demonstrating its potential to enhance the properties of expansive soil for construction applications.

The stabilization of expansive soil using bagasse ash, in combination with either cement or lime is one of the best additive combinations which can effectively reduce soil shrink-swell behaviour and enhance strength parameters such as MDD (maximum dry density), CBR (California Bearing Ratio), and UCC (unconfined compressive strength). From literature study, it is inferred that most of the studies focused on the partial substitution of lime, cement or bagasse. Few studies are carried out to evaluate the performance using all three materials on the mechanical properties of soil. Moreover, there are limited studies to analyse and compare mechanical properties with the heat of hydration for different combinations of bagasse, lime and cement. The research on crack formation analysis using image processing has not been widely explored. Thus, the objective of the current investigation is to evaluate the role of the cement, lime and bagasse on the subgrade soils with a CBR of 3%. The CBR and UCC of subgrade soils with different proportions of bagasse, lime and cement are determined and compared. The study further seeks to identify the effect of heat of hydration on different curing periods. A study on the evaluation of the crack development pattern on the UCC samples for different curing periods using Image processing is attempted. The study also provides detailed insights into the microstructural and morphological changes occurring in stabilized soils with the aid of SEM (Scanning Electron Microscopy). The optimum cement, lime and SBA based on the UCC and heat of hydration is attained using DoE software.

2 Materials and Methods

2.1 Materials

Soil

The soil as shown in Fig. 1 is collected for the study. The lumps in the soil were crushed. To identify the physical properties tests of the soil were conducted. The properties of the soil are shown in Table 1 and is classified as Clay soil with low compressibility (CL) as per ISI.S. 1498–1970.

Lime

Lime is a well-established additive for stabilizing fine-grained soils as shown in Fig. 2. To decrease plasticity, increase strength, durability, workability and fatigue strength and make it more friable than the original clay which makes it more suitable for subgrade. Hydrated lime is used in the study as per IRC SP 89—(2010) guidelines. The amount of CaO present is found to be 64%.

Cement

The Pozzolanic Portland cement (PPC) is used for the study as shown in Fig. 3. It interacts with the soil and reduces its affinity to water. The characteristics and features of the cement are highlighted in Table 2.

SBA

The bagasse used for this study was obtained from the sugarcane industry, in Coimbatore as shown in Fig. 4. The specific gravity of the bagasse obtained is 1.788.

Fig. 1 Soil sample

Table 1 Properties of the soil

Sl. no	Type of soil	Property	Result
1	CL	Liquid limit	30.876%
2		Plastic limit	17.678%
3		Plasticity index	13.198%
4		MDD	1.968 g/cc
5		OMC	15.35%
6		CBR	3%
7		UCC	3.15 kg/cm^2

Fig. 2 Lime

Fig. 3 PPC

Table 2 Properties of cement

Sl. no	Tests	Results
1	Fineness	4.5%
2	Consistency	32%
3	Specific gravity	2.765

Fig. 4 SBA

2.2 *Experimental Program*

2.2.1 Hydrometer Test

For determining the particle size analysis of fine-grained soils such as clay, silt, and fine sand hydrometer analysis is used as per IS 2720–4 (1985)as shown in Fig. 5. It is based on Stokes' Law, which states that the settling velocity of a particle in a fluid is proportional to its size, shape, and density. For this test, 50 g of soil passing through a 75-micron sieve is mixed with 100 mL of a 5% Calgon solution to disperse the particles. The mixture is then transferred to a 1000 mL sedimentation cylinder and diluted with water. A hydrometer is used to measure the density of the suspension at different depths and times, as particles settle at varying rates depending on their size. Corrections for temperature and meniscus effects are made to ensure accurate

Fig. 5 Hydrometer analysis

hydrometer readings. The test helps determine the particle size distribution, and Stokes' Law is applied to calculate the particle diameters based on their settling velocity.

2.2.2 California Bearing Ratio (CBR)

The California Bearing Ratio (CBR) test, conducted as per IS 2720–16 (1987), is conducted to evaluate the strength of raw soil and soil mixtures like SLCB (Soil–Lime-Cement-Bagasse) for use in pavement design and subgrade strength determination. In this test, a soil sample is prepared within 4 days of curing, with a surcharge weight placed on the base plate, as shown in Fig. 6a. The CBR value is determined by applying a load with the help of a cylindrical plunger at a constant rate of 1.25 mm/min and measuring the resistance to penetration at specified depths. The test involves compacting 5 kg of soil mixed with its optimum moisture content (OMC) into a mould in 5 layers, each receiving 56 blows from a 4.89 kg rammer. The load at 2.5 mm penetration is compared with the standard load to calculate the CBR percentage. A load versus penetration curve is then plotted, as shown in Fig. 7. to analyse the soil's resistance to penetration and derive its CBR value.

Fig. 6 **a** Sample of soil mixture SLCB kept for soaking. **b** CBR test for SLCB mixture

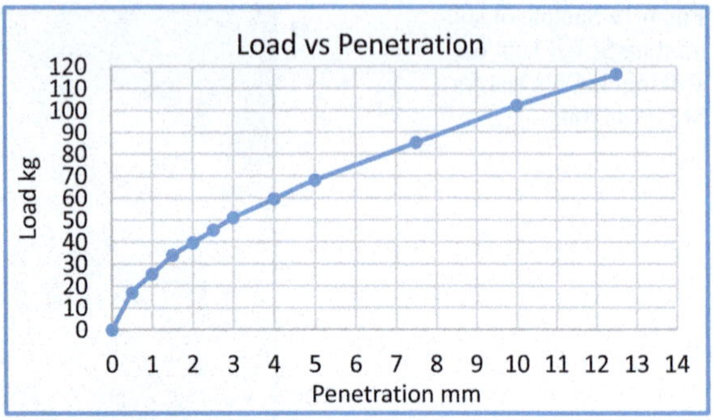

Fig. 7 Load versus penetration graph

2.2.3 Unconfined Compressive Strength (UCC)

As per IS 2720–10 (1991), the Unconfined Compressive Strength (UCS) is conducted to determine the compressive strength of raw soil and mixtures like SLCB (Soil–Lime-Cement-Bagasse) by testing the failure point under axial compression. For this test, the soil and SLCB mixture is prepared by mixing in a specific weight ratio. The samples are compacted in a mould at their MDD and OMC, ensuring uniform compaction by placing the soil in layers and tamping it. Once compacted, the samples are enclosed in an air-tight bag and cured for 3 and 7 days, as shown in Fig. 8 at atmospheric temperature. After curing, the samples are tested using an unconfined compressive testing machine. The UCS is determined as the load per unit area at which the sample fails in the axial compression, indicating the soil or mixture's strength when subjected to compressive forces.

Fig. 8 UCC samples kept for curing

2.2.4 Heat of Hydration

As per IS 4031 (part 9)—1988 heat of hydration test is conducted for the soil mixture with stabilizers in the ratios by weight for unhydrated samples and hydrated samples for 7 days which is stored in an air-tight container to prevent differences in moisture content. Figure 9a shows the heat of the hydration machine. A test is conducted to measure the heat capacity. The calorimeter is assembled and the stirrer for at least 5 min is allowed to run to allow the temperature to become uniform. For unhydrated samples, 3 g of mixture in the ratio per weight is added in the calorimeter and 7 g for loss of ignition is kept at the same time. For a hydrated sample of 4.2 g and 7 g of sample for the heat of hydration and loss of ignition respectively are tested. For the ignition loss shown in Fig. 9b, the sample is placed in the furnace and raised to 900 °C temperature and the sample is placed for 3 h at the temperature cooled in a desiccator and then weighed. It determines the heat generated when it is reacted with a mixture.

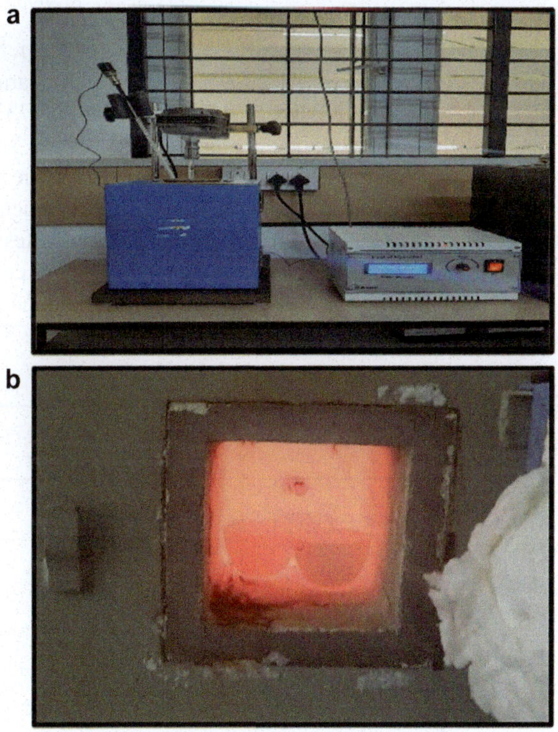

Fig. 9 **a** Heat of Hydration. **b** Ignition loss at 900 °C

2.2.5 Image Processing

A crack is the initial failure of a component. It is important to analyse these cracks to avoid accidents. So, a method for crack detection is required. The proper detection of cracks is analysed with the help of MATLAB software. Many models were proposed for image processing as the general outline for the process as it follows by capturing images and image processing by edge detection for extracting the cracks boundaries, crack detection and crack analysis (length, area, orientation, etc.). As the captured image is converted to a binary image by Sobel edge detection and following length, area, width and orientation will be calculated using MATLAB.

3 Results and Discussions

3.1 Basic Property of Soil

Hydrometer analysis is used to determine the particle size distribution of the soil, and represented in Fig. 10, shows the gradation characteristics, with the D10, D30, and D60 values calculated as 0.0056528 mm, 0.0737138 mm, and 0.209457 mm, respectively. Additionally, the liquid limit of the soil was determined using the Casagrande apparatus, as shown in Fig. 11, where the water content corresponding to 25 blows was found to be 30.876%. This value indicates the soil's moisture content at the boundary between its plastic and liquid states, crucial for understanding its consistency and behaviour under varying moisture conditions.

The graph shown in Fig. 12 illustrates the change in dry density for water content. The test was conducted in the laboratory using the Standard Proctor Test resulting

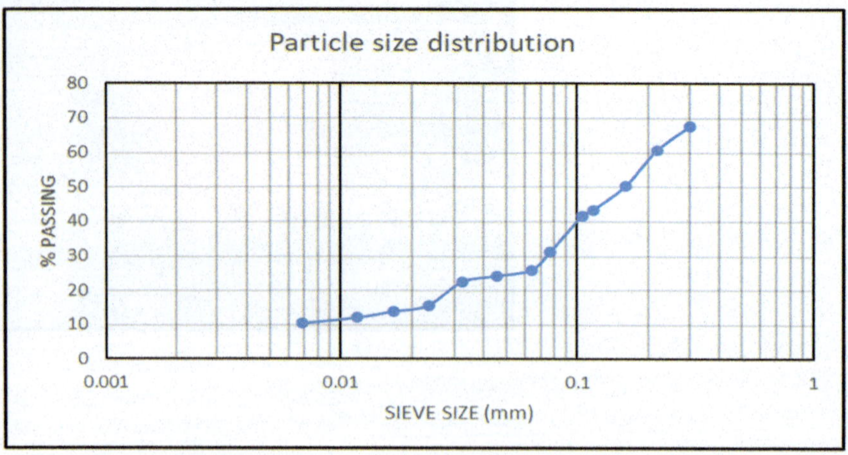

Fig. 10 Particle size distribution curve

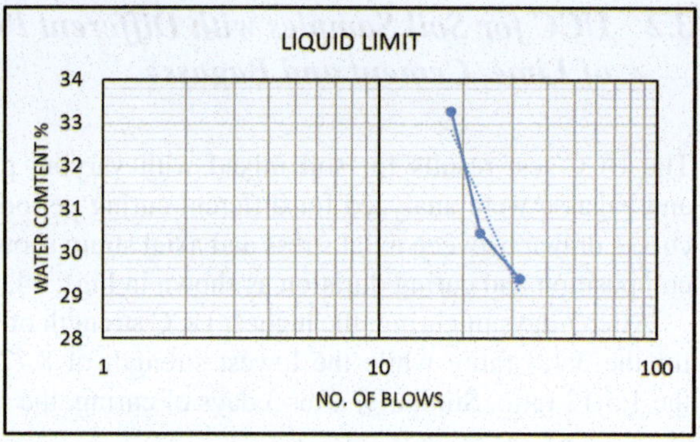

Fig. 11 Water content versus number of blows

in an optimum moisture content of 15.35% and with a maximum dry density of 1.968 g/cc. The curve drawn between axial stress and axial strain for the Unconfined compression test conducted in the laboratory for the soil is shown in Fig. 13. The maximum axial stress obtained was 3.15 kg/cm^2.

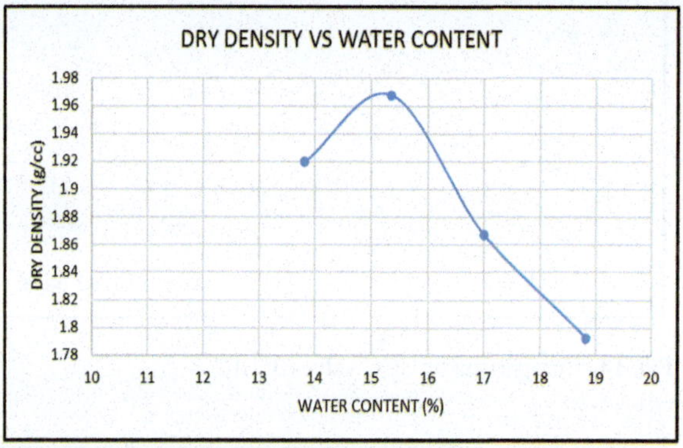

Fig. 12 Dry density versus water content

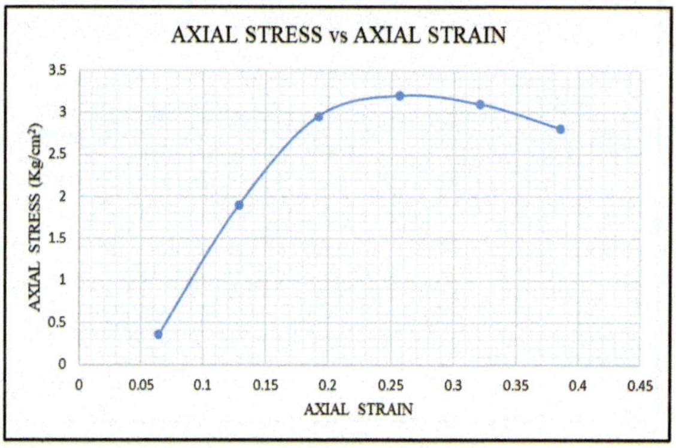

Fig. 13 Axial stress versus axial

3.2 *UCC for Soil Samples with Different Proportions of Lime, Cement and Bagasse*

The UCC test results for soil mixed with varying proportions of lime, cement, and bagasse were analyzed for different curing periods such as 3 and 7 days. The curves drawn between axial stress and axial strain show the strength variations with composition and curing duration as shown in Figs. 14, 15, 16 and 17.

After 7 days of curing, the highest UCC strength of 17.931 kg/cm^2 was recorded for the 3:7:8 ratio, while the lowest strength of 8.771 kg/cm^2 was observed for the 1:5:12 ratio. Similarly, after 3 days of curing, the maximum UCC strength was

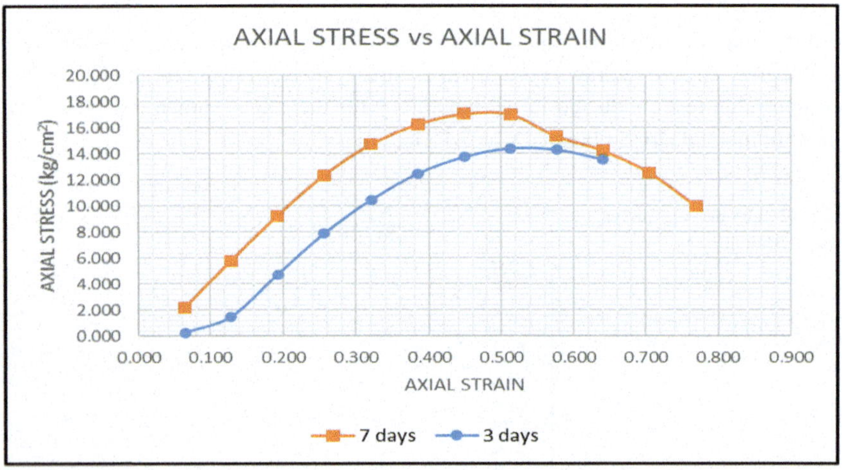

Fig. 14 Comparison of UCC values 3:7:12

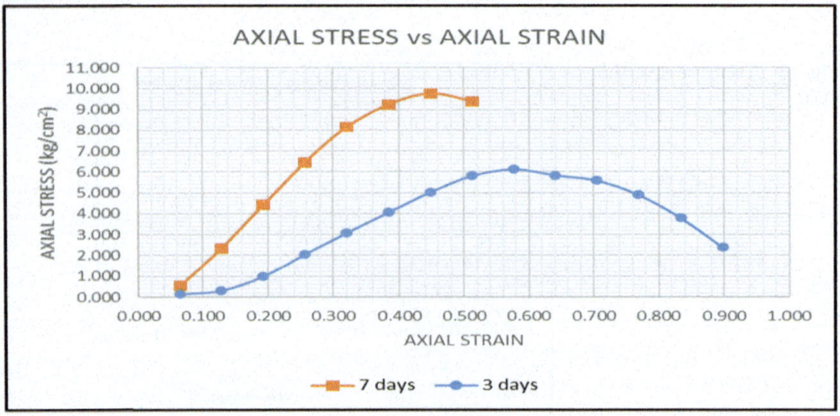

Fig. 15 Comparison of UCC values 1:3:10

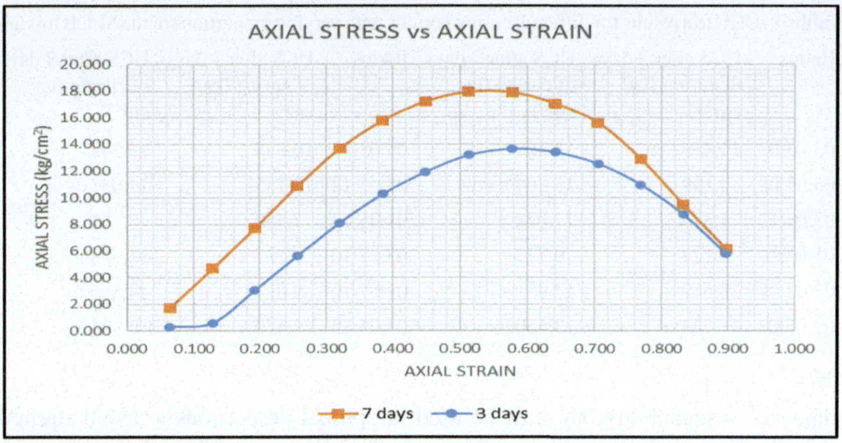

Fig. 16 Comparison of UCC values 3:7:8

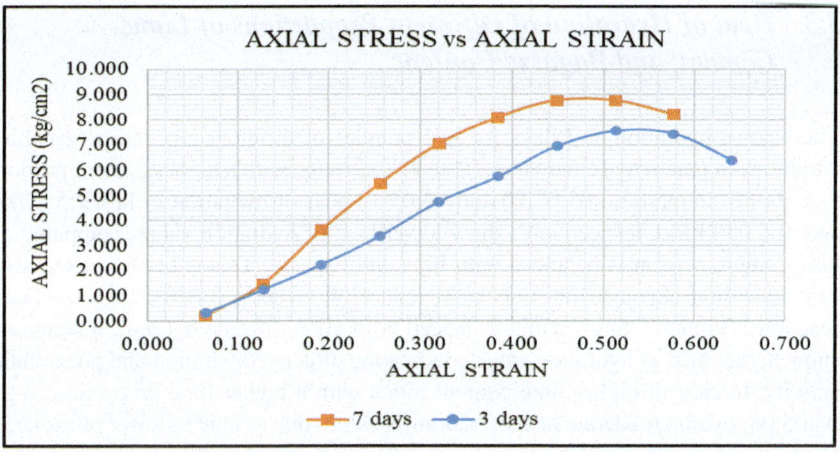

Fig. 17 Comparison of UCC values 1:5:12

14.375 kg/cm^2 for the 3:7:12 ratio, and the minimum was 6.122 kg/cm^2 for the 1:3:10 ratio. These results indicate that the UCC strength increases with curing time and is significantly influenced by the proportion of lime, cement, and bagasse in the mixture. Table 3. represents the Unconfined Compressive Strength (UCS) results of various mixes of cement, lime, and bagasse in different proportions after 3 and 7 days of curing. The proportions are expressed as ratios (cement: lime: bagasse). Cement as a primary binding material and higher proportions of cement significantly enhance UCS. Lime and Bagasse contribute to pozzolanic reactions, improving long-term strength and reducing swelling and shrinkage. As an agricultural byproduct, bagasse

Table 3 UCC test results for different curing periods with varying proportions of the SLCB mixture

Ratios	UCS after 3 days of curing (kg/cm^2)	UCS after 7 days of curing (kg/cm^2)	Ratios	UCS after 3 days of curing (kg/cm^2)	UCS after 7 days of curing (kg/cm^2)
03:03:08	10.439	12.398	05:03:10	10.898	16.240
05:05:08	14.199	15.264	05:05:12	10.155	11.031
03:07:12	14.375	17.074	01:05:12	7.550	8.771
01:03:10	6.122	9.777	05:07:10	9.237	14.588
03:07:08	13.669	17.931	01:07:10	9.677	14.506
01:05:08	11.733	13.560	03:03:12	8.470	9.394

enhances sustainability but must be used in optimal proportions to avoid strength reduction.

3.3 *Heat of Hydration of Different Proportions of Lime, Cement, and Bagasse Content*

The heat of hydration at 7 days for soil samples of different mix ratios revealing variations in reactivity is shown in Table 4. The ratios with a higher cement proportion (For example ratio 05:03:10 and 05:05:12 heat of hydration is 117.325 kJ/kg and 165.030 kJ/kg respectively) show lower heat of hydration values compared to ratios where cement is balanced with lime and bagasse. This is because excessive cement without adequate lime or bagasse reduces the synergy for efficient pozzolanic reactions. Similarly, ratios with a balanced proportion of cement, lime, and bagasse show higher heat of hydration values, indicating effective hydration and pozzolanic activity. In case of higher lime content ratios with a higher lime proportion (e.g., 05:05:08) exhibit moderate heat of hydration values due to lime's slower pozzolanic reaction, which releases less heat compared to cement. The ratio showing a higher proportion of baggage content shows a moderately high value of heat of hydration (For Example ratio 01:05:08 and 3:7:12 heat of hydration is 376.909 kJ/kg and 180.545 kJ/kg respectively).

The ratio 3:7:8 exhibits the highest heat of hydration at 405.200 kJ/kg, indicating rapid hydration and significant early strength development due to the optimal balance of components. In contrast, 1:5:12 records the lowest heat of hydration at 112.292 kJ/kg, reflecting slow reactivity and limited effectiveness for early strength applications. The ratio 1:3:10 shows a moderate heat of hydration of 231.225 kJ/kg, suggesting steady hydration and controlled strength gain over time. Similarly, 3:7:12 demonstrates a mid-range heat of hydration at 180.545 kJ/kg, offering a balance between early reactivity and sustained strength development. These results emphasize that 3:7:8 is the most effective for rapid strength gain, while 1:5:12 is the least reactive, with 1:3:10 and 3:7:12 providing moderate and balanced performance. The ratio

Table 4 Summary of the test results Heat of hydration for soil samples with various proportions of Lime, cement, and bagasse content

Ratios	Heat of hydration at 7 days (kJ/kg)	Ratios	Heat of hydration at 7 days (kJ/kg)
03:03:08	250.430	05:05:12	165.030
05:05:08	166.120	01:05:12	112.292
03:07:12	180.545	05:07:10	231.082
01:03:10	231.225	01:07:10	284.370
03:07:08	405.200	01:05:08	376.909
05:03:10	117.325	03:03:12	252.578

having maximum UCC strength produced maximum heat of hydration and the ratio having minimum UCC strength produced minimum heat of hydration.

3.4 *Image Processing*

The Image processing was performed on soil samples with maximum and minimum UCC strength after 3 and 7 days of curing to analyse crack pattern development as shown in Table 5.

The samples with maximum UCC strength exhibited narrower crack widths and shorter crack lengths, indicating improved structural integrity and resistance to deformation. In contrast, the samples with minimum UCC strength with wider cracks and longer crack lengths showed lower cohesion and strength. The ratio with 3:7:8 achieved a maximum UCC value of 17.931 kg/cm^2 after 7 days of curing reflecting the optimal balance of cement, lime, and bagasse for compressive strength and 3:7:12 achieved a maximum UCC value of 17.074 kg/cm^2 after 3 days curing indicating faster initial strength gain. For the optimum proportion 3:7:8: the orientation of the crack is 90°, indicating vertical cracks that occur due to uniform stress distribution and for the lower ratios 1:3:10 and 1:5:12 orientation of cracks are horizontal (0° orientation), which are more problematic for structural stability. The ratio with 3:7:8 has higher UCC values correlated with higher heat of hydration, as these ratios have a balance of cement, lime, and bagasse that ensures optimal pozzolanic activity and also shows narrow and short cracks making it ideal for soil stabilization. When comparing with ratio 1:5:12 has the lowest UCC, widest cracks, and the weakest structural performance due to poor balance in the composition.

Table 5 Image processing of various mix

Ratio	UCC (Kg/cm^2)	Number of days	Length (cm)	Width (cm)	Orientation	Original image	Binary image
3:7:12	max	3	2.85	0.02	90–0–90–4		
3:7:8	max	7	1.79	0.1	90		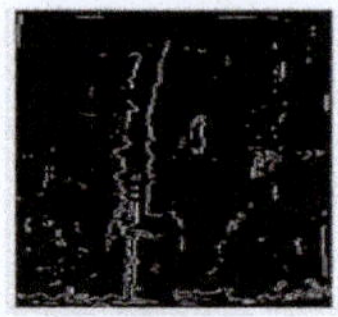
1:3:10	Min	3	3.99	0.238	0		
1:5:12	min	7	23.88	2.51	0	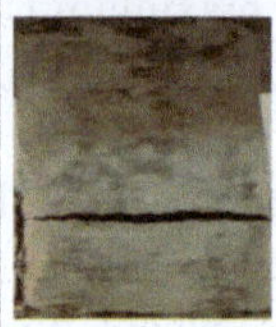	

3.5 Optimum Percentage of Lime, Cement, and Bagasse Content

The optimization of lime, cement, and bagasse content for soil stabilisation is carried out using Design of experiments software considering the UCC and heat of hydration., It is found that the optimal percentages of lime, cement and SBA is found to be 3:7:8. as shown in Fig. 18. The results indicate that after three days of curing, the Unconfined Compressive Strength (UCC) reached 13.109 kg/cm^2, nearing the maximum of 14.375 kg/cm^2, while after seven days, the UCC strength achieved 17.025 kg/cm^2, close to the maximum of 17.931 kg/cm^2. Additionally, the heat of hydration at seven days was 379.927 kJ/kg, approaching the highest value of 405.2 kJ/kg. These findings demonstrate that the optimized mix ratio effectively enhances strength development and hydration reactivity, achieving near-maximum performance within short curing durations and validating its suitability for soil stabilization applications.

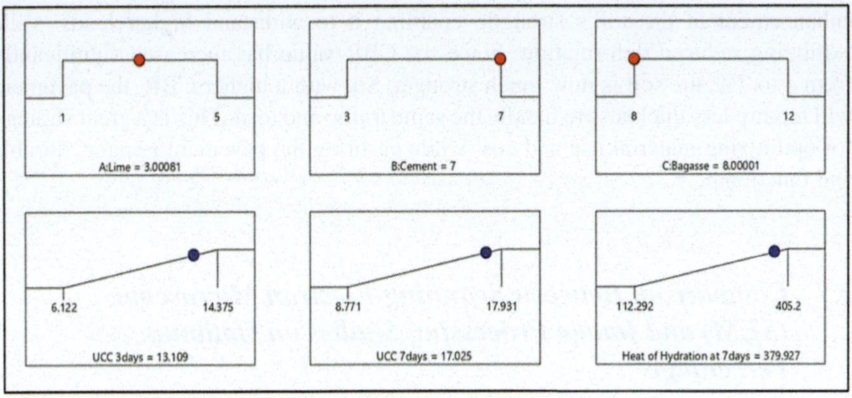

Fig. 18 Optimum percentage of lime, cement, and bagasse content

3.6 *Result for CBR Test on Soil with Optimum Percentage of Lime, Cement, and Bagasse*

The Load versus Penetration Curve in Fig. 19. illustrates the performance of soil stabilized with the optimum mix of 3% lime, 7% cement, and 8% bagasse under applied loads. The obtained CBR value is 7% of the optimum percentage of lime, cement, and bagasse. This behaviour demonstrates the enhanced stiffness and durability the stabilising agents provide, making the optimized soil mix effective for applications requiring high resistance to load and deformation.

The natural soil has a CBR of 3%, which indicates moderate strength. It requires a certain thickness of pavement to bear the load effectively. When the soil is blended with cement, lime, and bagasse ash in a 3:7:8 ratio, the CBR value improves to 7%, indicating enhanced strength and load-bearing capacity. This marks a notable

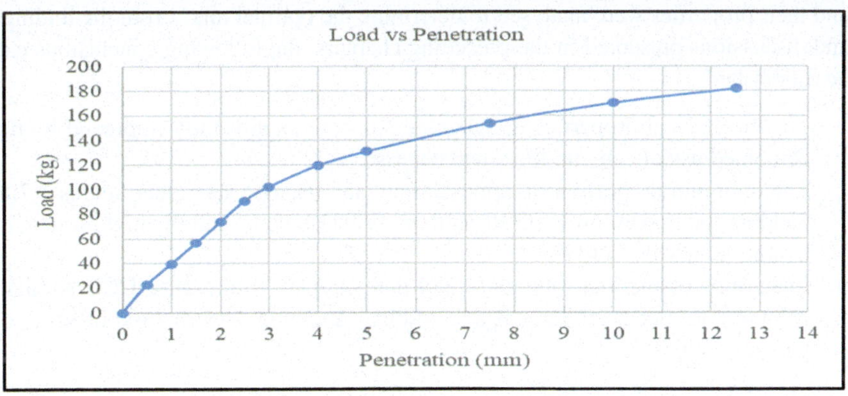

Fig. 19 Load versus Penetration curve for the optimum percentage

enhancement in the soil's strength, enabling it to withstand higher loads while exhibiting reduced deformation. Since the CBR value has increased significantly from 4 to 7%, the soil is now much stronger. So, with a higher CBR, the pavement will require less thickness to handle the same traffic and load. This is a great strategy for optimizing material use and cost while ensuring the pavement remains durable and functional.

3.7 Comparison Between Scanning Electron Microscope (SEM) and Image Processing Studies on Optimum Percentage

The SEM image is at a magnification of 250X, revealing the microstructure of the stabilized soil mix having a 3:7:8 ratio of cement, lime, and bagasse ash as shown in Fig. 20a. The texture shows agglomerated particles bound by cementitious compounds, indicating pozzolanic reactions between lime, cement, and bagasse ash. The Fine and coarse particles are visible, suggesting good particle bonding and improved soil stability. In terms of Cracks and Voids, small discontinuities and voids are visible, potentially caused by incomplete binding or shrinkage during curing as shown in Fig. 20b. But here in case of optimum proportion these cracks appear limited, reflecting improved structural integrity due to the stabilizers.

4 Conclusions

The primary objective of this study was to examine the effect of cement, lime, and bagasse content on the characteristics of subgrade soils. Different sets of soil samples were prepared by incorporating different proportions of cement, lime, and bagasse, and their properties were analysed to determine the optimal mix. From the findings and discussions presented in the preceding chapters, the following conclusions can be established:

1. The mechanical properties of subgrade soils are significantly improved by the combined use of cement, lime, and bagasse.
2. The optimum proportion of cement, lime, and bagasse was identified as 3:7:8, yielding the maximum UCC strength of 17.931 kg/cm^2 and the highest heat of hydration of 405.2 kJ/kg.
3. Image processing analysis confirmed that samples with higher UCC strength exhibited shorter crack lengths and widths, validating the structural integrity of the optimal mix.

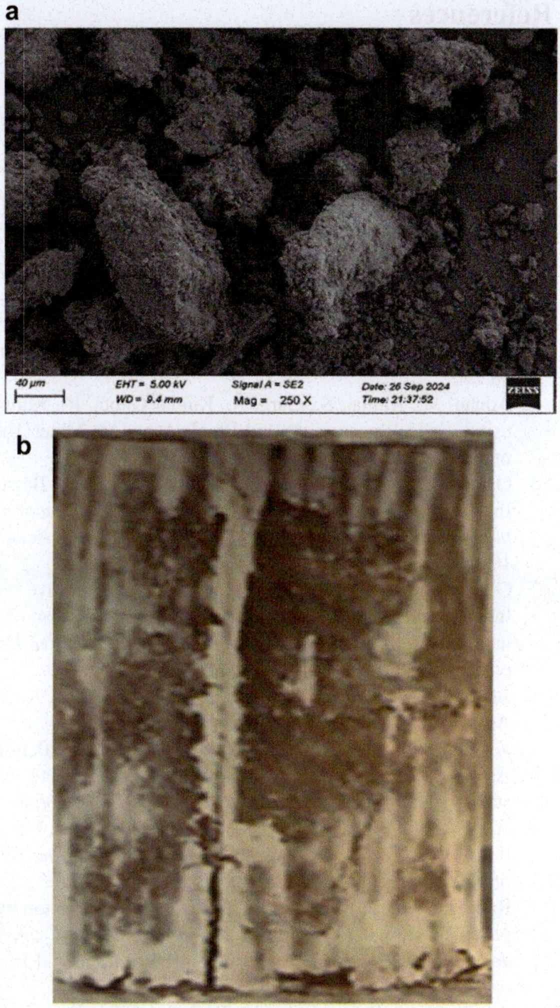

Fig. 20 **a** SEM Image and Original Image of optimum proportion 3:7:8. **b** SEM Image and Original Image of optimum proportion 3:7:8

4. This study concludes that incorporating cement, lime, and bagasse in the proportion of 3:7:8 is the most effective mix for stabilizing subgrade soils, ensuring better strength and durability while leveraging sustainable waste materials.
5. The SEM image reveals the bonding between soil particles and stabilizing agents such as lime, cement and baggage ash for the optimum percentage 3:7:8.It helps in identifying the reduction of void spaces and dense matrix formation due to stabilization.

References

1. Nik Daud, N. N., Jalil, F. N. A., Celik, S., Albayrak, Z. N. K. (2019). The important aspects of subgrade stabilization for road construction. In *IOP conference series: Materials science and engineering*. Institute of Physics Publishing. https://doi.org/10.1088/1757-899X/512/1/012005
2. Gidebo, F. A., Yasuhara, H., & Kinoshita, N. (2023). Stabilization of expansive soil with agricultural waste additives: a review. *International Journal of Geo-Engineering, 14*(1). https://doi.org/10.1186/s40703-023-00194-x
3. Khandelwal, A., Kishor, R., Singh, V.P. (2023). Sustainable utilization of sugarcane bagasse ash in highway subgrade-a critical review. In *Materials Today: Proceedings* (pp. 114–119). Elsevier Ltd. https://doi.org/10.1016/j.matpr.2022.11.310
4. Niraj, B., & Sharma, T. (2023). Experimental study on stabilization of clayey soil using agricultural and industrial waste. *Mater Today Proc.* https://doi.org/10.1016/j.matpr.2023.08.290
5. Adnan, M., Kumar, S., Garg, N., Kumar Gupta, K., Kumar Das, S. (2023). Soil stabilization using waste 'Bagasse ash and lime': A review. Mater Today Proc. https://doi.org/10.1016/j.matpr.2023.05.018
6. Osinubi, K. J., Bafyau, V., & Eberemu, A. O. (2009). Bagasse ash stabilization of lateritic soil. In *Appropriate technologies for environmental protection in the developing world—selected papers from ERTEP 2007* (pp. 271–280). Kluwer Academic Publishers. https://doi.org/10.1007/978-1-4020-9139-1_26
7. Oluwaseun, M. G., & Kennedy, C. (2024). Comparative evaluation of stabilized soil properties incorporating bagasse ash with cement versus lime on determination of best additive combinations. *International Journal of Mechanical and Civil Engineering, 7*(1), 26–42. https://doi.org/10.52589/ijmce-zsncxbzt
8. Jawad, I. T., Taha, M. R., Majeed, Z. H., & Khan, T. A. (2014). Soil stabilization using lime: Advantages, disadvantages and proposing a potential alternative. *Research Journal of Applied Sciences, Engineering and Technology, 8*(4), 510–520. https://doi.org/10.19026/rjaset.8.1000
9. Akula, P., & Little, D. N. (2020). Analytical tests to evaluate pozzolanic reaction in lime stabilized soils. *MethodsX, 7*. https://doi.org/10.1016/j.mex.2020.100928
10. Al-Mukhtar, M., Khattab, S., & Alcover, J. F. (2012). Microstructure and geotechnical properties of lime-treated expansive clayey soil. *Engineering Geology, 139–140*, 17–27. https://doi.org/10.1016/j.enggeo.2012.04.004
11. Rathod, R. G., & Ravande, K. (2018). Soil stabilization by using coarse grain crush sand and pavement design for same. In *2018 International Conference on Smart City and Emerging Technology, ICSCET 2018*, Institute of Electrical and Electronics Engineers Inc. https://doi.org/10.1109/ICSCET.2018.8537307
12. Pandey, A., & Rabbani, A. (2017). Soil stabilisation using cement soıl stabılısatıon. *Article in International Journal of Civil Engineering and Technology*, *8*(6). [Online]. https://iaeme.com/Home/search?string=SOIL+STABILISATION+USING+CEMENT+&all=1

Experimental Investigation on the Durability Properties of Concrete with Partial Replacement of Cement by Paper Sludge

Asmitha Raheem, D. Anupama Krishna, and Rama Mohan Rao Pannem

Abstract One of the main issue in the construction industry is the high cement usage since it produces an enormous amount of carbon dioxide. This leads to the greenhouse effect and global warming. Apart from this, another critical problem is waste management which accumulates day by day and leads to pollution. To overcome these issues different replacements such as industrial by-products and waste materials have been introduced. Papermill sludge also known as hypo sludge is an innovative material obtained from paper mills after treatments. In this study, concrete mixes were cast by partially replacing cement with paper sludge at 5, 10, 15 and 20% and were compared with normal concrete. Paper sludge used for the study was obtained as solid waste from the paper mill which was dried, ground, and sieved through 90 μm. This chapter focuses on the mechanical and durability properties of concrete with partial replacement of cement by paper sludge. The compressive strength exhibited a 24.55% enhancement at 10% paper sludge replacement relative to control mix and was considered optimum. The presence of SiO_2 (12.8%), CaO (17.60%) & Al_2O_3 (8.16%) as observed from XRF analysis in paper sludge may have contributed to the strength enhancement of the concrete. The results indicated that paper sludge-incorporated concrete exhibited enhanced strength and durability compared to the control mix, which can be attributed to reduced porosity and refined microstructural composition.

Keywords Papermill sludge · Hypo sludge · Cement replacement · Waste product · Sustainable material

---

A. Raheem
School of Civil Engineering, Vellore Institute of Technology, Vellore, India
e-mail: asmithanr9607@gmail.com

D. A. Krishna (✉)
Department of Civil Engineering, Mar Baselios College of Engineering & Technology, Trivandrum, India
e-mail: anupama.d@mbcet.ac.in

R. M. R. Pannem
Centre for Disaster Mitigation and Management, Vellore Institute of Technology, Vellore, India
e-mail: rao_pannem@vit.ac.in

P. Danish et al. (eds.), *Innovative Technologies for Smart and Sustainable Infrastructure*,
https://doi.org/10.1007/978-3-031-92801-7_12

1 Introduction

Concrete is known to be the most durable and frequently used construction material over decades. Concrete is good in compression but weak in tension, so it is usually reinforced with rebars. Also, it exhibits superior fire resistance compared to wooden construction. With developing technology and limitations in concrete, many smart building materials were developed from normal concrete overtime. Strength properties were improved than conventional concrete. But usage of high cement content in concrete was still a problem. Since it emits carbon dioxide, CO_2, which leads to greenhouse effect and global warming. These will have severe effect on biodiversity and natural ecosystem. With developing infrastructure and new structures, the demand for cement also increases industry. 10 billion tonnes of concrete are produced every year and majority of it is produced using Portland cement. CO_2 emissions occur during the calcination process. The chemical reaction taking place during calcination process is given by Eq. (1).

$$\mathrm{CaCO_3} + \mathrm{heat} \rightarrow CaO + CO_2 \tag{1}$$

Different kinds of cement replacements are developed which reduces the cement content as well as improves the strength compared to conventional concrete. Class F fly ash was used as cement replacement in recycled aggregateconcrete which reduced creep and drying shrinkage with enhanced chloride ion penetration resistance [1]. Recycled glass sand (RGS) and pozzolanic glass powder (PGP) were used as sand replacement and cement replacement respectively [2]. The usage of RGS as sand replacement affected the fresh concrete properties. The mix with 100% RGD showed inconsistency with bleeding and segregation due to their particle shape and inconsistent material properties. The workability and fresh properties were enhanced when PGP was used as cement replacement due to its refined shape and texture. Cement was replaced with fly ash at 0%, 10%, 15%, 20%, 25% and 30%. Mix 6 with 30% fly ash showed 65% strength gain at 7th day (42.7 MPa). The 28th day strength was obtained as 57.2 MPa which was higher than all other mixes except for 25% fly ash (57.9 MPa) [3]. A comparative study between silica fume, metakaolin, and GGBS as cement replacement was conducted by authors in [4]. The compressive strength of concrete incorporating various materials followed a similar range with highest observed in silica fume (57.5 MPa) > metakaolin (56 MPa) > GGBS (53.5 MPa). The same trend was observed for split tensile strength. Out of the three materials, 10% silica fume showed better strength due to high calcium and silica content. Authors in [5] studied the flexural performance of RC beams with rice husk ash (RHA) as partial cement replacement (0%, 10%, 20%) and crumb rubber (CR) as partial fine aggregate replacement (0%, 5%, 10%). The mechanical properties were enhanced when 10% RHA was incorporated due to higher silica content whereas CR affected the strength. The flexural strength was enhanced by 11.1% because of reduced porosity. The flexural strength got reduced for 20% RHA but was higher than that of the control mix.

The demand for sustainable materials is rising day by day globally due to environmental issues. One such material is paper sludge that has shown potential to enhance the performance of concrete. Previous studies have highlighted the environmental benefits and mechanical performance of using waste materials in concrete [6]. Thus by incorporating waste materials as alternatives, both waste disposal as well as carbon footprint can be reduced. Industrial by-products or waste materials are used as cement replacements since waste management is another critical problem world is facing nowadays. Wastes such as sludges gets accumulated day by day causing pollution. Therefore, to enhance performance and sustainability, various kinds of materials can be utilized as cement replacement and aggregate replacement in concrete. It contains some constituents which is present in cement at various percentages imparts the strength. Some of the waste materials used as cement replacements are fly ash [1, 3, 7–10], slag [4, 10–14], metakaolin [4, 15, 16], rice husk ash [5, 12, 17–19], paper sludge ash [11, 20–23] etc. which have binding properties to the concrete.

1.1 Paper Mill Sludge

Paper mill sludge (PMS) is one of the innovative waste materials which is obtained in the paper mill industry after various treatments. Transforming pulp and paper mill sludge (PPMS) into construction materials is a feasible way for sustainability [23–26]. In recent studies, concrete with cement partially replaced by paper sludge exhibited superior strength than conventional concrete. Million tons of solid paper sludge are produced per year. It has many environmental impacts since large amount of sludges are getting accumulated per day. Paper mill sludge has a wide range of applications such as soil conditioning, landfill material, cement replacement material as illustrated in Fig. 1.

The per capita paper consumption globally is estimated to be 57 kg. Also, it was predicted that paper consumption per person in India will become 17 kg by 2024–2025 [27]. India produces around 10.11 million tons of paper per year accounting for 2.6% of the total world's paper production. Around 50 kg of sludge are produced from 1 ton of paper in the paper industry [28]. Paper sludge (PS) is the final waste product and no further recycling process is possible. Paper sludge obtained from the paper industry is also known as hypo sludge. It is derived as solid waste from pulping, deinking & wastewater treatment. Black liquor sludge is also obtained as waste product from paper industry before producing solid paper sludge. Black liquor sludge (BLS) is used as admixtures in concrete. When cement is partially replaced with paper sludge, it improves the strength properties of concrete. Figure 2 shows the image of the paper mill sludge powder. These can be used in concrete production as cement substitutes because of its omparable pozzolanic properties [1]. Cement was replaced with paper sludge at various percentages such as %, 10%, 15%, 20% etc. It is used as a replacement since the chemical constituents of cement responsible for strength improvement is there in paper sludge also. These materials can improve mechanical properties and other properties. Up to 10% of portland cement, showed

Fig. 1 Uses of Paper Mill Sludge (PMS)

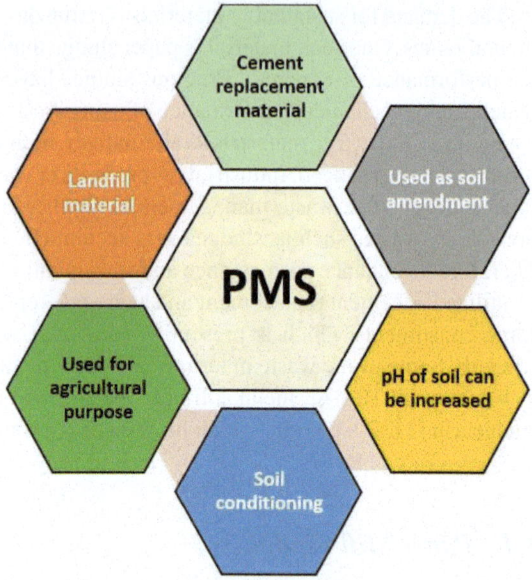

a positive effect on the mechanical performance of mortars [2]. When it is replaced with cement, the cement content can be reduced and also the cost can be reduced thereby it becomes economical in one way or other way.

Fig. 2 Paper sludge powder obtained from solid paper waste through drying, grinding and sieving

2 Literature Review

The increase in demand for eco-friendly and sustainable construction material has led researchers to explore industrial by-products as partial replacements for cement and aggregate in concrete. Among these, paper sludge and its derivatives have gained attention due to its superior performance while reducing environmental impact. The following literature provides key findings from recent research on the use of paper sludge-based materials in concrete.

Authors in [6] reviewed the utilization of plastic waste in concrete production and highlighted the benefits and limitations. The study shows that construction industries can contribute to the reduction of landfill waste, reduction of carbon emission by incorporating plastic waste into concrete. Also, it highlights the sustainability and environmental impact associated with it, mechanical and physical properties of plastic-incorporated concrete, thermal and acoustic properties & economic and practical feasibility. The study cites the research showing that replacement of 50% of fine or coarse with plastic can give lightweight concrete with superior insulation properties [22] studied the mechanical and durability performance of concrete with waste paper sludge ash as partial cement replacement (2.5, 5 and 7.5%).5% ash gave higher strength compared to control and was considered as optimum. The results from acid attack (H2SO4, HCL, H3PO4) exhibited that mass loss was higher for mix with 2.5% ash than 7.5%. Authors in [23] studied the impact of waste paper sludge ash (WPSA) in light weight foam concrete. The cement was partially replaced with WPSA at 0%, 5%, 10%, 15%, 20%, 25%. The enhanced mechanical properties were shown for 15% WPSA replacement. The porosity and water absorption were reduced by 37.75% and 17.4% respectively for maximum WPSA replacement. The impact of incinerated and non-incinerated waste paper on the properties of cement mortar was studied by [29]. They observed that incorporating 5% incinerated waste paper ash (IWPA) and 1% non-incinerated waste paper ash (NIWP) enhanced the compressive and tensile strength even though flow rate of fresh mortar was reduced. Authors in [20] examined the effect of replacing cement with waste paper ash at 0, 5, 10, 155 and 20%. It was observed that workability was reduced with ash increment and 10% was found to be optimum.

Authors in [30] explored the feasibility of paper sludge ash (PSA) as a hydrophobic powder in concrete surface coatings. The results demonstrated that 12% PSA reduced water absorption, conductivity, and sorptivity by 84%, 85%, and 86%, respectively. The samples coated with PSA exhibited superior self-cleaning and water-repelling properties. Authors in [31] investigated the performance of partially replaced paper pulp based (M-20) concrete. The cement was partially replaced at 0, 5, 10, 15 and 20%. The mechanical properties were enhanced at 5 and 10%. Therefore 10% was considered as optimum [32]. The performance of high-performance concrete with hypo sludge at 0–25% and jute fiber at 0.5–1.5% was studied. Compressive and flexural strength was enhanced when 15% hypo sludge and 1% fiber was incorporated into the mix. High split tensile strength was obtained for 10% hypo sludge +1% fiber. Cement was partially replaced with 10% hypo sludge with varying polypropylene fiber at 0, 15, 30 and 45%. Mix 2 with 10% hypo sludge and 0.15% fiber showed maximum strength. The strength decreased when the fiber content was increased and exceeded 0.15%. Authors in [33] studied the performance of silica fume and paper waste ash on improvement of concrete behavior. Cement was replaced with paper waste ash and silica fume at 0, 2.5, 5, 7.5 and 10%. The results showed that 10% silica fume increases the flexural tensile and indirect tensile strength. Tensile/compressive strength ratio of concrete with silica fume and ash was between 13.55 and 17.79 and that of ordinary concrete was about 10. The specimens with silica fume and waste paper ash showed higher compressive strength. The electrical resistance was increased upto 10% silica fume. Pores were reduced due to usage of paper waste ash thus surface became smooth and less porous. 10% silica fume enhanced the concrete integrity as observed form SEM images.

3 Methodology

Figure 3 exhibits the methodology which includes material testing, was done on cement, fine aggregate, coarse aggregate, paper sludge as per the IS codes. A M30 grade concrete was mix-designed and control cube specimens were cast to find the compressive strength. Dry components were first added and mixed in drum mixer. Water was added and mixed until a uniform paste-like mix was obtained. All mixtures were cast into cubes for evaluating 7th, 28th, 90th, and 120th day compressive strength. Specimens were kept at room temperature for 2 days and they were demoulded and water cured until 28 days. Control cube specimens were tested for 7th, 28th, 90th & 120th day strength and compressive strength was obtained. To determine the elemental composition of paper sludge, X-Ray Fluorescence Spectroscopy (XRF) were conducted. After determining the 28th day compressive strength, cubes were cast with partial replacement of cement by paper sludge at 5, 10, 15, and 20%. For each replacement 3 cubes were cast for different curing ages and the optimum paper sludge dosage was determined. Different durability tests were performed on specimens with optimum paper sludge and control specimens. Finally, results obtained from different tests were analysed and comprehensive discussions were conducted to derive conclusions.

4 Materials

The different materials procured for the experimental study were:

- Ordinary Portland Cement (OPC)—(IS:4031,2019)
- Fine aggregate (FA)—Manufactured sand (IS:2386,2021 & IS:383,2021)
- Coarse aggregate (CA) (IS:2386,2021 & IS:383,20,221)
- Paper sludge (PS)

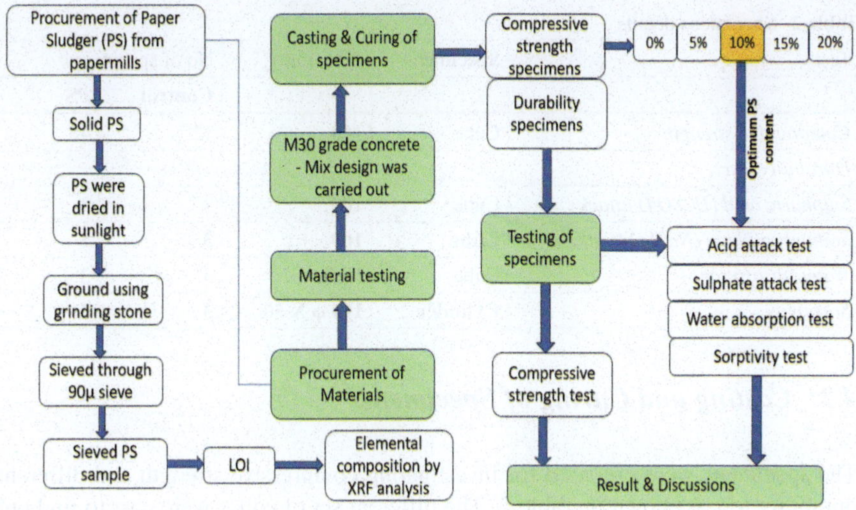

Fig. 3 Flow chart illustrating methodology followed in the experimental work

- Water.

The material tests were conducted on various materials and was conforming to respective IS codes. For the experiment, paper sludge was procured from RPC paper mills, Punalur, Kerala. The procured material was dried, grinded and was sieved through 90 microns. The sieved paper sludge was used to replace the cement.

4.1 Mix Proportion

The mix were designed as per the IS 10262: 2019 [34]. For control mix, mix proportion of M30 normal concrete is 1:1.8:3.38 with water-cement ratio of 0.54. Control cube specimens were cast for evaluating 7th, 28th, 90th and 120th day compressive strength. After evaluating 28th day strength, cement was partially replaced with paper sludge at 5%, 10%, 15%, 20%. The mix proportion after replacement is shown in the Table 1. And an optimum content of paper sludge was determined.

Table 1 Cement partially replaced with PS at different %

Materials (Kg/m^3)	Cement partially replaced with PS at				
	0% (PS0)	5% (PS5)	10% (PS10)	15% (PS15)	20% (PS20)
Cement	359	341.05	323.1	305.15	287.2
PS	*0*	*9.2*	*18.4*	*27.52*	*36.8*

Table 2 Specimens details

Test	Specimen	Size (mm)	No of specimens	
			Control	**PS**
Compressive strength	Cube	150	12	48
Durability test				
Sulphuric acid (H2SO4) attack	Cube	100	3	3
Sodium sulphate (Na2SO4) attack	**Cube**	**100**	**3**	**3**
Water absorption	Cube	100	3	3
Sorptivity	**Cylinder**	**100 φ X 50**	**3**	**3**

4.2 Casting and Curing of Specimens

The specimens were prepared for investigating compressive strength and different durability test as shown in Table 2. The different set of cubes were cast to find out the compressive strength of control mix, mix with paper sludge as partial cement replacement. The cast specimens were water cured in the water tank for 28 days. The different durability test was conducted on cube and cylinder specimens i.e., control and specimens with optimum PS corresponding to each test.

5 Results and Discussion

5.1 Loss on Ignition (LOI) (ASTM D7348-21)

The LOI test was conducted based on ASTM D7348-21[35]. The weight of the empty crucible was noted and 5 g of PS sample sieved through 90μ sieve was taken in the crucible (W1) as shown in Fig. 4a. Then the crucible was placed in a muffle furnace and exposed to a temperature of 950 °C for 1 h as shown in Fig. 4b. After cooling to room temperature, the weight of the crucible with PS was noted (W2). The percentage LOI was calculated by Eq. 2. The LOI value of the PS sample was obtained as 46.43%. The obtained value of LOI was required for XRF analysis to determine the total elemental composition. The LOI value obtained by [36] was 70.11% but a low LOI of 4.5% was observed by [37]. This significant variation in LOI can be attributed to variations in the components present and their sources.

$$\%\text{LOI} = \left(\frac{W_2}{W_1}\right) \times 100 \tag{2}$$

From Fig. 4 (C) it was observed that the colour of the sample changes from grey to cream after it has been subjected to a temperature of 950 °C for 1 h. The LOI value obtained for PS is much higher than that of OPC. This may be due to the moisture or

Fig. 4 **a** PS in crucible before LOI **b** PS in muffle furnace after 950 °C **c** PS after LOI

water content present in paper sludge since it is a raw material and also sludge will be washed during the treatment process in the paper industry.

5.2 Compressive Strength

The control cube specimens and cement partially replaced with paper sludge at 5, 10, 15, and 20% were cast to determine 7th, 28th, 90th and 120th day compressive strength. Figure 5a and b show the image of control cube specimens and cube specimens with 5% PS after testing after casting & testing. The graph plotted for compressive strength values corresponding to different replacements are shown in Fig. 6.

The cube compressive strength increased upto 10% replacement of cement with PS. Compressive strength increased by 7.43% and 25.44% for 5% and 10% addition of PS with reference to control specimens respectively. Beyond 10% the strength

Fig. 5 **a** Control specimens after testing **b** Specimens with 5% PS after testing

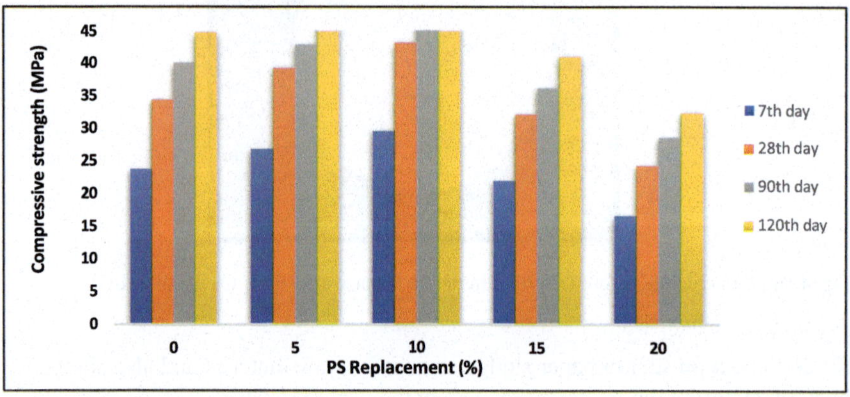

Fig. 6 Compressive strength obtained for different replacement of PS

starts to decrease. The maximum value was obtained for 10% addition of PS that is 43.20 MPa as the 28th day compressive strength with reference to control specimens. Compressive strength decreased by 6.53% and 29.33% for 15% and 20% addition of PS with reference to control specimens respectively. This may be due to lack of SiO_2 and CaO present in the PS as obtained from XRF analysis. Therefore, it can be concluded that 10% addition of PS can be used as the optimum content of paper sludge in the concrete.

5.3 Durability Properties

Resistance to chemical attack: The PS and NC specimens immersed in sulphuric acid and sodium sulphate solution are shown in Fig. 7. The results of the resistance to chemical attack test for both the control specimen and specimen with optimum paper sludge are shown in Fig. 9. From the test results, it was understood that weight loss and loss in compressive strength were more in control specimens than PS specimens after specimens were immersed in both Na_2SO_4 and H_2SO_4. The visual appearance of specimens immersed in sulphuric acid showed that they had undergone surface erosion which was more in control specimens than in PS specimens as shown in Fig. 8a white layer was formed over the control specimen's concrete surface. Sulphuric acid reacts with calcium hydroxide (CH) of cement in concrete and produces gypsum. This will react with calcium aluminate hydrate (C3A) to produce ettringite. The formation of ettringite causes internal pressure and leads to the formation of cracks [14]. The initial layer of the surface was spalled off from the surface as shown in Fig. 8a. It was also observed that there was no gypsum formation on PS specimens. This may be due to the limited availability of calcium hydroxide to react with sulphuric acid. The compressive strength of both the control and PS specimens were reduced by 37.77% and 25% respectively. The specimens immersed in sodium sulphate showed that there was no change noted in visual appearance compared to that before immersion. Also, there was no formation of white layer, cracking or spalling as shown in Fig. 8b. The strength of both control and PS specimens immersed in sodium sulphate was reduced by 20% and 6.25% respectively. So, it can be concluded that concrete replaced with paper sludge showed better performance in both acid and sulphate nature than normal concrete.

Water absorption test: The graph plotted with the results of the water absorption test for both the control specimen and specimen with optimum paper sludge are shown in Fig. 10. From the test results it is found that the water absorption value

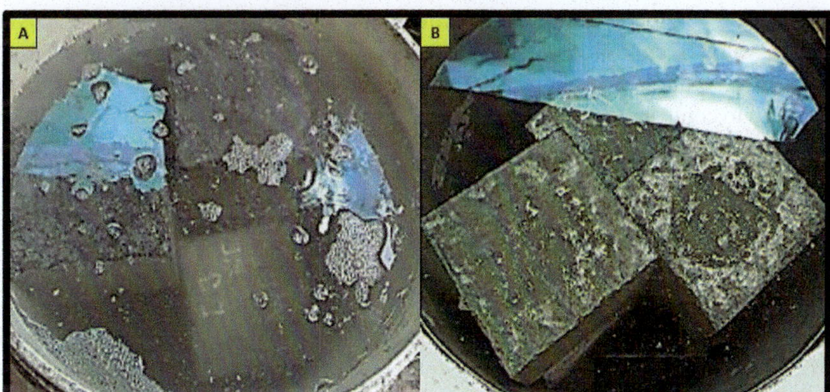

Fig. 7 Specimens immersed in **a** H_2SO_4 **b** Na_2SO_4

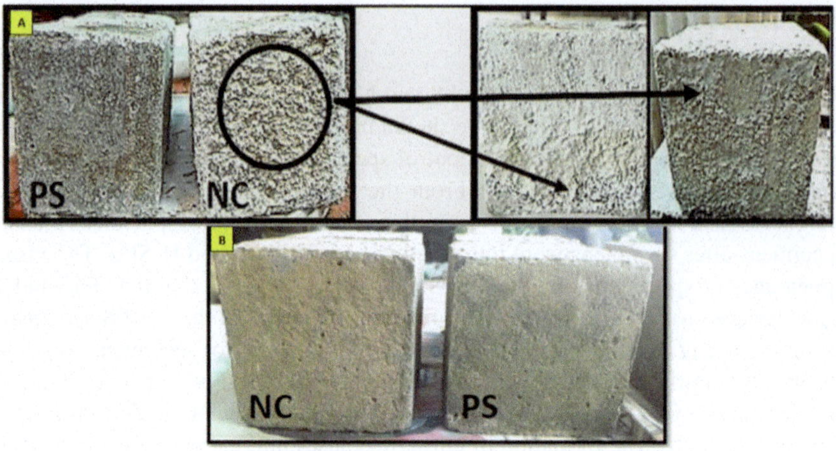

Fig. 8 Specimens after immersed in **a** sulphuric acid **b** sodium sulphate

Fig. 9 Loss in strength and weight loss after acid attack

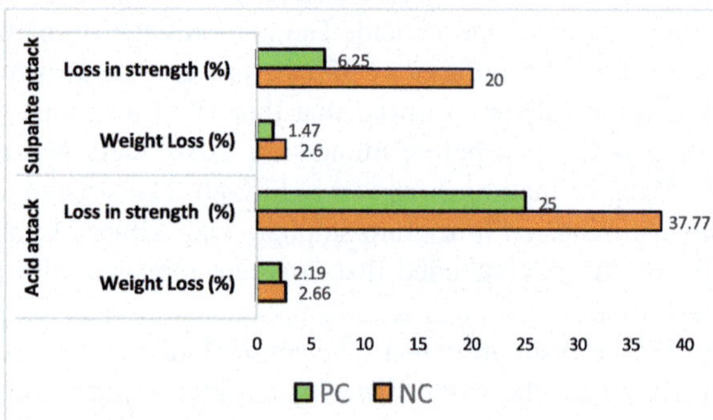

obtained for both control and PS specimens increased with respect to time. And also, the value of control is much higher than that of PS. This may be due to lesser pores and also the moisture present in the paper sludge in the PS specimens when compared to control specimens. The water absorption value was obtained as 1.398% and 1.713% for 10% and 20% waste paper sludge ash receptively [38]. The amount of water added while mixing may have some influences because workability of concrete decreases as soon as water is added to the concrete. It shows that paper sludge has high water absorbing capacity which leads to low workability of concrete. But after the durability test, it was observed that PS was having low water absorption value since it may have reached its saturation point. Also, water absorption value of PS after 1 hour, 2 hours and 24 hours shows only slight change whereas the value of control varies greatly after 2 hours. Water absorption was reduced by 83–84% compared to the control mix when cement was partially replaced with 12% super hydrophobic PSA [30].

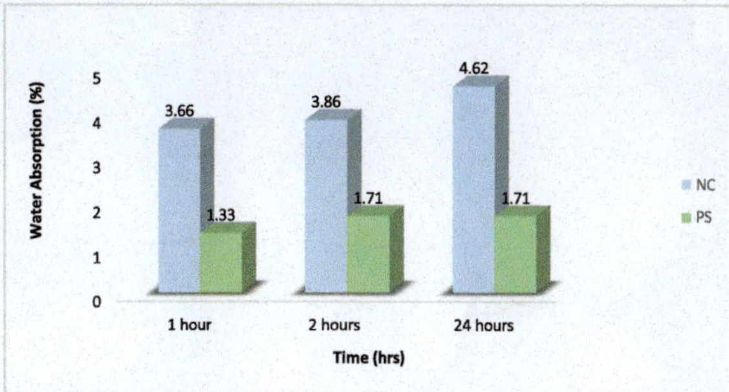

Fig. 10 Water absorption capacity of PS and NC

Sorptivity test: The sorptivity specimens before immersion and after immersed in water are shown in Fig. 11. The results of sorptivity test for both control specimen and specimen with optimum paper sludge are shown in Table 2. From the results it is found that the sorptivity value of NC is 0.00412 mm/s$^{1/2}$ higher than 0.0025 mm/s$^{1/2}$ that of PS. The sorptivity was reduced by 83 to 86% with respect to control specimens when 12% super hydrophobic PSA was used [30]. Sorptivity is the tendency of concrete to absorb and transmit water by capillary action. It is very important to minimize the number of pores present in the concrete to ensure its durability property in terms of penetration or permeability of liquid through the concrete. High value of sorptivity indicates high permeability of liquid into the concrete. This is due to the presence of large number of pores which can lead to ingress of chloride, sulphate or even water into the concrete. This in turn leads to the degradation of concrete. Higher the penetration or permeability rate, faster will be the degradation. Curing also influences the sorptivity rate. Proper curing is necessary for the concrete to achieve its potential performance. It is generally accepted that curing is even more important for concrete with mineral admixture than conventional concrete. This is due to the pozzolanic reaction between amorphous silica present in the admixture and CaOH2 from the cement hydration which requires water to continue. Additionally, water curing has a greater impact on the sorptivity than concrete strength [39]. During the hydration reaction in cement paste, consumed water develops pores. Lesser pores in PS specimens are due to water consumed from paper sludge thereby the possibility of pore formation will be lesser and fine particles of paper sludge fill the voids inside the concrete. So, from the results, it can be concluded that concrete made with optimum paper sludge has lesser pores than normal concrete.

5.4 *XRF Analysis*

The elemental composition of paper sludge was determined by XRF. The elements such as SiO2, TiO2, Al2O3, MnO, Fe2O3, CaO, P2O5, K2O, MgO, Na2O, SO3 were determined. The XRF analysis were tested from National Centre for Earth Science Studies (NCESS), Akkulam. The paper sludge powder sieved through the

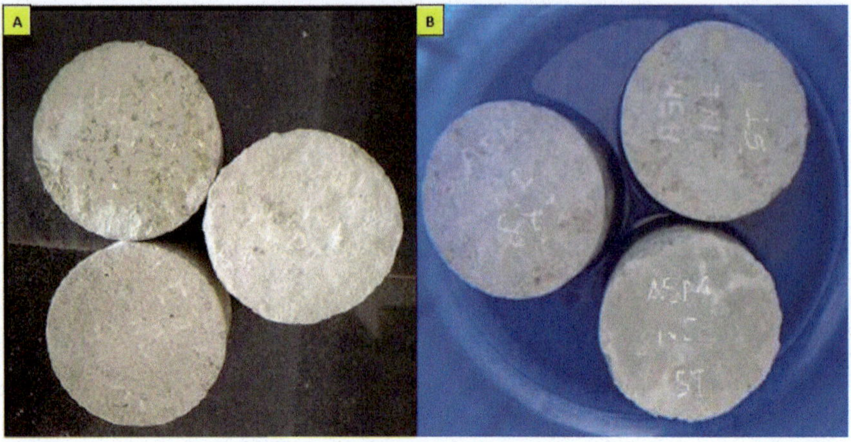

Fig. 11 **a** Specimens after oven dry **b** sorptivity test setup

Table 3 Sorptivity test results

Specimen designation	Sorptivity (mm/s$^{1/2}$)
NC	**0.00412**
PS	0.0025

90μ sieve was given as the sample for testing. The loss on ignition (LOI) value was provided for determining the total elemental composition of the sample. The elemental composition of PS compared with that of OPC are shown in Table 3.

The elements such as SiO2 and CaO impart the strength whereas Al2O3 imparts the setting property of cement. From the elemental composition it is found that % of SiO2 present in PS is 12.80% which is lesser than that of OPC. PS has got 8.16% Al2O3 which is higher than 4.4% as in OPC. The % CaO in OPC is 60.5% which is much higher than 17.60% as in PS. Therefore, it can be concluded that an increase in strength is due to the presence of different elements present is the same as that of cement but in different amounts due to which it can impart strength to the concrete when cement is replaced with PS. The CaO and SiO2 content in paper sludge ash (PSA) was found to be higher, while Al2O3 was absent, possibly due to variation in components present [40]. However, a composition of 78.60% CaO, 8.55% SiO2, and 8.89% Al2O3 was reported by [41] (Table 4).

6 Recommendations

The experimental work was done only with basic materials of concrete. Different replacements were adopted such as to find the optimum and found that 10% PS was giving better performance. Without using any other extra materials, with 10% PS

Table 4 XRF of PS & OPC

Sample	Elemental composition (%)	
	PS	OPC [42]
SiO2	12.80	19.5
TiO2	0.73	0.2
Al2O3	8.16	4.4
MnO	0.02	-
Fe2O3	1.53	2.6
CaO	17.60	60.5
MgO	2.30	2.9
Na2O	3.49	0.3
K2O	0.42	0.9
P2O5	0.30	0.9
SO3	3.97	4.0
LOI	46.43	2.7

a strength of 43.2 MPa for 28th day and 50.66 MPa for 120th day was achieved. The durability properties of PS specimens exhibited better performance than normal concrete. The workability was reduced with increasing PS which shows its absorbing capacity but from water absorption test it was clear that only up to a certain stage it can absorb after which it becomes saturated. The variations in papermill waste can vary the elements present in it and will change the performance of concrete. Also, the waste paper sludge will be washed during the treatment at the factory. So, after procurement, it should be dried completely otherwise it can affect both the grinding process and workability. Chemical and mineral admixtures can be used to overcome workability issues and to improve concrete performance. The performance of concrete with fiber & paper sludge can be done to enhance its properties.

7 Conclusions

In conclusion, it was observed that from XRF analysis, that paper sludge imparts strength due to the presence of 12.80% SiO2, 17.60% CaO and 8.16% Al2O3. The LOI value obtained for PS was 46.43% which is much higher than that of OPC due to moisture or water content present in the paper sludge. The optimum % of paper sludge replacement was found to be 10%. Compressive strength increased by 7.43% and 25.44% for 5% and 10% PS respectively with reference to control specimens. After water absorption test, it was observed that value of PS after 24 h was 1.71% which was less than 4.62% that of NC due to the presence of the moisture present in the paper sludge. Sorptivity of NC was 0.00412 mm/s$^{1/2}$ which was higher than that of PS specimens. This indicates that the ingression of chloride into the concrete was more due to a larger number of pores. The formation of pores in PS will be lesser because the fine particles of PS fill the voids. So, it can be concluded that the performance of concrete made with paper sludge as a replacement material for cement shows satisfactory results when compared to normal concrete. Therefore, it can be used as a cement replacement material in concrete due to the presence of elements like SiO2, CaO, and Al2O3. They can impart strength to the concrete. Also, it is durable as it gives better results from durability tests mainly due to the formation of fewer pores in PS specimens. Hence paper sludge can be used as a cement replacement material.

References

1. Kou, S. C., Poon, C. S., & Chan, D. (2007). Influence of fly ash as cement replacement on the properties of recycled aggregate concrete. *Journal of Materials in Civil Engineering, 19*, 709–717. https://doi.org/10.1061/(ASCE)0899-1561(2007)19:9(709)
2. Taha, B., & Nounu, G. (2009). Utilizing waste recycled glass as sand/cement replacement in concrete. *Journal of Materials in Civil Engineering, 21*, 709–721. https://doi.org/10.1061/(ASCE)0899-1561(2009)21:12(709)
3. Zachar, J., & Asce, M. (2010). More sustainable and economical precast and prestressed concrete using fly ash as a cement replacement. In *2nd international conference on sustainable construction materials and technologies* (pp. 117–123). https://doi.org/10.1061/(ASCE)MT.1943-5533.0000243
4. Naga Venkat, G., Chandramouli, K., Ahmed, E., & Nagendrababu, V. (2020). Comparative study on mechanical properties and quality of concrete by part replacement of cement with silica fume, metakaolin and GGBS by using MÀSand as fine aggregate. https://doi.org/10.1016/j.matpr.2020.10.819
5. Hussein, N., Elmaaty, M. A., Alturki, M., & Elsayed, M. (2025). *Enhancing flexural performance of rubberized concrete beams through incorporation of rice husk ash as cement replacement*. https://doi.org/10.1016/j.engstruct.2025.119958
6. Almohana, A. I., Abdulwahid, M. Y., Galobardes, I., et al. (2022). Producing sustainable concrete with plastic waste: A review. *Environmental Challenges, 9*. https://doi.org/10.1016/J.ENVC.2022.100626
7. Karumanchi, M., Bellum, R. R., Chennupati, M., et al. (2022). Influence on mechanical properties of concrete of cement replacement with fly ash and river sand replacement with foundry sand. *Mater Today Proc, 65*, 3547–3551. https://doi.org/10.1016/J.MATPR.2022.06.146
8. Padavala, S. S. A. B., Dey, S., Veerendra, G. T. N., & Phani Manoj, A. V. (2024). Experimental study on concrete by partial replacement of cement with fly ash and coarse aggregates with palm kernel shells (Pks) and with addition of hybrid fibers. *Chemistry of Inorganic Materials, 2*, Article 100033. https://doi.org/10.1016/J.CINORG.2024.100033
9. Islam, M. N., Noaman, M. A., Islam, K. S., Hanif, M. A. (2024). Mechanical properties and microstructure of brick aggregate concrete with raw flyash as a partial replacement of cement. *Heliyon, 10*, e28904. https://doi.org/10.1016/J.HELIYON.2024.E28904
10. Orozco, C., Babel, S., Tangtermsirikul, S., & Sugiyama, T. (2024). Comparison of environmental impacts of fly ash and slag as cement replacement materials for mass concrete and the impact of transportation. *Sustainable Materials and Technologies, 39*, Article e00796. https://doi.org/10.1016/J.SUSMAT.2023.E00796
11. Ingale, S. D., Nemade, P. D. An investigation on fresh state properties of cement paste and concrete with GGBS and waste paper sludge ash. https://doi.org/10.1016/j.matpr.2022.04.183
12. Ramakrishna, J., & Gopi, R. (2023). Experimental investigation on partial replacement of cement and coarse aggregate by rice husk ash and steel slag in concrete. *Mater Today Proc.* https://doi.org/10.1016/J.MATPR.2023.08.340

13. Ho, C. M., Doh, S. I., Chin, S. C., & Li, X. (2024). The effect of particle sizes of steel slag as cement replacement in high strength concrete under elevated temperatures. *Construction and Building Materials, 411*, Article 134531. https://doi.org/10.1016/J.CONBUILDMAT.2023.134531
14. Mo, K. H., Johnson Alengaram, U., Jumaat, M. Z., & Yap, S. P. (2015). Feasibility study of high volume slag as cement replacement for sustainable structural lightweight oil palm shell concrete. *Journal of Cleaner Production, 91*, 297–304. https://doi.org/10.1016/J.JCLEPRO.2014.12.021
15. San Nicolas, R., Cyr, M., & Escadeillas, G. (2014). Performance-based approach to durability of concrete containing flash-calcined metakaolin as cement replacement. *ConstrBuild Mater, 55*, 313–322. https://doi.org/10.1016/J.CONBUILDMAT.2014.01.063
16. Subash Koneru, V., Nitin Chowdary Kolli, S., Gopal Vemuri, V., et al. (2023). Development of optimum sustainable metakaolin replaced cement concrete based on homogeneity, compressive strength and rapid chloride ion penetration. *Mater Today Proc, 80*, 1306–1310. https://doi.org/10.1016/J.MATPR.2023.01.061
17. Noaman, M. A., Islam, M. N., Islam, M. R., & Karim, M. R. (2018). Mechanical properties of brick aggregate concrete containing rice husk ash as a partial replacement of cement. *Journal of Materials in Civil Engineering, 30*. https://doi.org/10.1061/(ASCE)MT.1943-5533.0002272
18. Ma, W., Lv, B., & Wang, Y., et al. (2024). Freeze-thaw, chloride penetration and carbonation resistance of natural and recycled aggregate concrete containing rice husk ash as replacement of cement. *Journal of Building Engineering, 86*. https://doi.org/10.1016/j.jobe.2024.108889
19. Jayaraman, A., Vasudevan, M., Aravind, R., et al. (2023). Optimization of partial replacement of cement by silica fume and rice husk ash for sustainable concrete. *Mater Today Proc*. https://doi.org/10.1016/j.matpr.2023.05.598
20. Meko Kejela, B. (2020). Waste paper ash as partial replacement of cement in concrete. *American Journal of Construction and Building Materials, 4*, 8. https://doi.org/10.11648/J.AJCBM.20200401.12
21. Chenniappan, A., Shanmugam, J., & Rajendran, V. (2020). Study on flexural strength of concrete with partial replacement of cement with hyposludge and crimped steel fibres. In: *AIP Conf Proc*, 2240. https://doi.org/10.1063/5.0011063/1024784
22. Keerthana Devi, P., Sanchaya, M., Harikaran, M., et al. (2023). Effective utilization of waste paper sludge ash as a supplementary material for cement. *Mater Today Proc*. https://doi.org/10.1016/j.matpr.2023.05.599
23. Alyami, M., Mydin, M. A. O., & Zeyad, A. M., et al. (2023). Influence of wastepaper sludge ash as partial cement replacement on the properties of lightweight foamed concrete. *Journal of Building Engineering, 79*. https://doi.org/10.1016/j.jobe.2023.107893
24. Maheswaran, R., Marichelvam, M. K., Asok, S. P. (2023). Development of value-added sustainable products from paper mill sludge: An experimental approach. *Heliyon, 9*. https://doi.org/10.1016/j.heliyon.2023.e17517
25. Monosi, S. (2012). Reuse of paper mill ash in plaster blends. *The Open Waste Management Journal, 5*, 5–18. https://doi.org/10.2174/1876400201205010005
26. Corinaldesi, V., Fava, G., Ruello, M. L. (2010). Paper mill sludge ash as supplementary cementitious material. In: 2nd international conference on sustainable construction materials and technologies (pp. 85–94). https://doi.org/10.1061/(ASCE)MT.1943-5533.0000218
27. Ingale, S. D., & Nemade, P. D. (2023). Effect of paper sludge ash on properties of cement concrete: A review. *Mater Today Proc*. https://doi.org/10.1016/J.MATPR.2023.03.492
28. Goel, G., Kalamdhad, A. S. (2018). Paper mill sludge (PMS) and degraded municipal solid waste (DMSW) blended fired bricks–a review. *MOJ Civil Engineering, 4*, 81–85. https://doi.org/10.15406/MOJCE.2018.04.00101
29. Kubba, H. Z., Salah Nasr, M., Al-Abdaly, N. M., et al. *Influence of incinerated and non-incinerated waste paper on properties of cement mortar*. https://doi.org/10.1088/1757-899X/671/1/012113
30. Wong, H. S., Barakat, R., Alhilali, A., et al. (2015). Hydrophobic concrete using waste paper sludge ash. *Cement and Concrete Research, 70*, 9–20. https://doi.org/10.1016/j.cemconres.2015.01.005

31. Nigar, A., Shukla, V. K. (2020). Study of paper pulp concrete partially replaced for M-25 grade of concrete. *International Journal of Research in Engineering, Science and Management, 3.*
32. Sahu, K., Rathore, M. (2020). Assessment on utilization of hyposludge and jute fiber in the production of high—performance concrete. *International Research Journal on Advanced Science Hub, 02.*
33. Azar, J. P., Najarchi, M., Sanaati, B., et al. (2019). The experimental assessment of the effect of paper waste ash and silica fume on improvement of concrete behavior. *KSCE Journal of Civil Engineering, 23*, 4503–4515. https://doi.org/10.1007/S12205-019-0678-X
34. Bureau of Indian Standards. (2019). *IS 10262:2019—concrete mix proportioning—guidelines.* New Delhi.
35. ASTM International. (2021). *ASTM D7348–21: Standard test methods for Loss on Ignition (LOI) of solid combustion residues.*
36. Liaw, C.-T., Chang, H.-L., Hsu, W.-C., & Huang, C.-R. (1998). A novel method to reuse paper sludge and co-generation ashes from paper mill. *Journal of Hazardous Materials, 58*, 93–102.
37. Shubbar, A. A, Sadique, M., Nasr, M. S., et al. (2020).The impact of grinding time on properties of cement mortar incorporated high volume waste paper sludge ash. *Karbala International Journal of Modern Science, 6*, 396–403. https://doi.org/10.33640/2405-609X.2149
38. Mirza, M. I., Malik, M. I., Ahmad, S., et al. (2015). Prospect of partial utilization of waste glass powder and waste paper sludge ash in concrete. *Article in International Journal of Civil Engineering and Technology, 6*, 160–174.
39. Shankar, K. G., Rama Rao, G. V. (2017). Variation of sorptivity in diverse grades of concrete proportioned with fly ash, wood husk ash, and rice husk ash. *International Research Journal of Engineering and Technology (IRJET)*
40. Niranjan, R., Rajendrakumar, N., & Saravanan, R. (2020). Evaluation on mechanical properties of Lm14 reinforced with paper sludge ash. *Mater Today Proc, 24*, 1210–1216. https://doi.org/10.1016/j.matpr.2020.04.435
41. de Azevedo A. R. G., Alexandre, J., Marvila, M. T., et al. (2020). Technological and environmental comparative of the processing of primary sludgewaste from paper industry for mortar. *J Clean Prod. 249.* https://doi.org/10.1016/j.jclepro.2019.119336
42. Mejdi, M., Saillio, M., & Chaussadent, T., et al. (2020). Hydration mechanisms of sewage sludge ashes used as cement replacement. *Cem Concr Res, 135.* https://doi.org/10.1016/J.CEMCONRES.2020.106115

Variables Influencing the Performance of 3D Concrete Printing for Sustainable Construction: Current Status and Future Prospects

Bishnu Kant Shukla, Gaurav Bharti, Bhupender Parashar, Pushpendra Kumar Sharma, Abhimanyu, and Shivam Verma

Abstract A type of additive manufacturing called three-dimensional concrete printing (3DCP) makes it possible to create intricate architectural details straight from digital three-dimensional design using CAD software (3D CAD) models eliminating the requirements of traditional molds or fixtures. This innovative technology is gaining traction due to its efficiency in producing precise prototypes with reduced material waste and minimal human intervention. Its applications range from manufacturing structural components to on-site construction of walls, addressing various design and technical challenges. With a focus on sustainable practices, such as the use of recycled aggregates, ecologically friendly binders, chemical admixtures, and nano-additives to enhance mix designs, this chapter critically examines developments in 3DCP. It also examines current studies on the rheological characteristics needed for efficient printing, highlighting its potential for environmentally beneficial structural and ornamental uses. The chapter ends with a list of research gaps, especially with regard to improving longevity and continued reliability of 3D-printed concrete buildings.

Keywords 3-Dimensional concrete printing · Sustainable practice · Rheology · Additive manufacturing · Recycled materials · Nano-additives · Durability

---

B. K. Shukla (✉) · G. Bharti
Department of Civil Engineering, JSS Academy of Technical Education, Noida, Uttar Pradesh 201301, India
e-mail: bishnukantshukla@gmail.com

B. K. Shukla · P. K. Sharma · Abhimanyu
School of Civil Engineering, Lovely Professional University, Phagwara, Punjab 144411, India

B. Parashar
Department of Mathematics, JSS Academy of Technical Education, Noida, Uttar Pradesh 201301, India

S. Verma
New Jersey Institute of Technology, Newark, NJ 07102, USA
e-mail: ssv8@njit.edu

P. Danish et al. (eds.), *Innovative Technologies for Smart and Sustainable Infrastructure*,
https://doi.org/10.1007/978-3-031-92801-7_13

1 Introduction

Concrete remains the predominantly used building material, and printing of 3D concrete (3DCP) has developed as a novel technique of construction. The process involves creating a three-dimensional model using computer-aided design (CAD) software, which is then converted into G-code to control the extrusion process [1]. 3DCP offers several advantages, including customized designs, reduced labor costs, minimized material waste, and the elimination of formwork. Removing formwork alone can reduce labor expenses by 50–80% and cut material waste by 30–60% at construction sites [2]. Despite these benefits, traditional concrete construction presents challenges such as high costs and limited design flexibility when formwork is required. The automation of concrete manufacturing has been explored since the nineteenth century, with Thomas Edison's 1917 patent for machine-built concrete houses being an early example, though it failed due to technological limitations. Achieving an optimal rheological balance for 3D printing remains a challenge, requiring the material to be workable during pumping but stable post-extrusion, balancing workability and thixotropy [3]. As 3DCP constructs structures layer by layer, it results in anisotropic mechanical properties. While it allows for mold-free and cost-effective fabrication, concerns about material strength and structural reliability remain [1]. Research continues to refine these aspects for greater efficiency and durability in 3D-printed structures of concrete. According to Llatas, the construction sector contributes nearly 80% of global waste, emphasizing the need for sustainable solutions. The high cost of specialized formwork limits architectural flexibility, as rectilinear structures are less stable than curved designs, where stress is better distributed [2–4]. Another challenge is the distinction in structural characteristics of printed elements of concrete. The central section of extruded concrete has more strength as compared to the interfaces between layers, whether horizontally or vertically aligned. The direction of applied loads significantly impacts structural performance (Fig. 1). The elimination of molds, cost savings, and fast fabrication make 3DCP a strong alternative to traditional construction. The ability to create complex geometries has further driven research and technological advancements in this field [5].

Early 3DCP research examined the printability and structural characteristics of fine aggregate-based concrete. Studies found that complex geometric designs improved cost and material efficiency. Recent advancements have introduced cement-derived substances and new printing techniques for large-scale structures. The crystallization and hydration within layers enhance strength over time. Due to microstructural differences, 3D-printed concrete shows strength variations, but the American Concrete Institute (ACI) sets a 28-day compressive strength minimum of 55 MPa [6]. Structural characteristics are influenced by premature drying, which traditional concrete avoids through mold protection. Layer interfaces also affect structural integrity, requiring strong interfacial bonding to reduce weaknesses. Ensuring proper adhesion between layers improves the robustness and stability of 3D-printed components [7].

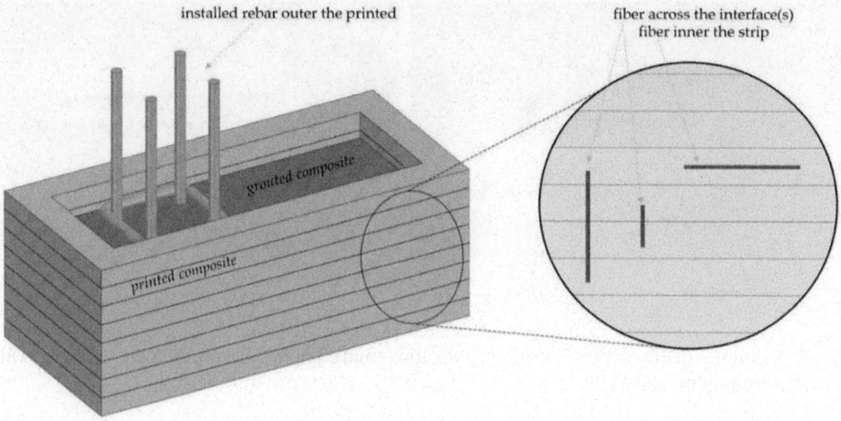

Fig. 1 Reinforcement placement in a 3D Printed concrete structure: inner strips, interface fibers, and external rebar [4]

2 Analysis of Existing Literature

2.1 Additive Manufacturing of Concrete

In traditional concrete, open time refers to the time interval when the material maintains its moldability before hydration acceleration begins. In 3D concrete printing (3DCP), it refers to how long the mix retains viscosity and yield stress, ensuring smooth extrusion and layer deposition. Since process operation dictates the required extrusion rate, open time is a critical factor. Material deposition per layer depends on the component size and working surface area, influencing process complexity. Each layer must be added within a specific cycle time, which determines the interval before subsequent layer placement [8, 9]. Extending open time remains challenging, as it requires admixtures while maintaining mix stability. Research continues to optimize formulations, but isolating individual factors remains complex, as noted by Reiter et al. [10]. The ability to model and simulate the entire 3DCP process, including material aggregation, extrusion, and solidification, is crucial for efficient implementation.

Extrusion nozzles determine print quality and material flow in 3D concrete printing. Common shapes include circular [11] and rectangular [12, 13]. Extrudability ensures continuous material deposition by allowing the mixture to pass through the nozzle with minimal resistance. Figure 2 illustrates 3DCP systems, such as a 4-axis gantry printer and layered printing of complex structures. Since no standard extrudability test exists, evaluation is done visually, influenced by nozzle shape, material properties, and deposition movement [13]. Nozzle shape affects extrusion consistency; round nozzles are widely used, while rectangular ones enhance layer alignment and interlayer bonding [14]. In spray-based 3DCP, wet-mixed mortars

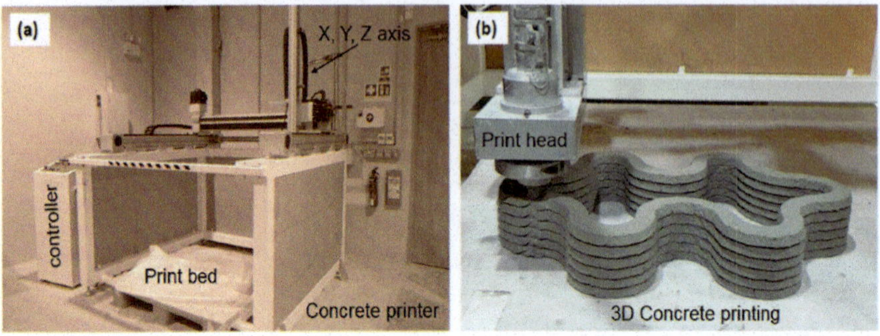

Fig. 2 Examples of 3D concrete printing applications and equipment: **a** 4-axis gantry printer, **b** 3D concrete printing process [16]

improve pumpability and shape retention. To define key material properties in 3D concrete printing, Camacho et al. [15] proposed:

- **Open Time**: Duration the material remains workable for extrusion and pumping.
- **Stabilization and Printing Cycle**: Time needed for a layer to stabilize before the next is applied.

Improving process efficiency, mathematical accuracy, and hardened properties requires strict control over material characteristics and design constraints, along with process complexity and operational adjustments. Print quality and structural performance depend on factors like workability of fresh concrete, material deformation, hardened properties, design adherence, and flexibility in execution [17]. Printability is significantly influenced by rheology and nozzle configuration, which determine the structural strength of printed elements. In most cases, printed components exhibit higher strength than required for structural and surface finishing applications. To minimize surface irregularities, automated troweling has been introduced, improving the final layer's smoothness [18, 19]. For doubly curved surfaces, conformal printing enhances strength. 4-axis and 5-axis systems improve nozzle positioning, ensuring better layer bonding and structural integrity by optimizing interaction between deposited layers and the working surface [20].

2.2 *Process of Printing*

Optimizing the printing of 3D concrete (3DCP) requires a comprehensive strategy that includes component design, finishing, process standardization, and material selection to enhance efficiency and structural integrity [21].

1. **Component Design**: Co-simulation integrates component modeling and tool path generation, ensuring optimized design. Understanding both fresh and hardened material properties, assessed through standardized testing, is essential for developing structural performance models that improve durability.
2. **Finishing**: Near net-shape CAD geometry minimizes post-printing refinements, reducing additional processing time while improving precision.
3. **Process Standardization**: Manufacturing-specific components will be standardized to incorporate reinforcements and composite materials effectively. Integrating these elements into design software enhances control over production.
4. **Material Selection**: Optimized aggregate formulations ensure consistent fresh-state properties. A uniform open-time supports continuous layer deposition, while precise setting time regulation aligns with printing process demands.

2.3 Advantages

3D printing in edifice offers benefits such as reduction in cost, efficient resource use, and increased structural durability [22–28].

- **Cost and Resource Efficiency**: Material waste is minimized, and resources are optimized, lowering overall expenses. Reduced labor demand also decreases costs related to wages and workplace injuries.
- **Faster Construction**: 3D printers convert CAD models into physical structures, enabling the completion of a single-story building in 12–24 h, significantly reducing construction time.
- **Improved Durability**: Printed structures have higher resilience and longer lifespans, with better resistance to temperature variations and extreme conditions.

2.4 Disadvantages

Despite its benefits, 3D printing in construction has challenges that must be addressed [26–28].

- **High Costs and Skilled Labour**: While 3D printing lowers construction costs, the printers require a large investment. Skilled labour is also necessary for operating CAD software and 3D printing systems.
- **Energy Consumption**: 3D printers consume nearly 100 times more electricity than traditional methods, raising efficiency and cost concerns.
- **Storage and Logistics**: Their large size complicates on-site storage and transportation, making space management crucial for large-scale use.

Researchers at the University of Southern California pioneered 3DCP technology, introducing “contour crafting”, which used a specialized concrete trowel head. Loughborough University later contributed by developing Energies, enabling direct nozzle printing without trowel blades [19, 20]. Using Fused Deposition Modeling (FDM) concepts, the D-Shape method was introduced to enhance printing efficiency and structural integrity. Meanwhile, the Institute for Advanced Architecture of Catalonia developed Mini Builders, a system with a distinct material deposition approach. As concrete extrudes, it gradually hardens, improving compressive strength and stability [21]. The layering process in 3DCP depends on printing speed and method. Figure 3 presents a laboratory-scale 3D concrete printer with an extruder for controlled material deposition. To assess printed concrete properties, researchers study rheological models to understand mix behaviour and flow characteristics. Some experts classify 3D-printed concrete as a non-Newtonian fluid, with buildability, as shown in Fig. 4, influenced by yield stress and thixotropy during printing.

The mechanical performance of additive-manufactured concrete depends on factors like yield stress, material properties, and structural stability. Studies have explored the conversion of non-Newtonian fluid characteristics into values representing printed concrete's strength and durability [22]. Gravity and yield stress influence extrusion path geometry, while surface tension, nozzle design, deposition speed, and material flow rate affect print quality and layer stability [23]. Ensuring stability in 3D-printed structures is essential for load-bearing capacity. A key challenge is structural buckling due to self-weight, addressed by lattice reinforcement or increased wall thickness. Using soil mechanics principles, researchers in Eindhoven studied freshly printed concrete's structural characteristics through uniaxial compression and

Fig. 3 Laboratory-scale 3D concrete printer with an extruder: (1) Electric Motor, (2) material hopper, (3) auger, (4) nozzle, (5) control panel, (6) frequency converter, (7) horizontal motion motor, (8) vertical motion manual drive, (9) supporting frame, (10) power cable [29]

Fig. 4 Examination of failure mechanisms in the squeezing test [29]

direct shear tests, applying a Coulomb-Mohr failure model. Casagrande et al. [25] analysed collapse mechanisms in printed structures based on past experiments. The environmental impact of waste arising from demolition and construction activities (CDW) is a growing concern. Recycled aggregates reduce resource depletion and waste, while transportation methods and distances affect global warming potential (GWP), energy use, and pollutant formation. Studies emphasize local recycling to cut transport emissions, with Life Cycle Assessment (LCA) used to optimize material substitution [26].

Replacing cement with zeolite by 10%, 20%, and 30% led to GWP reductions of 60.3%, 69.7%, and 64.3%, respectively. Lower substitution levels increased compressive strength, while 30% replacement weakened it. Efficient logistics in material production and site proximity reduce environmental impact, with studies highlighting the need for resource optimization in sustainable construction [27]. Automation in 3D construction will significantly impact the workforce. By 2030, computer-controlled 3D printing may automate 26% of construction jobs in Japan, 24% in UK, 41% in Germany and 35% in the US. While this reduces job opportunities, it enhances precision and efficiency. In developing nations, where construction employs a large workforce, automation could have major economic effects [28]. However, digital fabrication simplifies skill acquisition, reducing errors and improving efficiency. A life-cycle assessment was conducted, considering data uncertainty, impact factors, and sensitivity analysis. The study evaluated environmental and economic benefits, focusing on how concrete composition and geometric design influence construction sustainability. Given concrete's environmental footprint, 3D printing's ability to create complex designs makes it a key factor in sustainable development [29–32].

3 Digital Construction

A research team at MIT developed a digital construction platform designed for on-site monitoring, analysis, and fabrication. To enhance speed, accuracy, and accessibility, the system incorporates a small robotic arm, which is positioned using a large supporting structure for controlled material placement [33]. The printing material used in this system is polyurethane foam, chosen for its rapid curing time and high insulation properties. A similar approach to concrete printing has been developed at Loughborough University (UK), where a computer-controlled system extrudes substantial layers of concrete according to a digital model. However, challenges such as over-printing—where excess material accumulates at certain locations, causing unintended swelling—and under-printing, where insufficient material deposition results in printing defects, can affect print accuracy and stability [34–36]. This layered printing technique also introduces visual and structural issues. The staircase effect, a common problem in additive manufacturing, affects surface smoothness, while anisotropic mechanical behaviour arises due to variation in material properties

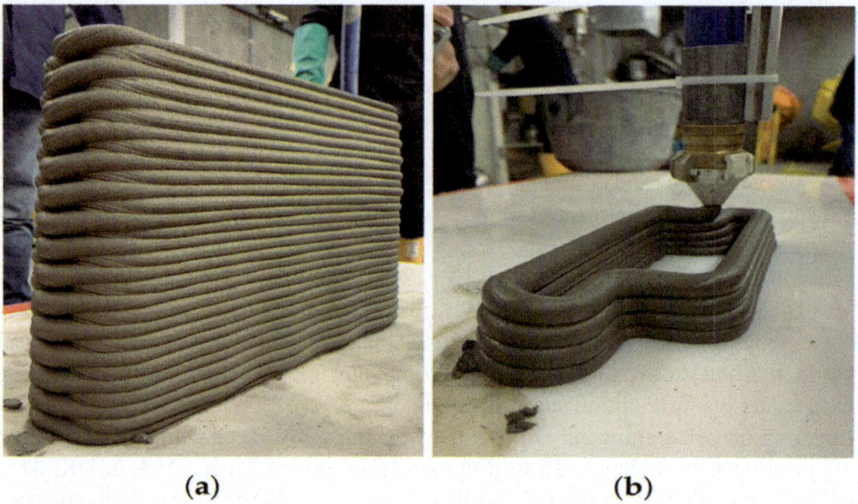

Fig. 5. 3D printing of Ultra-High-Performance Concrete (UHPC): **a** A 3D-printed wall, and **b** layered extrusion process [37]

between layers. Figure 5 illustrates the process of 3D printing of concrete demonstrating ultra-high-performance (UHPC), depicting layer-wise deposition and the gradual formation of a printed wall structure.

Planning, scheduling, cost estimation, and post-construction management of facility are all covered by the all-inclusive construction project management method known as Building Information Modelling (BIM). A BIM-integrated workflow for additive manufacturing can drastically cut down on project time by combining all phases into a single platform [8, 38, 39]. By storing and combining crucial production and equipment data, including printer management, material delivery, and post-processing processes, BIM also makes it possible to automate the process of 3D printing [40]. Many construction professionals consider BIM a leading design tool, likely to become the standard of industry. Since 3D printing technology is also seen as a transformational advancement in construction, adopting a BIM-based 3D printing approach could provide significant cost and labour savings for the sector. Before initiating the 3DCP process, it is advisable to conduct structural and material simulations to assess the performance of the final digital design [41–43]. Finite element analysis (FEA) and topology optimization software play a crucial role in evaluating structural integrity and material behaviour for newly developed designs [42]. Utilizing FEA simulations based on material properties can provide insights into how the material will perform under different conditions. Additionally, integrating sensors and actuators within 3D-printed structures allows for continuous monitoring of architectural elements, such as facades exposed to environmental factors. Beyond sensors, reinforcement can also be incorporated during the printing process, enhancing structural resilience against various loads [44].

One of the primary objectives of SC3DP is the optimization of material properties for 3D printing applications. To analyse the evolution of stress and viscosity, researchers conduct rheological testing at regular intervals. The use of structural optimization techniques supports the development of sophisticated geometries, which are otherwise challenging to construct. This approach has been widely adopted in structural mechanics, particularly when combined with finite element simulations [40]. The optimization process involves determining whether individual elements should be retained or removed during each stage of the design. The SIMP (Solid Isotropic Material with Penalization) method has been used at several scales, from the construction of micro- and nano-architectures to the enhancement of building structural design (Fig. 6, based on data from [40]). Even though these phenomena may be investigated independently, integrating them with process modeling to precisely forecast the hardened characteristics of a 3D-printed component is still a major obstacle and a continuous goal in 3DCP component design [45–47]. 3D printing in construction is a complicated process that calls for specific materials, hardware and software compatibility, accurate planning and control algorithms, creative formwork methods, altered structural designs, and conformity to industry norms [48, 49]. Improvements in rheology control techniques and material design strategies are required to guarantee printability, economic viability, and environmental sustainability for large-scale applications. Various cementitious materials and printing techniques have been used to construct large-scale structures in recent years. The integration of two advanced techniques—topology optimization for simulation-driven design and 3D concrete printing (3DCP) for manufacturing—presents a promising approach to achieving highly efficient concrete buildings. Numerical analysis results have been compared against girder deflection, providing insights into manufacturing challenges, such as printing setup, structural assembly, and reinforcement integration.

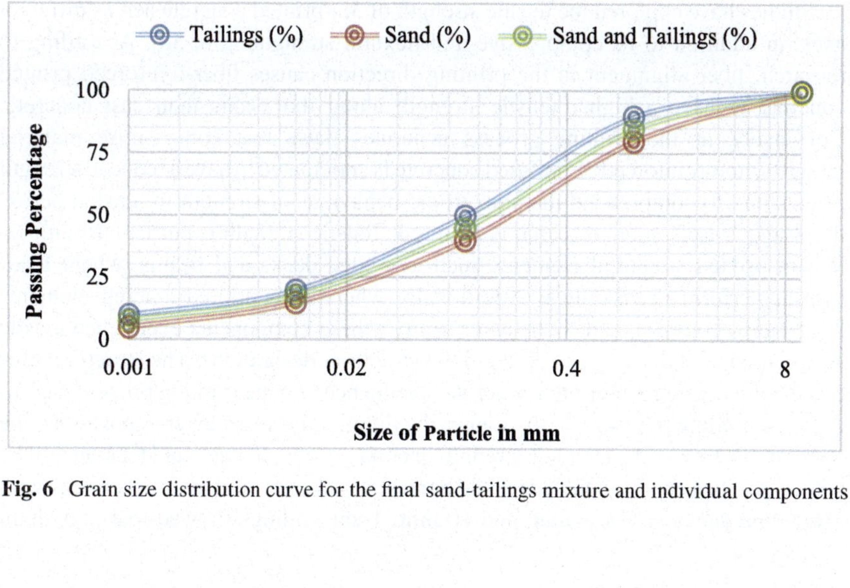

Fig. 6 Grain size distribution curve for the final sand-tailings mixture and individual components

4 Structural Characteristics

In 3D-printed concrete, the layered deposition process produces a unique structural configuration where the loading direction in relation to the print orientation determines the performance characteristics. Due to this, printed concrete elements exhibit anisotropic mechanical behaviour, with extruded layer orientations directly linked to the printing direction [1, 2, 40, 50, 51]. It is crucial to remember that the identical mix proportions were used to create both printed and cast concrete samples [52–54]. But according to a number of studies, printed concrete typically exhibits reduced strength in compression relative to ordinary cast concrete. This recommends that the process of printing itself contributes to the reduction in compressive strength. Therefore, a structural evaluation of printed elements is essential, considering anisotropic mechanical behaviour, load application direction, and strength variations resulting from the printing method [54]. Figure 7 displays 3D-printed specimens made from DETAX material, printed in three orientations: (a) 0° perpendicular alignment, (b) 45° alignment, and (c) 90° alignment. These orientations highlight the effect of deposition direction on anisotropic structural characteristics. Research indicates that fiber-reinforced printable mixes consistently demonstrate higher flexural strength along the Z-axis when compared to cast concrete of similar composition [55]. The alignment of fibers in the extrusion direction has been linked to enhanced flexural performance in fiber-reinforced printed mixtures [56]. Using smaller nozzle sizes and higher fiber content promotes greater fiber alignment with the printing direction, further improving structural characteristics. Bending resistance is an essential consideration in the construction of deck of bridges and structural beams. To guarantee optimum performance in load-bearing applications, these structures should be designed with a thorough awareness of the anisotropic behaviour of 3D-printed concrete.

Studies have explored the tensile strength of 3D-printed concrete across different axes, in addition to its compressive and flexural strengths [38, 58]. According to research, fiber alignment in the printing direction causes fiber-reinforced printed concrete to have a higher tensile strength along the x-axis than cast concrete. Conversely, as the printing process generates flaws that compromise material integrity, non-reinforced 3D-printed concrete is anticipated to have a reduced strength in tension. To comprehend the anisotropic behaviour of strength in tension across the Z and Y axes, more research is required. Tensile strength is particularly important in printed structural elements such as bridge decks and beams, where load-bearing performance is critical. Additionally, when lifting printed concrete elements like beams, columns, and shells using cranes, these components experience tensile stresses, which must be considered in the structural design [38]. The structural efficiency of hardened 3D-printed concrete is influenced by the printing process parameters. According to research, the longer the duration between layer depositions, the lower the 3D-printed concrete's strength in compression, flexure, and tension will be achieved. Chen et al. conducted tensile bond strength tests on printed specimens with layer time gaps of 20 s, 1 min, and 10 min. Their findings showed that maximum

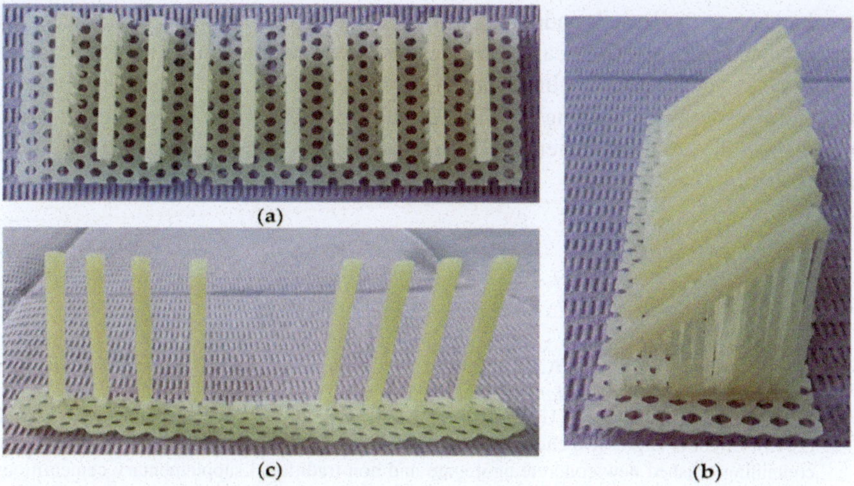

Fig. 7 Effect of printing orientation on 3D-printed specimens: **a** 0° alignment, **b** 45° alignment, and **c** 90° alignment [57]

strength in bond was achieved in specimens printed with a 20-s interval. In comparison to cast concrete samples, the bond strength dropped by 13% when the time interval was extended from 1 to 20 min. The time interval between layer depositions has a major impact on the failure mechanisms and failure strength of 3D-printed concrete [59].

5 Conclusion

In sustainable construction, three-dimensional concrete printing (3DCP) has become a game-changer, providing increased productivity, design flexibility, and less material waste. This study critically evaluated the key variables influencing 3DCP, including material composition, rheological properties, and structural performance. Sustainable practices, such as the incorporation of recycled aggregates, nano-additives, and alternative binders, were explored to optimize mix designs and improve environmental impact. Despite all of its benefits, there are still issues with guaranteeing the mechanical dependability and long-term durability of 3D-printed concrete buildings. Issues such as anisotropic mechanical behaviour, interlayer bonding, and process-induced defects require further investigation to reinforce the stability and functionality of printed structures. Additionally, advancements in digital fabrication techniques, such as topology optimization and integrated reinforcement strategies, are essential for improving the applicability of 3DCP in large-scale infrastructure projects. Future research should focus on refining printing parameters, developing standardized testing methodologies, and exploring automation-driven quality control mechanisms to facilitate broader adoption in the construction industry. The accuracy and dependability of 3DCP can be further improved by integrating real-time monitoring systems with Building Information Modeling (BIM). 3DCP has the potential to transform sustainable building by filling up these research gaps and providing structurally sound, resource-efficient, and reasonably priced solutions for contemporary infrastructure requirements.

References

1. Ismail, M. S., & Waliuddin, A. M. (1996). Effect of rice husk ash on high strength concrete. *Construction and Building Materials, 10*(7), 521–526.
2. Alnahhal, M. F., Alengaram, U. J., Jumaat, M. Z., Alsubari, B., Alqedra, M. A., & Mo, K. H. (2018). Effect of aggressive chemicals on durability and microstructure properties of concrete containing crushed new concrete aggregate and non-traditional supplementary cementitious materials. *Construction and Building Materials, 163*, 482–495.
3. Sikora, P., Chung, S. Y., Liard, M., Lootens, D., Dorn, T., Kamm, P. H., Stephan, D., & Abd Elrahman, M. (2021). The effects of nanosilica on the fresh and hardened properties of 3D printable mortars. *Construction and Building Materials, 281*, Article 122574.
4. Cao, X., Yu, S., & Cui, H. (2022). Experimental investigation on inner-and inter-strip reinforcements for 3D printed concrete via automatic staple inserting technique. *Applied Sciences, 12*(4), 2099.
5. Nadi, M., Majdoubi, H., Haddaji, Y., Bili, O., Chahid, M., Oumam, M., Tamraoui, Y. & Hannache, H. (2025). Digital fabrication processes for cementitious materials using three-dimensional 3D printing technologies. In *Recent developments and innovations in the sustainable production of concrete* (pp. 595–620). Woodhead Publishing.
6. Bardhan-Roy, B. K. (1996). Lightweight aggregate concrete in the UK. *Concrete, 30*(6).
7. Wangler, T., Lloret, E., Reiter, L., Hack, N., Gramazio, F., Kohler, M., & Flatt, R. J. (2017). Digital concrete: Opportunities and challenges. *RILEM Technical Letters, 1*(1), 67–75.
8. Lim, S., Buswell, R. A., Le, T. T., Austin, S. A., Gibb, A. G., & Thorpe, T. (2012). Developments in construction-scale additive manufacturing processes. *Automation in Construction, 21*, 262–268.
9. Le, T. T., Austin, S. A., Lim, S., Buswell, R. A., Gibb, A. G., & Thorpe, T. (2012). Mix design and fresh properties for high-performance printing concrete. *Materials and Structures, 45*, 1221–1232.
10. Reiter, L., Wangler, T., Roussel, N., & Flatt, R. J. (2018). The role of early age structural build-up in digital fabrication with concrete. *Cement and Concrete Research, 112*, 86–95.
11. Bos, F., Wolfs, R., Ahmed, Z., & Salet, T. (2016). Additive manufacturing of concrete in construction: Potentials and challenges of 3D concrete printing. *Virtual and Physical Prototyping, 11*(3), 209–225.
12. Da Silva, W. R. L. (2017). 3d concrete printing: from material design to extrusion. Presented slides. In *Annual Civil Engineering Workshop at Ecole Centrale de Lille* (ACE Workshop 2017).
13. Lim, S., Buswell, R. A., Valentine, P. J., Piker, D., Austin, S. A., & De Kestelier, X. (2016). Modelling curved-layered printing paths for fabricating large-scale construction components. *Additive Manufacturing, 12*, 216–230.
14. Khoshnevis, B., Hwang, D., Yao, K. T., & Yeh, Z. (2006). Mega-scale fabrication by contour crafting. *International Journal of Industrial and Systems Engineering, 1*(3), 301–320.
15. Camacho, D. D., Clayton, P., O'Brien, W., Ferron, R., Juenger, M., Salamone, S., & Seepersad, C. (2017). Applications of additive manufacturing in the construction industry–a prospective review. In *ISARC. Proceedings of the international symposium on automation and robotics in construction* (Vol. 34). IAARC Publications.

16. Panda, B., Noor Mohamed, N. A., Paul, S. C., Bhagath Singh, G. V. P., Tan, M. J., & Šavija, B. (2019). The effect of material fresh properties and process parameters on buildability and interlayer adhesion of 3D printed concrete. *Materials, 12*(13), 2149.
17. Meza, L., Singer, R., Vazquez, N., Keenan, R., Burgoyne, H., Hogstrom, K., & Tan, W. L. (2016). Concept for a fully in situ resource-derived habitat for Martian environment. In *15th Biennial ASCE conference on engineering, science, construction, and operations in challenging environments* (pp. 425–436). American Society of Civil Engineers.
18. El-Sayegh, S., Romdhane, L., & Manjikian, S. (2020). A critical review of 3D printing in construction: Benefits, challenges, and risks. *Archives of Civil and Mechanical Engineering, 20*, 1–25.
19. De Schutter, G., Lesage, K., Mechtcherine, V., Nerella, V. N., Habert, G., & Agusti-Juan, I. (2018). Vision of 3D printing with concrete—Technical, economic and environmental potentials. *Cement and Concrete Research, 112*, 25–36.
20. De Soto, B. G., Agustí-Juan, I., Hunhevicz, J., Joss, S., Graser, K., Habert, G., & Adey, B. T. (2018). Productivity of digital fabrication in construction: Cost and time analysis of a robotically built wall. *Automation in Construction, 92*, 297–311.
21. Mata-Falcón, J., Bischof, P., & Kaufmann, W. (2019). Exploiting the potential of digital fabrication for sustainable and economic concrete structures. In *First RILEM international conference on concrete and digital fabrication–digital concrete 2018* (pp. 157–166). Springer International Publishing.
22. Buswell, R. A., De Silva, W. L., Jones, S. Z., & Dirrenberger, J. (2018). 3D printing using concrete extrusion: A roadmap for research. *Cement and Concrete Research, 112*, 37–49.
23. Ferreiro-Cabello, J., Fraile-Garcia, E., Martinez-Camara, E., & Perez-de-La-Parte, M. (2017). Sensitivity analysis of Life Cycle Assessment to select reinforced concrete structures with one-way slabs. *Engineering Structures, 132*, 586–596.
24. Khoshnevis, B., & Dutton, R. (1998). Innovative rapid prototyping process makes large sized, smooth surfaced complex shapes in a wide variety of materials. *Materials Technology, 13*(2), 53–56.
25. Zhao, P., Rao, C., Gu, F., Sharmin, N., & Fu, J. (2018). Close-looped recycling of polylactic acid used in 3D printing: An experimental investigation and life cycle assessment. *Journal of Cleaner Production, 197*, 1046–1055.
26. Uppala, S. S., & Tadikamalla, M. R. (2017). A review on 3D printing of concrete-the future of sustainable construction. *i-Manager's Journal on Civil Engineering, 7*(3), 49.
27. Hack, N., Lauer, W., Langenberg, S., Gramazio, F., & Kohler, M. (2013). Overcoming repetition: Robotic fabrication processes at a large scale. *International Journal of Architectural Computing, 11*(3), 285–299.
28. Xiao, J., Li, W., Fan, Y., & Huang, X. (2012). An overview of study on recycled aggregate concrete in China (1996–2011). *Construction and Building Materials, 31*, 364–383.
29. Dvorkin, L., Konkol, J., Marchuk, V., & Huts, A. (2022). Effectiveness of polymer additives in concrete for 3d concrete printing using fly ash. *Polymers, 14*(24), 5467.
30. Truong-Thinh, N., Ngoc-Phuong, N., & Phuoc-Tho, T. (2011). A study of pipe-cleaning and inspection robot. In *2011 IEEE international conference on robotics and biomimetics* (pp. 2593–2598). IEEE.
31. Ford, S., & Despeisse, M. (2016). Additive manufacturing and sustainability: An exploratory study of the advantages and challenges. *Journal of Cleaner Production, 137*, 1573–1587.
32. Corcione, C. E., Palumbo, E., Masciullo, A., Montagna, F., & Torricelli, M. C. (2018). Fused Deposition Modeling (FDM): An innovative technique aimed at reusing Lecce stone waste for industrial design and building applications. *Construction and Building Materials, 158*, 276–284.
33. Paul, S. C., Van Zijl, G. P., Tan, M. J., & Gibson, I. (2018). A review of 3D concrete printing systems and materials properties: Current status and future research prospects. *Rapid Prototyping Journal, 24*(4), 784–798.
34. Kim, Y., Hanif, A., Kazmi, S. M., Munir, M. J., & Park, C. (2018). Properties enhancement of recycled aggregate concrete through pretreatment of coarse aggregates–Comparative assessment of assorted techniques. *Journal of Cleaner Production, 191*, 339–349.

35. Laiblová, L., Pešta, J., Kumar, A., Hájek, P., Fiala, C., Vlach, T., & Kočí, V. (2019). Environmental impact of textile reinforced concrete facades compared to conventional solutions—LCA case study. *Materials, 12*(19), 3194.
36. Tay, Y. W. D., Panda, B., Paul, S. C., Noor Mohamed, N. A., Tan, M. J., & Leong, K. F. (2017). 3D printing trends in building and construction industry: A review. *Virtual and Physical Prototyping, 12*(3), 261–276.
37. Medicis, C., Gonzalez, S., Alvarado, Y. A., Vacca, H. A., Mondragon, I. F., García, R., & Hernandez, G. (2022). Mechanical performance of commercially available premix UHPC-based 3D printable concrete. *Materials, 15*(18), 6326.
38. Martens, P., Mathot, M., Bos, F., & Coenders, J. (2018). Optimising 3D printed concrete structures using topology optimisation. In *High tech concrete: Where technology and engineering meet: Proceedings of the 2017 fib symposium, held in Maastricht, The Netherlands, June 12–14, 2017* (pp. 301–309). Springer International Publishing.
39. Zhou, S., & Li, Q. (2008). Design of graded two-phase microstructures for tailored elasticity gradients. *Journal of Materials Science, 43*, 5157–5167.
40. Feng, P., Meng, X., Chen, J. F., & Ye, L. (2015). Mechanical properties of structures 3D printed with cementitious powders. *Construction and Building Materials, 93*, 486–497.
41. Arunothayan, R., Nematollahi, B., Bong, S. H., Ranade, R., & Sanjayan, J. (2020). Hardened properties of 3d printable ultra-high performance fiber-reinforced concrete for digital construction applications. In *Rheology and Processing of Construction Materials: RheoCon2 & SCC9 2* (pp. 355–362). Springer International Publishing.
42. Ma, G., Li, Z., Wang, L., Wang, F., & Sanjayan, J. (2019). Mechanical anisotropy of aligned fiber reinforced composite for extrusion-based 3D printing. *Construction and Building Materials, 202*, 770–783.
43. Zhu, B., Pan, J., Nematollahi, B., Zhou, Z., Zhang, Y., & Sanjayan, J. (2019). Development of 3D printable engineered cementitious composites with ultra-high tensile ductility for digital construction. *Materials & Design, 181*, Article 108088.
44. Chen, Y., Jansen, K., Zhang, H., Rodriguez, C. R., Gan, Y., Çopuroğlu, O., & Schlangen, E. (2020). Effect of printing parameters on interlayer bond strength of 3D printed limestone-calcined clay-based cementitious materials: An experimental and numerical study. *Construction and Building Materials, 262*, Article 120094.
45. Álvarez-Fernández, M. I., Prendes-Gero, M. B., González-Nicieza, C., Guerrero-Miguel, D. J., & Martínez-Martínez, J. E. (2021). Optimum mix design for 3d concrete printing using mining tailings: A case study in Spain. *Sustainability, 13*(3), 1568.
46. Zdeb, T., & Śliwiński, J. (2011). The influence of selected material and technological factors on mechanical properties and microstructure of reactive powder concrete (RPC). *Archives of Civil Engineering, 57*(2), 227–246.
47. Matthäus, C., Back, D., Weger, D., Kränkel, T., Scheydt, J., & Gehlen, C. (2020). Effect of cement type and limestone powder content on extrudability of lightweight concrete. In *Second RILEM international conference on concrete and digital fabrication: Digital concrete 2020 2* (pp. 312–322). Springer International Publishing.
48. Sun, J., Huang, Y., Aslani, F., & Ma, G. (2020). Properties of a double-layer EMW-absorbing structure containing a graded nano-sized absorbent combing extruded and sprayed 3D printing. *Construction and Building Materials, 261*, Article 120031.
49. Ting, G. H. A., Tay, Y. W. D., Qian, Y., & Tan, M. J. (2019). Utilization of recycled glass for 3D concrete printing: Rheological and mechanical properties. *Journal of Material Cycles and Waste Management, 21*, 994–1003.
50. Hack, N., Wangler, T., Mata-Falcón, J., Dörfler, K., Kumar, N., Walzer, A. N., & Kohler, M. (2017). Mesh mould: an on site, robotically fabricated, functional formwork. In *Second Concrete Innovation Conference (2nd CIC)* (Vol. 19, pp. 1–10).
51. Bravo, M., De Brito, J., Pontes, J., & Evangelista, L. (2015). Mechanical performance of concrete made with aggregates from construction and demolition waste recycling plants. *Journal of Cleaner Production, 99*, 59–74.

52. Davies, D. R., & Kitchener, J. N. (1996). Massive use of pulverised fuel ash in concrete for the construction of a UK power station. *Waste Management, 16*(1–3), 169–180.
53. Shayan, A., Diggins, R., & Ivanusec, I. (1996). Effectiveness of fly ash in preventing deleterious expansion due to alkali-aggregate reaction in normal and steam-cured concrete. *Cement and Concrete Research, 26*(1), 153–164.
54. Mays, G. C., & Barnes, R. (1991). Performance of lightweight aggregate concrete structures in service. *Structural Engineer, 69*, 351–361.
55. Bardhan-Roy, B. K. (1996). Lightweight aggregate concrete in the UK. *Concrete, 30*(6)
56. Eastman, C. M. (2011). *BIM handbook: A guide to building information modeling for owners, managers, designers, engineers and contractors*. John Wiley & Sons.
57. Derban, P., Negrea, R., Rominu, M., & Marsavina, L. (2021). Influence of the printing angle and load direction on flexure strength in 3D printed materials for provisional dental restorations. *Materials, 14*(12), 3376.
58. Rehman, A. U., & Kim, J. H. (2021). 3D concrete printing: A systematic review of rheology, mix designs, mechanical, microstructural, and durability characteristics. *Materials, 14*(14), 3800.
59. Chen, Y., He, S., Gan, Y., Çopuroğlu, O., Veer, F., & Schlangen, E. (2022). A review of printing strategies, sustainable cementitious materials and characterization methods in the context of extrusion-based 3D concrete printing. *Journal of Building Engineering, 45*, Article 103599.

Revolutionizing Sustainable Construction with Circuit Board-Derived Concrete: A Comprehensive Review

Aditya Dudhal, Jagruti Patil, and Pravin Minde

Abstract The increasing amount of waste electronic circuit boards i.e. Printed Circuit Board (PCB) is a serious environmental issue. This chapter explores the ability to use PCB waste as a sustainable addition to concrete, transforming e-waste into a valuable building material. Based on experimental data, the research highlights how adding PCB waste can improve concrete properties, like flexural and compressive strength, while also offering environmental benefits like reduction in usage of natural resources, reduction in landfill waste, and minimizing the carbon footprint of traditional concrete production. The study examines how finely mixed PCB components enhance concrete's strength, durability, and environmental efficiency, showing its suitability for modern infrastructure. It also discusses the role of this innovative material in smart infrastructure, emphasizing its contribution to sustainable development and the promotion of a circular economy. Linking e-waste management with sustainable construction provides a practical solution aligned with global sustainability goals.

Keywords Electronic waste · Sustainable concrete · Sustainability · Printed circuit board · E-waste recycling · Smart infrastructure · Durability enhancement

1 Introduction

Sustainable construction is a critical approach aimed at minimizing the environmental footprint of building practices. This methodology states the adoption of energy-efficient techniques, the use of renewable and recyclable materials, and innovative strategies designed to reduce waste while conserving resources. By prioritizing these practices, the construction industry can significantly enhance the long-term sustainability and durability of its projects [1, 2]. The importance of sustainable construction becomes increasingly important in light of the environmental challenges associated with conventional building methods. These traditional practices contribute

A. Dudhal · J. Patil (✉) · P. Minde
Dr. Vishwanath Karad, MIT World Peace University, Pune, India
e-mail: jagruti.patil@mitwpu.edu.in

P. Danish et al. (eds.), *Innovative Technologies for Smart and Sustainable Infrastructure*,
https://doi.org/10.1007/978-3-031-92801-7_14

to increased carbon emissions, depletion of natural resources, and excessive waste generation [3]. By embracing sustainable construction techniques, the industry not only eliminates these dangerous effects but also fosters a healthier environment. Furthermore, the promotion of eco-friendly materials and technologies may lead to substantial cost reduction and improved performance in building operations over time [4].

Incorporating electronic waste into construction industry represents a pioneering step to sustainability. This practice involves repurposing outdated electronics as valuable components in building projects, effectively reducing landfill contributions while recycling materials that would otherwise be discarded.

The amount of electronic garbage produced worldwide increased to around 53.6 million metric tonnes in 2019, and it is expected to reach 74.7 million metric tonnes by 2030, making it a significant problem [5]. E-waste contains hazardous materials that can contaminate soil, water, and air if failed to manage properly [6] (Fig. 1).

Given the substantial volume of e-waste produced globally, utilizing PCBs as an alternative to either cement or sand in concrete emerges as an efficient solution for achieving sustainability. Non-metallic fractions of PCBs, which contain about 70% of their weight, can be repurposed as partial alternative for aggregates in concrete [8]. This not only helps in waste management but also enhances the mechanical properties of concrete [9].

Fig. 1 E-waste statistics [7]

2 Electronic Waste and Circuit Boards: An Overview

2.1 Compositions and Sources of Electronic Circuit Boards

The PCBs mainly consist of metals and non-metals. Mainly the metals are (copper, gold, lead, etc.) and the non-metals consisting of plastics, fibres, etc. [10]. The main source of the printed circuit boards is out of use and old smartphones, televisions, headphones, etc. [11].

2.2 Challenges in Managing E-waste

The management of electronic waste is challenging due to the hazardous materials present in electronic devices [12]. Improper disposal can lead to environmental degradation and health risks. The complexity of recycling PCBs, which contain valuable materials, adds to the difficulty of effective electronic waste management [11].

Managing e-waste presents various technical barriers like the presence of toxic substances, inadequate recycling infrastructure, and environmental pollution from improper disposal methods. The complexity of circuit board materials makes recycling difficult, leading to a huge amount of electronic waste dumped in landfills [13].

2.3 Existing Approaches to Recycling and Reusing Circuit Boards

The recycling and reuse of circuit boards have become critical components due to the growing environmental damage caused by electronic waste (e-waste). Various methodologies have been developed to recover valuable materials from discarded circuit boards, with each one having its own set of advantages and limitations [14]. This section reviews the existing approaches to recycling circuit boards, focusing on their efficiency, environmental safety, and potential for sustainable development.

Current recycling methods for circuit boards primarily include mechanical processing, pyrometallurgical recovery, and chemical extraction [15]. Mechanical processing involves shredding and grinding circuit boards to separate components based on size and density [16]. While this method is relatively straightforward and cost-effective, it often fails to recover certain valuable metals fully and can generate significant amounts of waste material. Pyrometallurgical recovery, on the other hand, employs high-temperature processes to obtain metals such as gold, silver, and copper [15]. Although this method is effective in recovering these precious metals, it poses substantial environmental risks due to emissions of toxic gases and the generation of hazardous by-products [17].

Chemical extraction methods utilize solvents or acids to dissolve metals from circuit boards. These techniques can achieve high recovery rates for specific metals; however, they often involve hazardous chemicals that require careful handling and disposal. The environmental implications of chemical extraction processes necessitate the development of safer alternatives that minimize ecological impact while maximizing resource recovery. Despite the advancements in these recycling techniques, there remains a pressing need for more sustainable practices that can effectively manage e-waste. The complexity of circuit board materials—comprising various metals, plastics, and composites—complicates recycling efforts and results in filling the landfills with e-waste. This highlights the importance of finding alternative uses for e-waste, particularly in construction applications where repurposed materials can contribute to sustainability goals.

2.4 Process of Extraction of PCB Waste

The process of recycling begins with the collection and sorting of discarded PCBs, which can be accomplished either manually or through automated systems [18]. Figure 2 shows a detailed flowchart of the process. Manual sorting involves laborers removing non-PCB materials like plastic casings and batteries, while automated systems enhance efficiency and accuracy in the sorting process [19]. Following sorting, pre-processing is conducted to remove electronic components through heat or chemical treatments. The PCBs are then shredded into smaller pieces (1–2 cm) using industrial shredders before undergoing further crushing with mechanical crushers such as hammer mills or ball mills [18].

The goal is to reduce PCB fragments to sizes between 1 and 5 mm for optimal separation. Mechanical process like using magnetic separators to extract ferrous materials like iron is used, while electrostatic techniques facilitate the separation of non-metals from metals [20]. Once the desired separation is achieved, a pulverization process grinds the remaining materials into a fine powder (below 100 μ), followed by refinement to ensure uniform particle size through fine mesh sieving. This meticulous processing not only enhances the quality of the recycled material but also prepares it for effective integration into concrete mixtures [18].

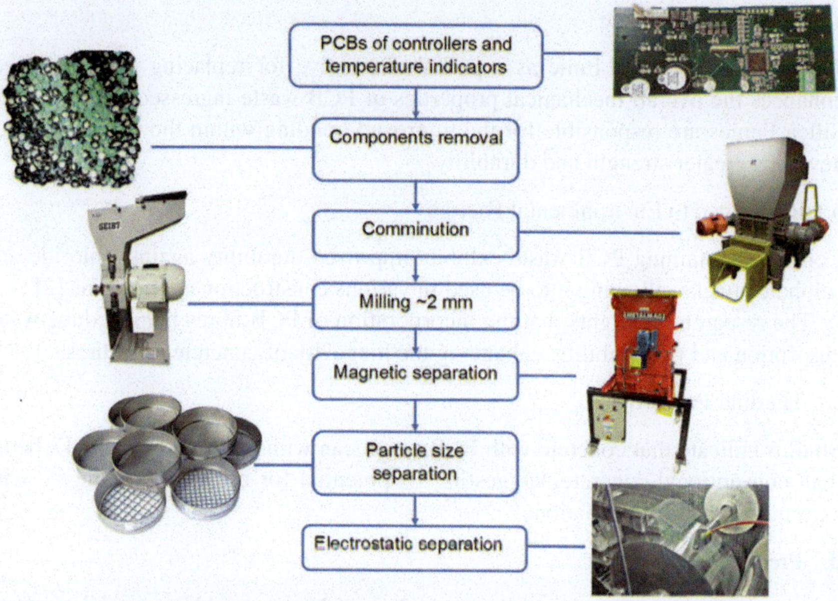

Fig. 2 Flowchart of the processing steps (20)

3 Circuit Board Waste in Construction Materials

3.1 Properties of Circuit Board Derived Materials Relevant to Concrete

The PCB powder improves various properties of concrete like workability and strength [21]. The specific gravity of OPC is generally greater than that of PCB powder, hence using PCB powder can improve the density and stability of concrete mixtures [22]. The chemical composition of PCB dust includes significant amounts of silica, which contributes to the pozzolanic activity in concrete [13].

3.2 Current Research on Incorporating Circuit Board Waste into Concrete

The integration of waste PCB fibres in concrete has been shown to significantly improve mechanical properties like compressive, tensile, and flexural strengths [23]. Research indicates that optimal replacement percentages of PCB fibres (typically around 5% by weight of cement) yield the best performance, enhancing the overall strength of the concrete mix [22].

a. Role of Silica Fume

The addition of silica fume as a partial alternative for replacing cement further enhances the overall mechanical properties of PCB waste ingressed concrete [22]. Silica Fumes are responsible for the improved bonding within the concrete which develops greater strength and durability.

b. Resistance to Environmental Factors

Concrete containing PCB waste exhibits improved durability against chloride and sulphate attacks, allowing it to be used in various construction applications [2].

The research highlights that the incorporation of PCB fibres helps reduce water absorption and permeability, enhancing the longevity of concrete structures.

c. Thermal Properties

Studies indicate that concrete with PCB waste can withstand thermal shocks better than conventional concrete, suggesting its potential for use in environments with extreme temperature variations.

d. Predictive Modelling

Recent studies have employed Artificial Neural Network (ANN) models to forecast the mechanical properties of PCB fibre-reinforced concrete, demonstrating high accuracy ($R^2 > 0.99$) [22].

These models can assist in optimizing concrete formulations by providing insights into the effects of different percent of PCB waste content on performance characteristics.

3.3 *Mechanisms by Which Circuit Board Additives Affect Concrete Performance*

The addition of PCB waste improves the microstructure of concrete, because when PCB waste is grinded into fine particles and used in cement matrix, they fill the voids. Due to this the porosity is reduced and the density is increased which results in improvement in compressive strength, reduced water permeability and better durability [23].

Circuit board materials often contain materials like metals, fiberglass, epoxy resins and other components. Silica based materials e.g. fiberglass contains amorphous silica which reacts with calcium hydroxide (Ca (OH)2) which is produced during hydration to form calcium silicate hydrate (CSH) gel [22]. This pozzolanic reaction minimizes free calcium hydroxide leading the concrete to be chemically more stable and resistant to aggressive environments.

Heavy metals in PCB waste like lead, cadmium produce leaching overtime when used in concrete leading to reduction in the overall performance of concrete [23].

Table 1 Mechanical characteristics of concrete samples after 28 days [25]

Sr no	Percentage of E-waste	Compressive strength (MPa)	Split tensile strength (MPa)	Flexural strength test (MPa)
1	0	19.50	1.35	2.86
2	3	19.10	1.33	2.72
3	5	18.35	1.28	2.55
4	10	18.10	1.19	2.37
5	15	21.50	1.51	2.91
6	20	19.90	1.45	3.01
7	25	17.40	1.15	2.84

4 Impact on Concrete Properties

4.1 Strength and Mechanical Performance

Research has shown that replacing OPC with PCB dust can lead to improved compressive strength and durability. The study conducted by [13] found that SCC mixtures with PCB powder replacement levels of 5% to 30% showed different mechanical properties. The compressive strength was highest at a replacement level of 5%.

Another study conducted by authors in [24] found that when coarse aggregate is replaced at 15% the compressive strength of concrete is maximized. At the mentioned optimal replacement level, the compressive strength is observed to be 20% higher compared to conventional concrete (Table 1).

The research conducted by authors in [25] represents that those fine aggregates replaced with PCB powder (at percentages of 3, 5, 10, 15, 20, and 25%) can enhance the compressive strength of concrete. This study shows a 10.25% increase in compressive strength at a 15% replacement compared to conventional concrete.

In another research paper by Kakria and Priya [26] the PCB powder was used in rigid pavement with dosages varying from (0, 3, 6, 9, 12) % With the addition of 3% powder, the material became 3.39% stronger in compression. The maximum strength was achieved by the addition of 9% of the non-metallic powder. However, by adding more than 9% of the powder, the strength started to go down (Fig. 3).

4.2 Durability and Resistance to Environmental Factors

Concrete which is made of PCB additives shows increased resistance to environmental degradation, it also includes chemical attacks and moisture absorption.

By using PCB waste in concrete, we can ensure longevity of the structures.

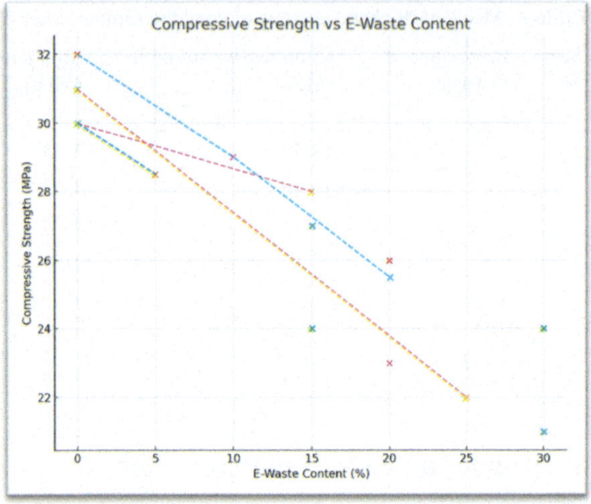

Fig. 3 Graph of compressive strength versus E-waste content [27]

a. Resistance to Moisture and Water Permeability: We can observe the hydrophobic nature of the concrete as PCB waste is rich in resin leading to reduced water absorption [23]. Capillary porosity of concrete may be lowered leading it lowered water ingress.
b. Chloride and Sulfate Resistance: The steel reinforcement can be delayed to corrosion as PCB additives might form shield against chloride ions. The sulfate ingress can cause expansive reactions which can be lowered by the inert components of PCB waste.
c. Freeze and Thaw Resistance: The ability of PCB waste to minimize pore connectivity contributes to the improvement of freeze–thaw durability.
d. Acid test

Physical effects on concrete can be observed due to chemical attacks. The physical effects include are cracking, spalling, increase of porosity and permeability and reduction in overall strength [23]. Most of the time, the chemical and physical effects on concrete structure work together at a same time, and one process may contribute to accelerate the other processes. An acid attack test was conducted by authors in [28] on WPCB fiber-reinforced concrete. Focus was to determine its chemical resistance. The acid attack test aimed to evaluate the chemical resistance of WPCB fiber-reinforced concrete. In this test, concrete cubes cured for 28 days were subjected to 5% hydrochloric (HCl) acid solution for 30 days, with the solution being changed every 15 days to maintain its concentration. The results indicated that mixes reinforced with 3, 4, and 5% WPCB fibers performed better in terms of weight loss compared to those with lower fiber content, as the presence of more fibers contributed to reduced deterioration [23]. Additionally, the compressive strength of the WPCB fiber-reinforced concrete showed minimal loss after exposure to the acidic environment, with the deterioration primarily attributed to bond failure between the

WPCB fibers and the cement matrix. Overall, the findings suggest that WPCB fiber-reinforced concrete can effectively withstand acidic environments, with higher fiber content leading to improved resistance against acid attacks.

4.3 Workability, Setting Time and Mix Design Considerations

Workability is an important factor in concrete and it can be evaluated by the slump cone test. As the amount of pcb waste in the concrete increases the workability in the concrete tends to decrease as observed by Suleman and Needhidasan [29]. It can be increased by adding admixtures like superplasticizers. As evaluated by Ebin and Isik [15] if 0% replacement is their workability is high but as the percentage of pcb waste increases to 16.5% the workability is low. In case of replacement of aggregates the workability also reduces due to irregular shapes and non-uniform sizes of the pcb waste.

The initial setting time of the concrete may be delayed due to the presence of non-reactive materials like resins.

5 Environmental Benefits

The following PCB waste practices in the current construction industry have various environmental benefits. The recycling of PCB waste reduces construction activities impact on the environment by minimizing landfill waste, conserving resources, and reducing emissions. These initiatives help sustain circular economy models that promote reusing materials.

1. Reduction in Landfill Waste
    a. E-Waste Issues: PCB waste is one of that type of e-waste that needs significant attention, unlike other forms of electronic waste. The decomposition rate is very slow, and it can stay in landfills for an estimated 100 years [3].
    b. Hazards: The damaged PCBs contain highly toxic materials such as mercury, cadmium and lead, which causes pollution of soil and groundwater. Hence the damaged PCBs are considered hazardous.
    c. Mitigation: Instead of disposing PCB waste in landfills it can be used as aggregate for concrete, asphalt, and composites [30].
    d. Volume Reduction: By recycling bulk amounts of PCB waste and reusing it in the construction industry we intend to lower the vast volume of waste that is taken to landfills.
    e. Prevention: By using construction grade products which contain concrete that effectively seals the toxic chemicals to avoid them from leaking and polluting the environment.

2. Conservation of Raw Materials and Natural Resources

    a. Resource Over exploitation: The construction industry demands the extraction of natural materials such as sand, stone, and limestone, which results in over exploitation of resources and results in damage of the ecosystem [31].
    b. Substantial Demand: The demand for raw materials from the field of construction industry is at relatively higher scale as compared to other major industries; this increases a strain on natural reserves.
    c. Replacement of Aggregates: During the preparation of concrete mixes some amount of sand and stone is used which can be substituted with crushed PCB waste.

        Cementless Alternative: Silica and alumina, which are found in PCB waste, can function as supplementary cementitious materials and reduce the amount of cement required.

d. Decrease in Mining Activities: There is a lesser need to mine new materials when PCB waste is used, hence preserving ecosystems [31].

3. Reducing the Carbon imprint

Of the total worlds carbon dioxide emissions approximately 7–8% of carbon dioxide is emitted by the cement production.

Burning E Waste: Burning of PCB waste emits poisonous gases that give rise to pollution.

Using Less Cement: Emissions caused by the cement operator can be reduced by replacing partial amount of cement with PCB waste.

Energy Conservation: Building insulation based on PCB improves energy efficiency which in return decreases the consumption levels of carbon emission related to energy.

4. Moving Towards a Sustainability

Recycling E-Waste: Rather than using the general approach of take make and dispose which causes harm to the environment a sustainability alternative of reusing, recycling and repurposing can be used.

By this the PCB waste which would then have been discarded is used in construction practices to mostly enhance its factors thereby making it a valuable construction material [30].

Extending Material Lifespan: By repurposing PCB waste, its lifecycle is extended, reducing the need for new raw materials.

Closed-Loop System: Construction projects using recycled PCB waste create a closed-loop system where materials are continuously reused, minimizing waste generation.

6 Economic Feasibility

At first it may cost high to start using e-waste from PCBs in concrete, but considering the long run it can save money by reducing cost of materials and waste disposal. In today's construction industry the demand of eco-friendly materials is increasing, PCB based concrete can be used in buildings and infrastructure projects.

1. Cost Implications of Processing and Utilizing PCB Waste

Cheaper Waste Processing: Instead of dumping PCB waste in landfills or burning it, turning it into concrete is often cheaper because it avoids expensive disposal fees [30].

Less Energy Needed: Preparing PCB waste for concrete (like shredding and sorting it) uses less energy compared to extracting metals or full recycling.

Lower Transport Costs: When PCB waste will come from nearby recycling centers, transportation costs will be reduced, especially for local construction.

2. Difference in Costs with Regular Concrete

Replacing Expensive Materials Materials like sand or coment in concrete are somewhat costlier, replacing it with PCB waste can produce cheaper concrete: Saving on Disposal Fees: Companies that produce electronic waste can avoid paying high fees to dispose of it by selling it for concrete production. More Affordable Option. PCB-based concrete can be marketed as a low-cost and environmentally friendly alternative to regular concrete.

3. Upcoming Market Opportunities Demand for Green Materials

PCB-based concrete has a lot of market potential as more people search for environmentally friendly building materials.

Government Assistance: The production cost can be lowered if the government starts giving incentives to the companies that use recycled materials. Large-scale projects that prioritize sustainaibility, such as buildings or highways, can make use of PCB-based concrete.

4. Economic Benefits for E-Waste Management

Recycling Businesses can profit from Waste: Recycling businesses can profit by selling PCB waste as a raw ingredient for concrete. Creation of jobs in garbage collection, its processing, and usage in construction may be created by recycling E-waste PCBs and making PCB ingressed concrete. Additional revenue can be generated by the companies which manage e-waste by selling PCB waste as a raw material for concrete production, which will create a circular economy within the waste management sector sector.

5. Challenges

Setting up facilities to process PCB waste involves significant initial costs. Nevertheless, with government support, policy incentives, and the potential for long-term economic benefits, these challenges can be addressed. Collaboration between recycling and construction sectors, alongside resource sharing and process optimization, can make using PCB waste in concrete both cost-effective and sustainable.

7 Applications in Sustainable Construction

Use in Infrastructure and Building Projects

Circuit board-derived concrete can be utilized in roads, bridges, and buildings, contributing to sustainable infrastructure development. It can be used in buildings with partial replacement in coarse aggregate, fine aggregate, cement thereby improving various factors like compressive strength, durability, freeze thaw resistance, etc. [21]. It can also be used in road infrastructure in rigid pavement taking reference of research done by Kakria and Priya [26] thereby creating an eco-friendly pavement.

In case of flexible pavement, it can be used in combination with asphalt or bitumen which then increases or improves the stiffness, rutting resistance, and high temperature stability of asphalt materials particularly beneficial for road surfaces that experience high traffic loads and temperatures [30]. It can also be used in prefabricated modular construction by incorporating pcb waste into prefabricated bricks, blocks, panels, etc. These prefabricated components can then be used in sheds, warehouses, partition walls, lightweight structures, etc.

Role in Smart and Green Buildings

The PCB waste can be used in smart and green buildings as these are intended to improve sustainability and energy efficiency. This can be achieved by using glass fibers recycled from WPCB which are used to improve sound absorption and thermal insulation. It is proved that the sound absorption coefficient of RGF is superior to those of some traditional absorbers, such as Expanded Perlite (EP), Expanded Vermiculite (EV). It can also achieve the II rating of sound absorption according to national standard [32].

Also, PCB waste-based composites can be used to create lightweight, energy efficient panels, facades and cladding systems for both residential and commercial buildings.

Contributions to achieving sustainable development goals (SDGs)

Incorporating PCB waste in construction practices directly supports multiple sustainable development goals by addressing waste management, resource conservation, etc. [33].

The various SDGs that support using PCB waste are as follows:

(a) SDG 9: Industry, Innovation and Infrastructure: It contributes by using PCB waste in infrastructure projects.

(b) SDG 11: Sustainable cities and communities: It contributes by using PCB waste in improving sound and thermal insulation.
(c) SDG 12: Responsible Consumption and Production: It contributes by using it in prefabricated bricks, blocks and panels.
(d) SDG 13: Climate Action: It contributes by partially replacing cement with PCB powder which then reduces demand of cement and lowers greenhouse emissions.
(e) SDG 14 and SDG 15: Life below water and Life on Land: By using PCB waste in construction practices it saves the ecosystems from polluting.

8 Challenges and Barriers

1. Technical Hurdles in Processing and Incorporating Circuit Boards

Printed Circuit Boards (PCBs) contain a mix of materials such as metals, glass fibers, and polymers, making separation and processing intricate and resource-intensive [18]. Crushing and grinding PCBs into fine particles suitable for concrete applications require specialized equipment and considerable energy input. Ensuring the compatibility of PCB-derived materials with cement and aggregates is critical to maintaining the structural and durability properties of concrete. Hazardous components, including heavy metals and residual chemicals in PCBs, pose risks of leaching, potentially affecting both environmental safety and the concrete's long-term performance. Identifying the appropriate proportions of PCB material to replace traditional components in concrete while balancing performance and sustainability remains a technical challenge.

2. Regulatory and Policy Challenges

Due to the presence of toxic substances, PCBs are often classified as hazardous waste, which imposes strict regulations on their handling, transport, and utilization [34]. There is currently an absence of universally accepted guidelines or standards for using PCB waste in construction materials, creating a barrier to widespread adoption. Stringent environmental laws necessitate adherence to emission control measures during processing and the safe management of waste materials. Securing regulatory clearance for incorporating PCB waste into building materials can be a lengthy and demanding process.

Building trust among stakeholders and the public regarding the safety and environmental impact of using PCB waste in concrete requires well-designed awareness campaigns and transparent policies.

3. Safety and health risks in Handling and Application of e-waste

During the processing and handling of the e-waste, it may release harmful chemicals and fine particulates, which might cause significant health risk to workers. During the processing of e-waste materials on recycling company it is mandatory to use appropriate personal protective equipment (PPE) to reduce exposure to hazardous substances. During the process of construction it is mandatory to train personnel regarding safe handling practices and risk mitigation strategies should be taught for ensuring workplace safety. Some of the components of PCBs can be flammable, giving rise to potential fire hazards that might occur during processing or storage [34].

More research is needed to be done to evaluate the long term health impacts on person exposed to concrete containing PCB waste, both during construction and after installation.

9 Future Perspectives

1. Opportunities for Innovation and Research

Material Engineering: Researching how PCB waste behaves chemically, physically, and structurally when mixed with cement can lead to better material performance.

Eco-Friendly Processing: Developing sustainable ways to process PCB waste, like low-energy crushing, advanced separation methods, or chemical treatments, can make the process more practical and environmentally friendly.

Functional Additives: PCB waste could provide unique additives to concrete, improving properties like thermal resistance, electromagnetic shielding, or durability.

Environmental Impact: Conducting studies on long-term safety, like life cycle assessments and leachate analysis, can ensure the sustainable use of PCB waste in construction.

Collaborative Research: Bringing together experts from different fields—materials, engineering, and environmental science—can help solve challenges and unlock the full potential of PCB waste in construction.

2. Integration with Other Recycled Materials for Hybrid Solutions

Hybrid Blends: If the other recycled materials like fly ash, slag, or plastic waste are combined with PCB waste it can develop into a more stronger and ecofriendly building material [35].

Synergistic Effects: Studying how PCB waste interacts with alternative binders like geopolymer cement could lead to the development of innovative and low-carbon construction materials.

Circular Economy Models: Using PCB waste in circular economy plans can help the construction industry recover resources efficiently and adopt more sustainable practices.

Waste Utilization Networks: Connecting different waste streams to construction material production can encourage widespread use and improve waste management efficiency.

3. Standardization and Large-Scale Implementation Strategies

Developing Standards: Creating clear guidelines for processing, testing, and using PCB waste in concrete can increase trust and adoption in the construction industry.

Pilot Projects: Launching large-scale projects that showcase PCB-infused concrete in real-world applications can provide useful insights and build public confidence.

Incentive Programs: Governments may promote the use of recycled materials through providing tax exemptions or subsidies to businesses using sustainable practices.

Automation and Scale-Up: Processing and integrating PCB waste through automation can reduce expenses and make large-scale manufacturing more feasible.

International Cooperation: Research, policy, and technological collaboration across borders can hasten the implementation of sustainable PCB waste management strategies.

10 Conclusion

The integration of circuit board derived materials into concrete production represents a transformative step towards sustainable construction practices. Something important in improvement with sustainable building practice is adding raw materials from circuit boards into the concrete production. This approach not only addresses environmental challenges by making use of e-waste for a different purpose but also enhances the mechanical performance of concrete. Such approaches are crucial for encouraging a circular economy in the face of the presently booming trend of urbanization and the growing trend of green building. Ongoing studies should focus on optimizing material compositions and exploring broader applicability in the building industry for enhanced benefits from this new technology.

References

1. Seif, R., Salem, F. Z., & Allam, N. K. (2024). *E-waste recycled materials as efficient catalysts for renewable energy technologies and better environmental sustainability*. https://doi.org/10.1007/s10668-023-02925-7
2. Luhar, S., & Luhar, I. (2019). *Potential application of E-wastes in construction industry: A review*. https://doi.org/10.1016/j.conbuildmat.2019.01.080
3. Liu, K., Tan, Q., Yu, J., & Wang, M. (2023). *A global perspective on e-waste recycling*. https://doi.org/10.1016/j.cec.2023.100028
4. Vishnu Priyan, M., Ravella, D. P., & Alaneme, G. U. (2024). *Transforming electronic waste into sustainable building materials for a cleaner environment: A review*. https://doi.org/10.1007/s43939-024-00174-7
5. Vishnupriyan, M., Annadurai, R., Onyelowe, K. C., & Ganasen, N. (2023). *Review on electronic waste used as construction materials—a scientometric analysis*. https://doi.org/10.1080/23311916.2023.2283307
6. Santhanam, N., Ramesh, B., & Pohsnem, F. K. (2020). Concrete blend with E-waste plastic for sustainable future. In *Materials today: Proceedings* (pp. 959–965). Elsevier Ltd. https://doi.org/10.1016/j.matpr.2019.11.204
7. *E-waste statistics: electronic waste generation & consequences in 2025*.

8. Schneider, E. L., Veit, H. M., Hartmann, W. L., Stolz, C. M., Robinson, L. C., Oliveira, C. T., & Vargas, A. S. (2021). Cement matrix containing lightweight aggregate based on Non-Metallic Fraction Printed Circuit Boards (NMFPCB'S). *Journal of Materials Research and Technology., 14*, 2992–2997. https://doi.org/10.1016/j.jmrt.2021.08.090
9. Janardhanan, T., & Ramasamy, V. (2017). Improvements in the microstructural and mechanical properties of geopolymer concrete containing NMF's of E-wastes as partial replacement of aggregates. *European Journal of Environmental and Civil Engineering*, 1–12. https://doi.org/10.1080/19648189.2017.1357869
10. Kaya, M. *Electronic waste and printed circuit board recycling technologies the minerals, metals & materials series.*
11. Sua-iam, G., & Chatveera, B. (2022). Effect of printed circuit board dust on the workability and mechanical properties of self-compacting concrete: A preliminary study. *Case Studies in Construction Materials, 16*, (2022). https://doi.org/10.1016/j.cscm.2021.e00862
12. Kiddee, P., Naidu, R., & Wong, M. H. (2013). Electronic waste management approaches: An overview. *Waste Management., 33*, 1237–1250. https://doi.org/10.1016/j.wasman.2013.01.006
13. Oriby, A., Salem, S., & Salem, S. (2024). 5 far. *Resourceedings, 4*, 06–12. https://doi.org/10.21625/resourceedings.v4i2.1074
14. Chen, Y., Liu, J., Zhang, M., Feng, B., Wang, L., Wei, J., Fu, W., & Tan, X. (2022). Effects of waste printed circuit board powder on strength, durability and microstructure of cement-based materials. *Journal of Building Engineering., 61*, Article 105255. https://doi.org/10.1016/j.jobe.2022.105255
15. Ebin, B., & Isik, M. I. (2016). Pyrometallurgical processes for the recovery of metals from WEEE. In *WEEE Recycling* (pp. 107–137). Elsevier. https://doi.org/10.1016/B978-0-12-803363-0.00005-5
16. Van Yken, J., Cheng, K. Y., Boxall, N. J., Nikoloski, A. N., Moheimani, N., Valix, M., & Kaksonen, A. H. (2023). An integrated biohydrometallurgical approach for the extraction of base metals from printed circuit boards. *Hydrometallurgy, 216*. https://doi.org/10.1016/j.hydromet.2022.105998
17. Ban, B. C., Song, J. Y., Lim, J. Y., Wang, S. K., An, K. G., & Kim, D. S. (2005). Studies on the reuse of waste printed circuit board as an additive for cement mortar. *Journal of Environmental Science and Health. Part A, Toxic/Hazardous Substances & Environmental Engineering, 40*, 645–656. https://doi.org/10.1081/ESE-200046618
18. Hadi, P., Xu, M., Lin, C. S. K., Hui, C. W., & McKay, G. (2015). *Waste printed circuit board recycling techniques and product utilization.* https://doi.org/10.1016/j.jhazmat.2014.09.032
19. Goosey, M., & Kellner, R. (2003). Recycling technologies for the treatment of end of life printed circuit boards (PCBs). *Circuit World., 29*, 33–37. https://doi.org/10.1108/03056120310460801
20. Li, J., Xu, Z., & Zhou, Y. (2007). Application of corona discharge and electrostatic force to separate metals and nonmetals from crushed particles of waste printed circuit boards. *Journal of Electrostatics, 65*, 233–238. https://doi.org/10.1016/j.elstat.2006.08.004
21. Needhidasan, S., Ramesh, B., & Joshua Richard Prabu, S. (2020). Experimental study on use of E-waste plastics as coarse aggregate in concrete with manufactured sand. In *Materials today: Proceedings* (pp. 715–721). Elsevier Ltd. https://doi.org/10.1016/j.matpr.2019.10.006
22. Priyan, M. V., Annadurai, R., Alaneme, G. U., Ravella, D. P., Pradeepkumar, S., & Olaiya, B. C. (2023). A study on waste PCB fibres reinforced concrete with and without silica fume made from electronic waste. *Sci Rep, 13*. https://doi.org/10.1038/s41598-023-50312-z
23. Fresh and mechanical properties of concrete with replaced coarse aggregate by e-waste. *Eurasian Journal of Science and Engineering, 10*. https://doi.org/10.23918/eajse.v10i1p5
24. Rathore, V., & Rawat, A. (2019). Effective utilization of electronic waste in concrete mixture as a partial replacement to coarse aggregates. In *AIP conference proceedings*. American Institute of Physics Inc. https://doi.org/10.1063/1.5127161
25. Ganesh, S., Danish, P., & Bhat, K. A. (2020). Utilization of waste printed circuit board powder in concrete over conventional concrete. In *Materials today: Proceedings* (pp. 745–749). Elsevier Ltd. https://doi.org/10.1016/j.matpr.2020.11.161

26. Kakria, K., & Priya, S. (2021). Use of non-metallic powder reclaimed from waste printed circuit boards in rigid concrete pavement. *IOP Conf Ser Mater Sci Eng., 1075*, Article 012032. https://doi.org/10.1088/1757-899x/1075/1/012032
27. Kumar, G., Bansal, T., Haq, M., Sharma, U., Kumar, A., Jha, P., Sharma, D., Kamyab, H., & Valencia, E. A. V. (2024). Utilizing e-waste as a sustainable aggregate in concrete production: A review. *Buildings, 14*, 2495. https://doi.org/10.3390/buildings14082495
28. Priyan, M. V., Annadurai, R., Onyelowe, K. C., Alaneme, G. U., & Giri, N. C. (2023). Recycling and sustainable applications of waste printed circuit board in concrete application and validation using response surface methodology. *Sci Rep, 13*. https://doi.org/10.1038/s41598-023-43919-9
29. Suleman, S., & Needhidasan, S. (2020). Utilization of manufactured sand as fine aggregates in electronic plastic waste concrete of M30 mix. In *Materials today: Proceedings* (pp. 1192–1197). Elsevier Ltd. https://doi.org/10.1016/j.matpr.2020.08.043
30. Li, S., Sun, Y., Fang, S., Huang, Y., Yu, H., & Ye, J. (2022). Recycling non-metallic powder of waste printed circuit boards to ımprove the performance of asphalt material. *Materials, 15*. https://doi.org/10.3390/ma15124172
31. Ortiz, O., Castells, F., & Sonnemann, G. (2009). Sustainability in the construction industry: A review of recent developments based on LCA. *Construction and Building Materials, 23*, 28–39. https://doi.org/10.1016/j.conbuildmat.2007.11.012
32. Sun, Z., Shen, Z., Ma, S., & Zhang, X. (2013). Novel application of glass fibers recovered from waste printed circuit boards as sound and thermal insulation material. *Journal of Materials Engineering and Performance, 22*, 3140–3146. https://doi.org/10.1007/s11665-013-0587-y
33. Huisman, J., Stevels, A., Baldé, K., Magalini, F., & Kuehr, R. (2019). The e-waste development cycle—part i, introduction and country status. In *Waste Electrical and Electronic Equipment (WEEE) handbook* (pp. 17–55). Elsevier. https://doi.org/10.1016/B978-0-08-102158-3.00002-1
34. LaDou, J. (2006). Printed circuit board industry. *International Journal of Hygiene and Environmental Health, 209*, 211–219. https://doi.org/10.1016/j.ijheh.2006.02.001
35. Selvaraj, T., Kaliyavaradhan, S. K., Kakria, K., & Malladi, R. C. (2022). Use of e-waste in metakaolin blended cement concrete for sustainable construction. *Sustainability, 14*, 16661. https://doi.org/10.3390/su142416661

Reviving Heritage Structures: Sustainable Retrofitting of Dhajji Deewari Buildings

Mohd Hanief Dar, Muzamil Mushtaq Khanday, and Riyaz Ahmad Qasab

Abstract The restoration/retrofitting of heritage structures is necessary to safeguard the cultural heritage, ensuring safety and functionality to the modern environment. This chapter delves in dhajji deewari heritage buildings—a traditional form of timber-laced masonry construction key to earthquake-prone areas of South Asia, mainly Kashmir and northern Pakistan. Although dhajji deewari structures are globally celebrated for their natural resilience to seismic events, they can serve as vital facets of sustainable construction ideas relevant even today. This chapter discusses retrofitting alternatives to fortify these cultural establishments, without undermining their heritage. Important strategies consist of edge replacement of decayed timber, lightweight strengthening, and the logical incorporation of a modern structural support system that maintains the architecture of the structure and will preserve the structure for many more years. The approach highlights the use of the right materials and the application of minimum intervention techniques to preserve cultural value and, at the same time, enhance the buildings' capacity for survival in case of seismic disasters. By integrating these retrofitting methods into sustainable building principles, this chapter illustrates how dhajji deewari constructions can meet current safety standards while also being environmentally friendly. These measures are intended to preserve the embodied energy used within such historic structures, yielding results that are both environmentally and culturally critical. Such retrofitting of dhajji deewari buildings gives the best practices and practical consequences are also illustrated by case examples. It also focuses on the significance of policy frameworks and the presence of stakeholders leading to grateful and prosperous heritage protection. The retrofitting of heritage structures must be multidisciplinary and involves engineers, architects, policy-makers and the community at large working together. Finally, through this know-how, the present chapter integrates this

M. H. Dar · M. M. Khanday (✉) · R. A. Qasab
Department of Civil Engineering, National Institute of Technology Srinagar, Srinagar, India
e-mail: muzamil_2022phsciv002@nitsri.ac.in; muzamil4609@gmail.com

M. H. Dar
e-mail: hanief_2022phaciv006@nitsri.ac.in

R. A. Qasab
e-mail: riyaz_2022phsciv005@nitsri.ac.in

P. Danish et al. (eds.), *Innovative Technologies for Smart and Sustainable Infrastructure*,
https://doi.org/10.1007/978-3-031-92801-7_15

experience in order to further contribute into the dialogue of heritage conservation and sustainable retrofitting by demonstrating how the old can meet the modern approach and engineering. Practices such as dhajji deewari can be revived, which serves as a means of building resilient civic infrastructure that respects cultural heritage as well as addressing the challenges of contemporary disaster resilience.

Keywords Heritage conservation · Dhajji deewari · Sustainable retrofitting · Earthquake resilience · Cultural preservation

1 Introduction

The chapter covers the comprehensive idea of preserving traditional Dhajji Deewari architecture, a unique timber-framed masonry infill construction technique in earthquake-prone regions like Kashmir. It explores retrofitting methods with sustainable alternatives that balance structural resilience with cultural integrity, ensuring these heritage structures remain in existence for future generations. By combining traditional wisdom with modern sustainable engineering solutions, the chapter highlights innovative approaches of conserving these buildings while addressing present-day challenges of climate change, development, and seismic safety.

1.1 Cultural and Historic Importance of Dhajji Deewari

The term Dhajji Deewari is derived from the Persian term "patchwork quilt wall," which refers to a traditional architectural style in the western Himalayas. These structures are commonly found in Pakistan and Indian administered regions of Kashmir. This construction method is also documented in the Indian Standard Codes as brick-infilled timber frame construction. Dhajji, most frequently, comprises of a braced timber framework. Typically, the spaces between the bracing elements or frames are occupied by a narrow wall, known as single wythe, which is made from stone or brick masonry and usually constructed with mud mortar. Such structures are frequently found in regions including Britain, France, Germany, Central America, South America, Turkey, Greece, Portugal, Italy, and likely other Eastern European countries [1, 2]. The timber frame can absorb seismic energy due to its natural flexibility and non-rigid connections, providing significant resistance against earthquake forces [3, 4]. This method showcases the creativity of traditional builders and highlights the cultural and historical identity of the regions that utilize this technique.

As shown in Fig. 1, the Dhajji Deewari technique merges modern construction in Kashmir J&K India. It demonstrates how traditional methods can be adapted to present-day needs while preserving cultural and historical significance. The image illustrates a contemporary rendition of Dhajji Deewari and displays detailed timber-framed masonry, a distinctive feature of this construction style for centuries.

Fig. 1 A modern interpretation of Dhajji Deewari construction: Showcasing the traditional timber-laced masonry technique in new construction

Because of its deep cultural significance, the Dhajji Deewari technique is embedded in the cultural practice of Kashmir and is used for residential houses, places of worship, and palaces. The patterns of the woodwork of all constructions with the masonry not only support the structure's functionality but also contribute to the region's artistic traditions [5, 6]. In northern Pakistan, particularly Hunza Valley, Dhajji Deewari structures are integrated into this culture and are the main component of social gatherings and worship [7].

1.2 The Imperative for Sustainable Retrofitting

While Dhajji Deewari structures are naturally resilient, they experience multiple challenges due to aging, environmental degradation, and urbanization. Most structures are partially or not maintained regularly, resulting in structural instability [8, 9]. Moreover, swift urbanization and the arrival of modern construction materials have resulted in the neglect and demolition of several vertical heritage buildings, such as Dhajji Deewari structures [10, 11]. Adapting these buildings sustainably is vital to preserving their cultural heritage, safety, and function in modern environments.

Overall, reshaping Dhajji Deewari constructions requires blending organization and building attitude with structural safety. New eco-friendly, lightweight, and minimally invasive materials must be employed to retrofit these structures. To withstand earthquakes and preserve the environment modern engineering practices have to be blended with traditional construction methods [1, 12, 13].

2 Historical and Cultural Context of Dhajji Deewari

The Dhajji Deewari style combines elegant beauty with practicality. It has been built to withstand earthquakes and foster local artisanship. This building technique, created by a wooden structure filled with masonry, offers durability and is now an emblem of cultural identity while showcasing the ingenuity and adaptability of its makers through the ages. Figure 2 portrays a rural dhajji deewari setting of Kashmir, where the terraced fields and local craftsmanship reflect the beauty and creativity behind this earthquake-resistant building method.

2.1 Origins and Evolution of Dhajji Deewari

The Dhajji Deewari, a traditional construction method, has been practiced in the Kashmir Valley for centuries to erect residential and religious buildings [2]. The technique also spread to other seismically active areas like northern Pakistan, Afghanistan, and Iran, adapting to locally available materials and conditions. The construction of the Dhajji Deewari using timber and masonry is a result of the availability of local materials and the need for structures that could withstand seismic activity.

The method of construction continued to evolve over time in Kashmir as one where the builders kept on testing new combinations of wood and masonry that were toned to gain more strength in their built forms. Deodar wood, from various cedar, became the wood for practically all constructions due to its durability and decayed resistance [14]. The method was naturally practiced in northern Pakistan through stone masonry with timber gathered from the region's forests.

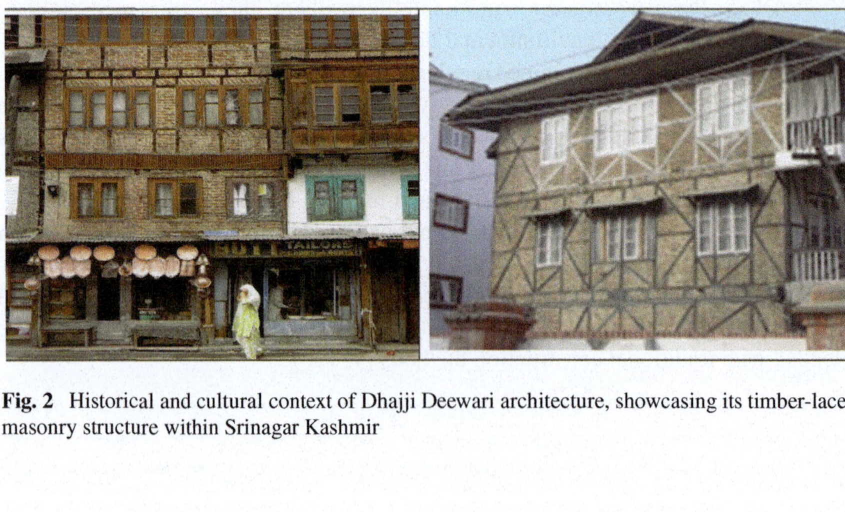

Fig. 2 Historical and cultural context of Dhajji Deewari architecture, showcasing its timber-laced masonry structure within Srinagar Kashmir

2.2 *Cultural Significance of Dhajji Deewari in Kashmir and Pakistan*

Not merely a manifestation of the architectural heritage of the valley, the Dhajji Deewari structures are also a representation of the cultural identity of Kashmir. Most are related to historical events, religious beliefs, and traditional lifestyles. For example, the Pari Mahal in Srinagar (a historic Dhajji Deewari building) is a true testament to the locality's cultural richness and architectural brilliance.

In northern Pakistan, especially in the Hunza Valley, the Dhajji Deewari structures are part of local culture and commonly utilized for communal gatherings and religious ceremonies. A classic example of a Dhajji Deewari structure preserved and retrofitted to contemporary safety standards is the Baltit Fort, listed as a UNESCO World Heritage [15].

2.3 *Dhajji Deewari in Other Countries*

The construction technique of Dhajji Deewari has been used in many countries across the globe like Britain, France, Germany, Central America, South America, Turkey, Greece, Portugal and Italy and other European countries under different names but with similar construction principles [16, 17]. Both the functional adaptability and durability are dictated by local materials and specific climatic conditions. This technique, called "Bundah," has recently been employed in the earthquake-resistant construction of residential houses in severe seismic zones of Afghanistan.

3 Issues with Dhajji Deewari Structures

3.1 *Problems Related to the Dhajji Deewari Fabrics*

Despite having great cultural significance and being remarkably seismic-resistant, the primary concern when it comes to the Dhajji Deewari structures is the lack of care that is showered on them. These structures endure poor maintenance and changes that are not in line with contemporary requirements. Deterioration, moisture exposure, and poor alterations are all issues that have an impact on these structures. It is important to mention that urbanization and the decreasing number of skilled artisans have made Dhajji Deewari structures hard to maintain. These facts pose a danger and place a limitation on the methods of the conservation of these elements making it harder to maintain.

3.2 *Decay of Timber Elements*

The main focus with regards to the durability of Dhajji Deewari is the breakdown of wooden portions because of moisture, termites, and fungus. After some time, the frames of the wooden elements begin to perish. In certain parts of the world, wooden frames may even entirely disintegrate. For places that experience earthquakes like Kashmir, where there is a lot of humidity, the weakened wooden frames magnify the issue. The remaining wooden structures end up amplifying the negative consequences.

3.3 *Urbanization and Neglect*

The rapid urbanization that many countries are experiencing has left untouched structures like the Dhajji Deewari ones at risk of demolishment. The only available solution is replacing these frames of buildings with modern construction that some regions lack. Many Dhajji Deewari buildings not maintained well because people do not know their cultural and historical value. Some owners lack the resources or knowledge needed for repairs. This lack of maintenance has caused many Dhajji Deewari buildings to deteriorate, with some nearly collapsing.

4 Techniques of Sustainable Retrofitting of Dhajji Deewari Buildings

The section "Techniques of Sustainable Retrofitting of Dhajji Deewari Buildings" focuses on integrating modern engineering solutions with traditional craftsmanship to enhance these heritage structures' structural resilience and longevity. Key sustainable retrofitting measures include:

4.1 *Timber Frame Reinforcement and Connection Enhancements*

The timber framework is the main lateral load-resisting feature of the Dhajji Deewari construction system. Over time, it can weaken due to decay/rot or insect damage and the weakening of connections between the timber members, which are mostly nailed. Reinforcing timber members include:

- **Steel plates or brackets:** These can be used to strengthen critical joints, improving the overall stability of the structure. Studies have shown that steel reinforcements can significantly enhance the seismic performance of timber frames without altering their traditional appearance [18].
- **Epoxy resin injection:** This technique is used to repair cracks in timber without replacing the original material, preserving the historical integrity of the structure [19].
- **Engineered timber replacement:** Severely damaged timber can be replaced with engineered timber, which offers improved strength and durability while maintaining the aesthetic qualities of the original structure [20].

4.2 Improved Masonry Infill

Traditional masonry infills, such as stone or mud, often weaken or add unnecessary weight, compromising structural safety. The necessary alterations may include:

- **Aerated concrete blocks:** These lightweight materials reduce the overall weight of the structure while maintaining its structural integrity. Research has shown that aerated concrete blocks can improve the seismic performance of masonry-infilled timber frames [1].
- **Stabilized mud bricks:** These bricks offer a sustainable alternative to traditional masonry, providing better binding strength and durability while preserving the architectural aesthetics of the structure [2].

4.3 Seismic Retrofitting

In earthquake-prone regions like Kashmir, enhancing the lateral resistance of Dhajji Deewari structures is crucial. Techniques include:

- **Diagonal bracings:** Installing diagonal bracings in the timber frame can help distribute lateral forces more effectively, reducing the risk of structural failure during seismic events [4].
- **Shear walls:** Adding shear walls in discreet locations can strengthen the building without altering its appearance. Studies have demonstrated that shear walls can significantly improve the seismic resilience of traditional timber-framed structures [8].

4.4 Moisture Protection

Moisture infiltration is a major concern for Dhajji Deewari buildings, as it can accelerate timber rot and masonry degradation. Preventative measures include:

Waterproof coatings: Applying waterproof coatings to exterior walls can prevent moisture penetration, protecting the structure from decay [21].

Breathable membranes: These membranes can manage internal moisture levels, preventing condensation and further damage to the timber and masonry [12].

4.5 Integration of Modern Materials and Community Involvement

The use of modern materials, such as fiber-reinforced polymers (FRPs), has gained traction in the retrofitting of heritage structures. FRPs are lightweight, strong, and can be bonded to complex shapes without altering the traditional appearance of the structure. Research has shown that FRPs can significantly improve the earthquake resistance of timber-framed buildings [13].

Community involvement is also crucial for the successful retrofitting of Dhajji Deewari buildings. Workshops and training programs can help local artisans learn traditional building techniques and modern retrofitting methods, ensuring the continuity of local heritage [1].

5 Policy Framework and Stakeholder Involvement

5.1 The Role of Policy in Heritage Conservation

Effective policy frameworks are essential for the sustainable retrofitting of heritage structures. Policies should focus on integrating heritage conservation into urban planning and providing financial support for the preservation of cultural sites. For example, the UNESCO World Heritage Centre has developed guidelines for the conservation of heritage structures in seismic zones, emphasizing the importance of sustainable retrofitting techniques [22].

Statistical data from the Indian National Trust for Art and Cultural Heritage (INTACH) indicates that only 30% of heritage structures in India receive regular maintenance, highlighting the need for stronger policy interventions to protect cultural heritage [23].

5.2 Stakeholder Participation in Heritage Conservation

The retrofitting of heritage structures requires a multidisciplinary approach involving engineers, architects, policymakers, and the local community. A case study from the Baltit Fort in Pakistan demonstrates the successful integration of stakeholder involvement in heritage conservation. The fort, a UNESCO World Heritage Site, was retrofitted using traditional techniques combined with modern materials, ensuring its preservation for future generations.

Engaging local communities in the retrofitting process fosters a sense of ownership and ensures the continuity of cultural practices. For example, in the Hunza Valley, community-led initiatives have been instrumental in preserving Dhajji Deewari structures, with over 70% of retrofitting projects being community-driven [7].

6 Conclusion

The sustainable retrofitting of Dhajji Deewari buildings presents a unique opportunity to preserve cultural heritage while enhancing disaster resilience. This chapter has highlighted the major issues facing these structures, including timber decay, moisture infiltration, and the challenges posed by rapid urbanization. To address these issues, a combination of traditional craftsmanship and modern engineering techniques has been proposed, including timber frame reinforcement, improved masonry infill, seismic retrofitting, and moisture protection.

The integration of modern materials, such as fiber-reinforced polymers (FRPs), offers a sustainable solution for enhancing the structural resilience of Dhajji Deewari buildings without compromising their cultural integrity. Furthermore, community involvement and stakeholder collaboration are essential for the successful implementation of retrofitting projects.

Policy frameworks play a critical role in heritage conservation, as evidenced by statistical data from organizations like INTACH and UNESCO. By integrating heritage conservation into urban planning and providing financial support for retrofitting projects, policymakers can ensure the preservation of Dhajji Deewari structures for future generations.

In conclusion, the sustainable retrofitting of Dhajji Deewari buildings not only safeguards cultural heritage but also contributes to the development of resilient civic infrastructure that can withstand the challenges of contemporary disaster resilience. Through a multidisciplinary approach that combines traditional wisdom with modern engineering, these heritage structures can continue to serve as enduring symbols of South Asia's rich history and identity.

References

1. Langenbach, R. (2008). Learning from the past to protect the future: Armature Crosswalls. *Engineering Structures, 30*(8), 2096–2100.
2. Dar, M. A. et al. (2014). Traditional earthquake resistant systems of Kashmir. *International Journal of Civil and Structural Engineering Research*, 2, 2.

3. Riyaz, N., et al. (2023). Housing in Kashmir: Traditional Dhajji Dewari housing as a seismic-resistant alternative. *In: 3rd International Conference on the Built Environment and Engineering (iconbee) 2022: Fostering a Transpicuous Construction Industry, 2881*, 1 5.
4. Qaisar Ali., et al. (2017) Shake table tests on single-story Dhajji Dewari traditional buildings. *International Journal of Architectural Heritage, 11* 7.
5. Himanshi Sharma. (2022). DHAJJI DEWARI-Traditional earthquake resistant construction of Kashmir. Thehimalayanarchitect.
6. Jokilehto, J. (2005). Definition of cultural heritage. *ICCROM*.
7. Ahmad, N., et al. (2012). Simplified engineering tools for seismic analysis and design of traditional Dhajji-Dewari structures. *Bull Earthquake Eng, 10*, 1503–1534.
8. Eva Frühwald Hansson. (2011). Analysis of structural failures in timber structures: Typical causes for failure and failure modes. *Engineering Structures, 33*, 2978–2982.
9. Sean Jackson. (2024). Types of structural failures and common causes. *Jackson and Sands Engineering Inc*.
10. ICOMOS. (2002). Principles for the conservation of heritage sites in China. *ICOMOS China*. https://www.getty.edu/conservation/publications_resources/pdf_publications/pdf/china_prin_heritage_sites.pdf.
11. Qazi Waleed, et al. (2022) A case for revival of traditional Earthquake resistant structures in Kashmir. *International Journal of Innovative Research in Engineering & Management*, ISSN: 2350–0557, *9* 5.
12. Hassan, M. A., et al. (2020). Improving structural performance of timber wall panels by inexpensive FRP retrofitting techniques. *Journal of Building Engineering, 27*, Article 101004.
13. Schober, K.-U., et al. (2015). FRP reinforcement of timber structures. *Construction and Building Materials, 97*, 106–118.
14. Dr. Suman Bala Sharma. Major timber yielding Plants of J&K State—Morphology, botanical name and family. www.scribd.com/document/522201320.
15. Syed Hamid Akbar et al. Preserving the Tangible and Intangible Values of the Baltit Fort in Gilgit-Baltistan Pakistan, through Adaptive Reuse. *In: SPACE International Journal of Conference Proceedings*, *2*(1), 1–9. https://doi.org/10.51596/sijocp.v2i1.20.
16. DF Laefer. Emergence, development, and prevalence of brick nogging in American vernacular structures. *Structural Analysis of Historical Constructions—2 Volume*.
17. Davis, H. The culture of building. https://www.amazon.in/Living-Over-Store-Architecture-Local/dp/041578316X.
18. Brain Ridout. (2000) Timber decay in buildings: the conservation approach to treatment.
19. Saad, K., & Lengyel, A. (2022). Strengthening Timber Structural Members with CFRP and GFRP: A State-of-the-Art Review. *Polymers, 14*(12), 2381. https://doi.org/10.3390/polym14122381
20. J. Custódio, S., Cabral-Fonseca. (2023). Advanced fiber-reinforced polymer (FRP) composites for the rehabilitation of timber and concrete structures. *Woodhead Publishing Series in Civil and Structural Engineering*. https://www.sciencedirect.com/science/article/pii/B9780128235508000210).
21. Sean Jackson. (2024). Types of structural failures and common causes. *Jackson and Sands Engineering Inc*. https://www.jacksonandsands.com/types-of-structural-failures-and-common-causes.
22. UNESCO World Heritage Centre. (2015). Guidelines for the conservation of heritage structures in seismic zones. https://whc.unesco.org/en/guidelines.
23. Indian National Trust for Art and Cultural Heritage (INTACH). (2020). Heritage conservation in india: challenges and opportunities. https://www.intach.org.

Materials Innovation

Investigating the Impact of Unmodified Sugarcane Bagasse Ash and Alccofine on Cement Paste: Combined Effects and Ideal Chemical Admixture Proportions

Amala Joseph, S. Bala Murugan, and Anusree S. Anirudhan

Abstract Cement production contributes significantly to CO_2 emissions, which can be mitigated using supplementary cementitious materials (SCMs). Agro-industrial byproducts can serve as SCMs to reduce CO_2 emissions. One such material is sugarcane bagasse ash (SCBA), known for its high silica content. The efficiency of SCBA can be further improved by combining it with alccofine (AF), but their interactions in cement paste are not fully understood. This research investigates the physical and mechanical characteristics of cement paste containing various proportions of unmodified SCBA (USBA) and AF. The study examined 28 mixtures using two chemical admixture dosages. USBA was refined by sieving through a No. 30 sieve. Paste formulations included binary mixtures with partial cement substitution by USBA or AF, and ternary blends with both materials. Higher chemical admixture content improved compressive strength, especially in ternary pastes with 20% and 30% USBA. Mixtures with 0.6% admixture exhibited lower consistency values than those with 0.4% dosage. USBA mixtures benefited more from 0.4% admixture dosage.

Keywords Unmodified sugarcane bagasse ash (USBA) · Alccofine (AF) · Consistency · Chemical admixture dosage · Water reducing effect

A. Joseph (✉) · S. B. Murugan
School of Civil Engineering, Vellore Institute of Technology, Vellore, India
e-mail: amala.lullu@gmail.com; amala.joseph2020@vitstudent.ac.in

S. B. Murugan
e-mail: balamurugan.s@vit.ac.in

A. S. Anirudhan
Department of Structural and Geotechnical Engineering, School of Civil Engineering, Vellore Institute of Technology, Vellore, India

P. Danish et al. (eds.), *Innovative Technologies for Smart and Sustainable Infrastructure*,
https://doi.org/10.1007/978-3-031-92801-7_16

1 Introduction

The scientific community's quest for more eco-friendly binders began many years ago, driven by the need to curtail carbon dioxide emissions from cement production. Recently, the construction sector has faced growing demands to implement sustainable methods and reduce their ecological impact. The integration of waste materials from agricultural and industrial sectors into cement-based products represents a viable solution. The successful integration of these byproducts necessitates enhanced fluidity for proper compaction and adequate strength development. Chemical admixtures, notably superplasticizers, are essential for this process. Determining the ideal admixture dosage is crucial, as insufficient amounts can lead to reduced flowability, while excess quantities may cause material separation, delayed setting and wasteful use of superplasticizers [1]. Chemical admixtures based on polycarboxylate ether have shown remarkable effectiveness in enhancing the mechanical characteristics of cementitious composites, especially during their initial stages of development [2]. The widespread adoption of polycarboxylate superplasticizer can be attributed to its minimal required dosage and superior water reduction capabilities [3].

In India, sugarcane bagasse ash (SCBA) is a plentiful agricultural waste product, often discarded on open land. SCBA is produced in the sugar industry cogeneration power plants that generate electricity [4]. This material shows potential as a renewable supplementary cementitious material. With a silica composition surpassing 70%, including both crystalline and amorphous forms, sugarcane bagasse ash exhibits significant pozzolanic properties [5]. The presence of amorphous silica in this ash contributes to the enhancement of strength over time [6]. The high silicon dioxide content in SCBA [7] contributes to its enhanced compressive strength and durability. The incorporation of SCBA as a substitute material offers dual benefits: it diminishes cement consumption and addresses environmental concerns associated with SCBA disposal [8]. In contrast, alccofine, a byproduct of steel manufacturing, consists of ultrafine particles [9]. Significantly higher levels of calcium oxide and aluminium oxide in alccofine [6] contribute to enhanced pozzolanic activity, improved workability, increased resistance to chemical attacks and reduced permeability. Substituting with Alccofine may result in a substantial enhancement of strength through the development of a more compact microstructure [10]. The exceptional fineness and high levels of pozzolanic oxides in Alccofine distinguish it from other supplementary cementitious materials. These characteristics make it particularly effective in enhancing the strength and durability properties of concrete, offering superior performance in improving the material's mechanical attributes and long-term resilience [9]. The water-absorbing characteristics and the expansive surface area of sugarcane bagasse ash necessitate higher water content [11]. Conversely, alccofine particles, with their unique chemical composition and glassy surface properties, enhance concrete's workability while decreasing segregation and bleeding by lowering water requirements [12].

Table 1 Properties of different materials

Materials	Specific gravity	Setting time (min.)		Specific surface area (cm^2/g)	Bulk density (kg/m^3)	Colour	LOI (%)	PO (%)
		Initial	*Final*					
OPC	3.13	96	187	3100	1440	Grey	1.88	27.47
USBA	1.8	–	–	2680	594	Cloudy grey	18	89.29
AF	2.8	–	–	12,000	700	Brownish grey	0.78	56.22
Chemical admixture	1.05	–	–	–	–	Light brown	–	–

LOI-Loss on ignition, PO-Pozzolanic Oxides ($SiO_2 + Al_2O_3 + Fe_2O_3$)

This study investigated the potential of unmodified sugarcane bagasse ash (USBA) and Alccofine as supplementary cementitious materials (SCMs) in cement paste formulations. Both sugarcane bagasse ash, an abundant agricultural waste, and Alccofine, a finely ground industrial by-product, exhibit properties that can enhance the performance of cement paste while mitigating its environmental impact. This study aims to explore the synergistic effects of combining these two materials and to determine the optimal chemical admixture dosages for achieving desirable physical and mechanical properties.

2 Experimental Program

2.1 Materials Used

2.1.1 Ordinary Portland Cement

For this investigation, 53-grade Ordinary Portland cement (OPC) was selected, which adhered to the standard IS:269:2020 [13] guidelines. The material characteristics of cement are listed in Table 1.

2.1.2 Supplementary Cementitious Materials

This study incorporated unmodified sugarcane bagasse ash (USBA) and alccofine (AF) as pozzolonic materials. The properties of different materials used for the study and their grain size distributions are shown in Tables 1 and 2 respectively. The sugarcane bagasse ash (SCBA) utilized in this research was obtained from a waste disposal site of a sugar manufacturing plant located in Tamil Nadu, India. The SCBA preparation process involved two distinct steps. Initially, samples underwent a 24-h drying

Table 2 Distribution of grain size

	OPC	USBA	AF
D_{10} (μm)	5.38	3.43	0.51
D_{50} (μm)	10.85	9.77	4.38
D_{90} (μm)	16.94	16.8	6.22

(a) Sugarcane bagasse after juice extraction

(b) Crushed sugarcane bagasse before burning

(c) Sugarcane bagasse ash

(d) Sieved sugarcane bagasse ash used for the study and its residues

Fig. 1 Various steps involved in producing USBA

period in an oven set at 110 ± 5 °C to eliminate moisture. Subsequently, the dried material was sieved through a No. 30 mesh to remove any coarser fibrous components. The presence of incompletely combusted carbon results in the low specific gravity of the USBA. Figure 1 illustrates the various steps involved in producing the USBA employed in this study. ASTM C618 [14] classifies natural pozzolans with a total pozzolanic oxide content (SiO_2, Al_2O_3 and Fe_2O_3) exceeding 70% as Class F. According to the total pozzolanic oxide content, the USBA used for the study qualifies for classification as a Class F pozzolanic substance.

The research incorporated Alccofine (AF) to improve the USBA's performance, adhering to the IS: 12,089:2013 [15] specifications. Compared to cement, Alccofine demonstrates superior characteristics, featuring a greater specific surface area of 12,000 cm^2/g and a reduced specific gravity of 2.8. The controlled granulation technique used for AF production yielded an exceptionally low Loss on Ignition (LOI). Figures 2 and 3 display SEM images of the USBA and AF, respectively. USBA comprises a mix of prismatic, irregular, spherical and irregular particles, whereas the AF consists of irregularly shaped particles with sharp edges.

2.1.3 Chemical Admixture and Water

The study utilized a high-performance superplasticizer derived from polycarboxylate ether, devoid of lignosulfonate, naphthalene salts and melamine–formaldehyde. This admixture, conforming to ASTM C494 (Type G) [16] requirements, was characterized by its light brown colour and specific gravity of 1.05. The paste specimens were cast using potable water according to the guidelines outlined in IS 456:2021 [17].

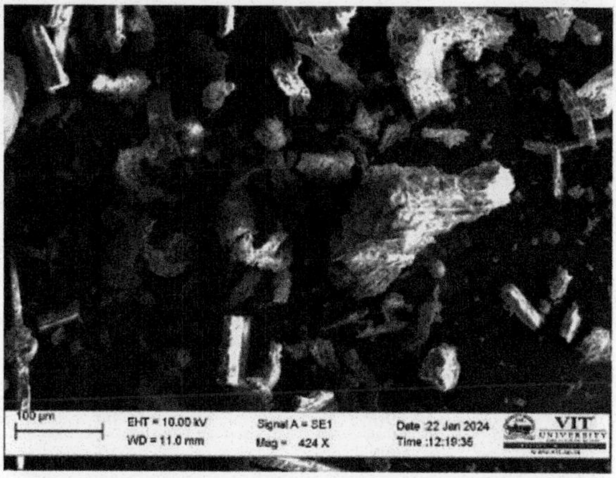

Fig. 2 Scanning electron microscopic images of USBA

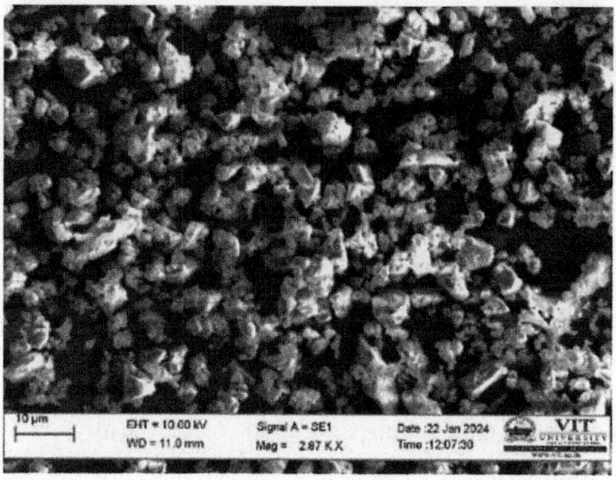

Fig. 3 Scanning electron microscopic images of Alccofine

2.2 *Mixture Combinations*

The research investigated 28 distinct blend formulations, with cement substitution rates reaching up to 60%, as outlined in Table 3. A mixture containing only cement was used as the reference standard. The study included six binary blends with USBA, six binary blends with AF and 15 mixtures with ternary blends incorporating both USBA and AF. The substitution percentage can be calculated by taking the designated number and multiplying it by 10. The water content was calculated based on standard consistency values (IS 4031 (Part 4): 2019 [18]). Figure 4a shows the Vicat apparatus used to measure standard consistency. Two different chemical admixture dosages (0.4 and 0.6) were employed to evaluate the interaction of the chemical admixtures and supplementary cementitious materials.

Table 3 Mixture combinations used for the study

Mixture	Replacement rates (%)			Standard consistency (%)	
	C	USBA	AF	0.4	0.6
Control	100	–	–	22	19.5
1USBA	90	10	–	27	25
2USBA	80	20	–	32.5	29
3USBA	70	30	–	37.5	33.5
4USBA	60	40	–	44	40.5
5USBA	50	50	–	51	44
6USBA	40	60	–	58.5	51.5
1AF	90	–	10	21.5	18.5
2AF	80	–	20	22	20
3AF	70	–	30	23	22.5
4AF	60	–	40	24	25
5AF	50	–	50	26.5	27
6AF	40	–	60	31.5	28
1USBA1AF	80	10	10	28	22
1USBA2AF	70	10	20	25	23
1USBA3AF	60	10	30	27	25.5
1USBA4AF	50	10	40	28	27
1USBA5AF	40	10	50	30.5	28.5
2USBA1AF	70	20	10	31.5	30
2USBA2AF	60	20	20	33	31
2USBA3AF	50	20	30	33	31
2USBA4AF	40	20	40	33.5	31.5
3USBA1AF	60	30	10	39.5	33.5
3USBA2AF	50	30	20	39	33
3USBA3AF	40	30	30	37.5	35.5
4USBA1AF	50	40	10	42.5	38
4USBA2AF	40	40	20	43	39.5
5USBA1AF	40	50	10	49.5	43

2.3 *Specimen Preparation and Curing*

A Hobart mixer was used for paste preparation. The process began with a one-minute low-speed mixing of the binder materials. Subsequently, 70% of the necessary water was added to the binder. The mixing process continued with an additional one minute of low-speed mixing, concluding with a brief 30 s of high-speed mixing. The remaining water with polycarboxylic ether-based superplasticizer was added to the cement paste with an additional 30 s of high-speed mixing. After pausing to

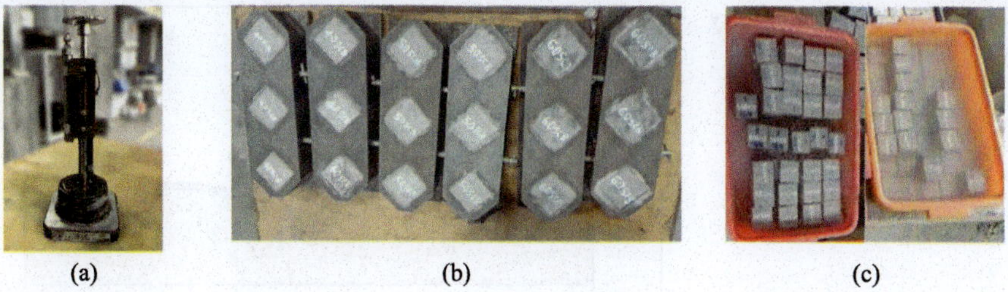

Fig. 4 **a** Vicat apparatus used to measure the consistency, **b** Casted specimens, **c** Specimen curing

scrape the sides of the mixer bowl, the paste underwent a final two-minute high-speed mixing. The prepared mixture was cast into cubic moulds measuring 50 mm on each side. The specimens (Fig. 4b) were demoulded 24 h after casting and then underwent water curing until the time of testing (Fig. 4c). After 7 and 28 days of curing, the specimens underwent compressive strength evaluations.

3 Results and Discussions

3.1 Standard Consistency

Figure 5 displays the standard consistency of various mixture combinations. The consistency of the control mixture decreases from 22 to 19.5 as the chemical admixture dosage rises from 0.4 to 0.6, suggesting that a higher dosage reduces consistency. For both chemical admixture dosages, the consistency increases with the USBA percentage, with 0.4 dosage yielding higher consistency values than 0.6 across all USBA mixtures. This implies that USBA enhances consistency, and higher USBA percentages lead to greater water demand, particularly at the 0.4 dosage, due to the porous nature of USBA particles [19]. A slight increase in consistency is observed with increasing AF percentage, with 0.4 dosage generally resulting in higher consistency values than 0.6 for all AF mixes. This indicates that AF contributes to increased consistency, especially at the 0.4 dosage, despite the AF mixture's ability to achieve desired workability at lower chemical admixture dosages owing to the glassy surface properties of AF particles [9, 20].

The combination of USBA and AF has a variable impact on standard consistency. In general, higher USBA percentages are associated with enhanced consistency, even when AF is present. At a chemical admixture dosage of 0.4, mixtures with larger proportions of USBA demonstrate the highest consistency values. When 10% USBA is combined with different percentages of AF, the overall trend shows higher consistency at 0.4 dosage compared to 0.6. Similarly, combinations of 20 and 30% USBA with varying AF percentages exhibit slightly higher consistency values at 0.4 dosage. For 40% USBA combined with different AF percentages, consistency values

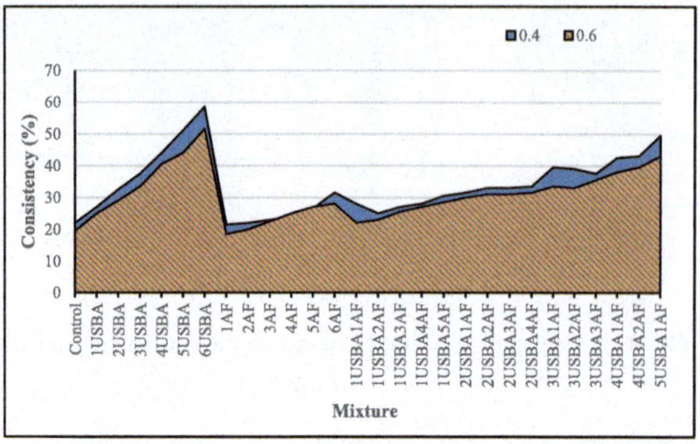

Fig. 5 Standard consistency of different mixture combinations

are also higher at 0.4 dosage. However, when 50% USBA is mixed with 10% AF, both consistency and compressive strength decrease at higher chemical admixture dosage. Most mixture combinations display elevated consistency levels at 0.4 compared to 0.6, indicating that USBA and AF have a more pronounced impact on consistency at the lower dosage. Generally, it was determined that introducing AF to USBA mixtures typically leads to reduced consistency, suggesting that AF may lower water demand when used in combination with USBA.

3.2 *Compressive Strength*

The control mixture's compressive strength is superior with a 0.4 dosage compared to 0.6, as observed after both 7 and 28 days. At 28 days, the USBA binary blended cement pastes exhibited higher compressive strength values than those with a 0.6 chemical admixture dosage. The compressive strength of both the control mixtures and those blended with USBA is displayed in Fig. 6. The binary blended mixtures containing AF, at a 0.6 dosage, demonstrate enhanced compressive strength after 7 and 28 days of curing. In binary blended mixtures containing AF, increased AF percentages lead to modest improvements in consistency but substantial gains in compressive strength. The 0.6 dosage produces higher compressive strengths than 0.4, indicating that AF effectively enhances strength even at elevated dosages. The addition of AF enhances the production of C-S–H, leading to improved compressive strength [21]. This enhancement is attributed to AF's strong pozzolanic properties and its ability to fill voids, resulting in better strength development [22]. The compressive strength of both the control mixture and the AF blended mixtures is depicted in Fig. 7.

For ternary blended mixtures, the strength gain from 7 to 28 days is more noticeable at 0.4 dosages compared to 0.6 dosages. The compressive strengths of control samples and ternary blended mixtures are presented in Fig. 8. The 1USBA1AF mixture exhibited the highest compressive strength for both chemical admixture dosages, while 5USBA1AF showed the lowest. Overall, using a 0.4 dosage leads to

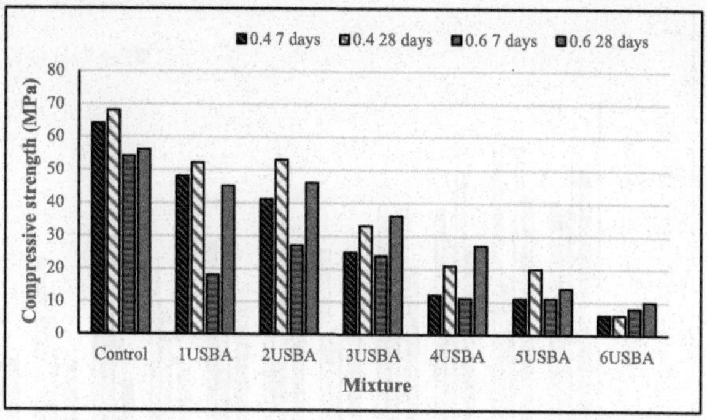

Fig. 6 Compressive strength of control and USBA blended mixtures

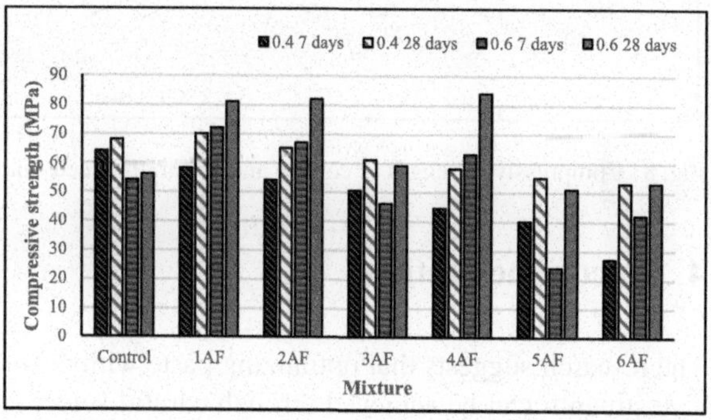

Fig. 7 Compressive strength of control and AF blended mixtures

greater compressive strength, with 1USBA1AF showing exceptional performance in terms of both consistency and compressive strength. The 2USBA1AF mixture demonstrates a good balance between these properties. Mixtures with high replacement levels display relatively low compressive strength at both dosages, indicating that increased USBA content may negatively impact overall strength when combined with AF, though this pattern may change with extended curing time [23]. For replacement percentages exceeding 30%, particularly with fixed 20% USBA content, the 0.6% chemical admixture dosage proves more effective than 0.4%. However, this trend becomes ineffective when the total replacement percentage surpasses 40%. The 0.4 dosage is generally more effective in maintaining higher consistency values across all mixes, while the 0.6 dosage tends to yield higher compressive strengths with increased AF percentages. For 10% fixed USBA percentages, the 0.4 dosage typically produces better compressive strengths. In fixed 30% USBA percentages, 3USBA1AF shows the highest compressive strength, suggesting an ideal mix for both consistency and strength. The combinations of 1USBA1AF, 2USCBA1AF and 3USBA1AF demonstrate a good balance of consistency and compressive strength, making them promising candidates for optimized mixes.

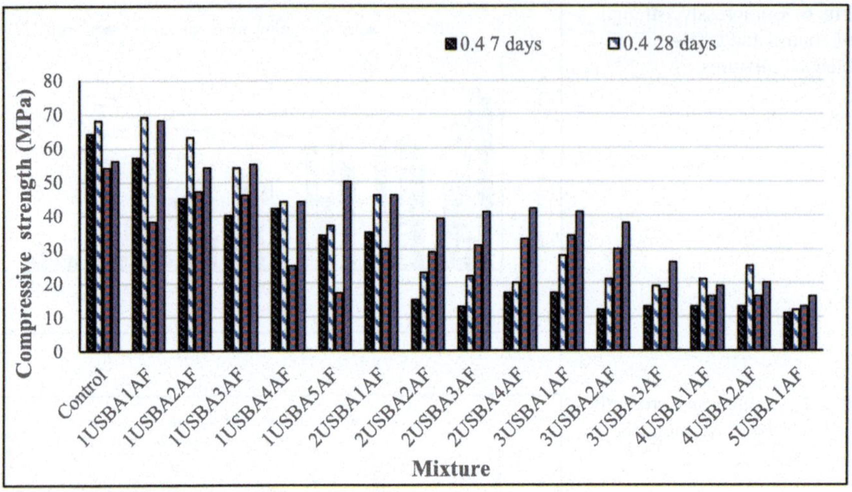

Fig. 8 Compressive strength of control and ternary blended mixtures

4 Recommendations

The research suggests that optimizing paste with desired consistency and compressive strength can be achieved through careful selection of USBA and AF percentages, combined with appropriate chemical admixture dosages. The identified optimal combinations serve as a valuable foundation for further investigation and potential real-world applications in cementitious composites. The following additional studies are recommended:

- Investigate the possible impacts of different curing durations and techniques on the final characteristics of the mixtures.
- Conduct long-term studies to assess the durability and performance of USBA and Alccofine mixtures under various environmental conditions, including exposure to aggressive chemicals, freeze–thaw cycles and wet-dry cycles.
- Perform detailed microstructural analyses using techniques such as XRD, SEM and TGA to understand the chemical interactions and hydration products formed in USBA and Alccofine blended mixtures.
- Explore the use of other agricultural wastes or industrial by-products as supplementary cementitious materials in combination with USBA and Alccofine to enhance the sustainability of concrete mixtures.
- Apply machine learning techniques to predict the performance of USBA and Alccofine mixtures based on various input parameters, such as admixture dosage, curing conditions and mixture proportions.

5 Conclusions

This research explored the use of unmodified sugarcane bagasse ash (USBA) and Alccofine as supplementary cementitious materials in cement paste formulations. Additionally, it examined the combined effects of these materials and aimed to identify optimal chemical admixture dosages to achieve desired physical and mechanical properties. The following conclusions were derived from the comprehensive experimental investigation conducted in this study.

As the chemical admixture dosage increases from 0.4 to 0.6%, the consistency of all mixtures generally decreases. The control mixture shows better compressive strength at 0.4 dosage compared to 0.6, while USBA mixtures achieve improved 28-day strength with 0.4 dosage. In contrast, AF blended mixtures perform best with 0.6 dosage, significantly improving both strength and consistency. For ternary blended mixtures, higher USBA percentages typically result in lower compressive strength, but 0.4 dosage maintains better consistency. On the other hand, 0.6 dosage yields higher strengths as AF percentages increase. The 3USBA1AF mixture demonstrates the most favourable combination of consistency and compressive strength.

Acknowledgements The authors feel highly indebted to Vellore Institute of Technology, Vellore, Tamil Nadu, India, for providing the opportunity to pursue the research.

References

1. Jayasree, C., & Gettu, R. (2008). Experimental study of the flow behaviour of superplasticized cement paste. *Materials and Structures/Materiaux et Constructions., 41*, 1581–1593. https://doi.org/10.1617/s11527-008-9350-5
2. Liu, J., C. Y. X. S. Q. R. Y. Y. (2019). Recent advance of chemical admixtures in concrete. *Cement and Concrete Research. 124*. https://doi.org/10.1016/j.cemconres.2019.105834.
3. Ma, R., Wang, Y., Li, H., Bai, Y. (2023). Progress in the polycarboxylate superplasticizer and their structure-activity relationship—A review. https://doi.org/10.1016/j.mtcomm.2023.105838.
4. Kannur, B., Chore, H. S. (2021) Utilization of sugarcane bagasse ash as cement-replacing materials for concrete pavement: an overview. *Innovative Infrastructure Solutions. 6*. https://doi.org/10.1007/s41062-021-00539-4.
5. Kirthiga, R., & Elavenil, S. (2023). Potential utilization of sugarcane bagasse ash in cementitious composites for developing inorganic binder. *Ain Shams Engineering Journal., 14*, Article 102560. https://doi.org/10.1016/j.asej.2023.102560
6. Joseph, A., Murugan, S. B. (2024) Investigating the sustainable application of unprocessed sugarcane bagasse ash and alccofine in cementitious mortar from a physical, mechanical, and microstructural perspective. *Innovative Infrastructure Solutions. 9*. https://doi.org/10.1007/s41062-024-01783-0.
7. Batool, F., Masood, A., & Ali, M. (2020). Characterization of Sugarcane Bagasse Ash as Pozzolan and Influence on Concrete Properties. *Arabian Journal for Science and Engineering, 45*, 3891–3900. https://doi.org/10.1007/s13369-019-04301-y
8. Dineshkumar, R., Balamurugan, P. (2021) Behavior of high-strength concrete with sugarcane bagasse ash as replacement for cement. *Innovative Infrastructure Solutions. 6*. https://doi.org/10.1007/s41062-020-00450-4.

9. Prasad Yadav, K., & Samanta, A. K. (2023). Characterization and development of alccofine based sustainable concrete—A review. *Mater Today Proc.* https://doi.org/10.1016/j.matpr.2023.05.331
10. Tinesh Kumaar, G., Gunasekar, S., & Ramesh, N. (2023). Impact of Alccofine on the strength and durability properties of concrete. *Materials Today: Proceedings.* https://doi.org/10.1016/j.matpr.2023.11.036
11. Ganesan, K., Rajagopal, K., & Thangavel, K. (2007). Evaluation of bagasse ash as supplementary cementitious material. *Cement and Concrete Composites, 29*, 515–524. https://doi.org/10.1016/j.cemconcomp.2007.03.001
12. Sagar, B., & Sivakumar, M. V. N. (2021). Use of alccofine-1203 in concrete: Review on mechanical and durability properties. *International Journal of Sustainable Engineering., 14*, 2060–2073. https://doi.org/10.1080/19397038.2021.1970275
13. IS 269, Ordinary Portland Cement—Specification. , New Delhi, India (2020).
14. ASTM C618, Standard Specification for Coal Fly Ash and Raw or Calcined Natural Pozzolan for Use in Concrete, West Conshohocken, PA 19428–2959 (2022). https://doi.org/10.1520/C0618-22.
15. IS 12089, Specification for Granulated Slag for the Manufacture of Portland Slag Cement. New Delhi, India (2018).
16. ASTM C494, Standard Specification for Chemical Admixtures for Concrete. West Conshohocken, PA 19428–2959 (2024). https://doi.org/10.1520/C0494_C0494M-24.
17. IS 456, Plain and Reinforced Concrete- Code of Practice. New Delhi, India (2021).
18. IS 4031 (Part 4), Methods of Physical Tests for Hydraulic Cement: Determination of Consistency of Standard Cement Paste. New Delhi, India (2019).
19. Gupta, C. K., Sachan, A. K., & Kumar, R. (2021). Examination of microstructure of sugar cane bagasse Ash and sugar cane bagasse ash blended cement mortar. *Sugar Tech., 23*, 651–660. https://doi.org/10.1007/s12355-020-00934-8
20. Srinath, B. L. N. S., Patnaikuni, C. K., Balaji, K. V. G. D., Kumar, B. S., & Manjunatha, M. (2021). A prospective review of alccofine as supplementary cementitious material. *Materials Today: Proceedings, 47*, 3953–3959. https://doi.org/10.1016/j.matpr.2021.03.719
21. Gautham Kishore Reddy, G., Ramadoss, P. (2020) Influence of alccofine incorporation on the mechanical behavior of ultrahigh performance concrete (UHPC). *Materials Today: Proceedings 33*, 789–797. https://doi.org/10.1016/j.matpr.2020.06.180.
22. Balamuralikrishnan, R., Saravanan, J. (2019) Effect of Alccofine and GGBS addition on the durability of concrete. *Civil Engineering Journal (Iran). 5*, 1273–1288. https://doi.org/10.28991/cej-2019-03091331.
23. Hernandez Toledo, U. I., Montes García, P., Caballero Aquino, T. (2012) Efecto de la ceniza de bagazo de cana y el tiempo de curado en las propiedades de morteros. 4 Congreso Nacional ALCONPAT, Xalapa, Veracruz, México (In Spanish).

Recent Developments in the Production of Synthetic Aggregates for Sustainable Infrastructure

R. Vignesh and A. Abdul Rahim

Abstract The process of rapid urbanization and industrialization requires an enormous amount of natural aggregates to produce concrete. To meet the current demand, the construction industry has focused on using synthetic aggregates, produced from eco-friendly raw materials, as a sustainable alternative to conventional aggregates. This chapter reviews the different origins of these eco-friendly alternative materials, their production methods, the various shapes of aggregates and the types of mortar used in the production process. Hardening methods such as sintering, cold-bonding, accelerated carbonation and alkaline activation may significantly affect the final properties of the produced synthetic aggregates. The chapter also outlines the significance of employing recycled concrete, waste plastic, industrial byproducts and agricultural waste as eco-friendly raw materials for the production of aggregates. The shapes of produced synthetic aggregates play a crucial role in the workability, strength and durability of the produced synthetic aggregate concrete. This chapter examines the differences among the angular, rounded, cubical and custom-shaped aggregates, which can influence the performance of produced concrete. The chapter also addresses the environmental benefits of using synthetic aggregates, such as reducing CO_2 emissions, conserving natural resources and utilizing industrial agro wastes. The objective of this chapter is to present innovative strategies and valuable insights about the production of synthetic aggregates using eco-friendly raw materials.

Keywords Synthetic aggregates · Sustainability · Concrete · Shapes · Life-cycle assessment

---

R. Vignesh
Department of Civil Engineering, Easwari Engineering College, Chennai, India

A. Abdul Rahim (✉)
Department of Structural and Geotechnical Engineering, School of Civil Engineering, Vellore Institute of Technology, Vellore, India
e-mail: abdulrahim.a@vit.ac.in

P. Danish et al. (eds.), *Innovative Technologies for Smart and Sustainable Infrastructure*,
https://doi.org/10.1007/978-3-031-92801-7_17

1 Introduction

The rapid growth in global infrastructure development has significantly increased the demand for construction materials, particularly aggregates. Natural aggregates, such as sand, gravel and crushed stone, are among the most consumed resources worldwide. However, their extraction has led to serious environmental consequences, including habitat destruction, depletion of resources and high carbon emissions. Comprehensive research has been conducted to explore the potential application of recycled materials, such as recycled concrete aggregates, crumb rubber, recycled bricks and refuse glass, as alternatives for natural aggregates in concrete. But the direct utilization of refuse ultimately impacted the properties of the concrete. For example, concrete composed entirely of 100% recycled coarse aggregates demonstrated inferior mechanical performance in comparison to conventional concrete [1]. Therefore, the production of synthetic aggregates (SAs) has emerged as a viable substitute for natural aggregates, contributing to sustainable infrastructure by reducing resource depletion, recycling waste materials and lowering carbon footprints. This chapter explores recent advancements in the production of synthetic aggregates, their properties, applications and their role in promoting sustainable infrastructure.

2 Need of Synthetic Aggregates

Natural aggregates have traditionally dominated construction due to their availability and cost-effectiveness. However, over-reliance on these resources has raised concerns about environmental degradation and resource scarcity. Hence, synthetic aggregates have emerged as a transformative solution addressing critical challenges in the dynamic landscape of modern concrete construction. These engineered materials represent a sophisticated approach to addressing resource scarcity, environmental sustainability and performance optimization in the built environment. The increasing focus on sustainability in construction has driven the need for innovative materials that align with global goals such as carbon neutrality and circular economy principles. Synthetic aggregates, made from industrial by-products, recycled materials and waste, have gained attention as a viable solution. This section discusses the historical development of synthetic aggregates and their alignment with sustainability objectives.

An increased demand for innovative building materials has resulted from population expansion and the acceleration of the construction industry to satisfy the requirements of global infrastructure development. Globally, the rate of depletion of natural resources is accelerating drastically. The rapid growth in industrialization results in the production of vast quantities of harmful industrial by-products, which causes significant challenges to handling and managing. Utilizing industrial by-products as construction materials is a secure and environmentally sustainable approach to eliminating waste that conserves natural resources for future generations [2]. Due to its numerous advantages, concrete has been the prevalent construction material worldwide for a few decades. Furthermore, it is a significantly more cost-effective material that exhibits extended durability and requires relatively little maintenance. Aggregates are crucial in providing structure to concrete, making up about 60–80% of its volume [3]. They also substantially impact the fresh and mechanical properties of the concrete. Hence, due to rapid urban growth, aggregate scarcity is emerging as a significant concern within the construction sector. On the other hand, constant aggregate exploitation results in the depletion of natural resources. Furthermore, the quarrying and natural aggregate processing industries are significant contributors to the emission of pollution, noise and greenhouse gases, all of which negatively impact the environment. Hence, there was an increased focus on developing synthetic aggregates derived from industrial byproducts and municipal wastes to manufacture robust concrete aggregates [4]. According to prior research, synthetic aggregates find extensive applications in fabricating prefabricated structural elements, producing structural concrete, acoustic insulation, thermal protection, landscaping, hydroponics and soil examination operations [5]. Therefore, developing novel sources for coarse aggregates is a critical issue in modern society [6]. Over the past few decades, synthetic aggregates have received increasing interest due to their unique characteristics, including low bulk density, high resistance to heat, excellent reliability, and notable thermal and acoustic insulation characteristics.

3 Synthetic Aggregates: Overview

Synthetic aggregates are man-made materials designed to replicate the properties of natural aggregates. Common types include lightweight aggregates (e.g. expanded clay, shale and fly ash-based aggregates), recycled aggregates (e.g. aggregates from construction and demolition waste), Geopolymer aggregates (e.g. alkali-activated materials). Synthetic aggregates are typically produced using waste materials such as industrial by-products (e.g. fly ash, slag, silica fume), municipal solid waste (e.g. incinerator bottom ash), plastics and polymers, recycled concrete and demolition debris.

4 Production Methods of Synthetic Aggregates

Several methods are utilized to fabricate synthetic aggregates derived from different sources. Specifically, three different methods are used to produce artificial aggregates: sintering, autoclaving and cold-bonding pelletization. One effective method for utilizing substantial quantities of different refuse materials is through the artificial production of aggregates through pelletization, a process also referred to as agglomeration [7]. The primary objective of all methods is to agglomerate the pulverized waste raw material into a particular shape and size suitable for use as aggregate in the intended application.

4.1 Sintering

High-temperature sintering is the standard method employed in the production of synthetic aggregates. LECA, or lightweight expanded clay aggregate, is a synthetic aggregate primarily produced at calcination temperatures ranging from 900 °C to 1200 °C. [8]. Furthermore, this sintering procedure emits hazardous substances and pollutants during the production of synthetic aggregates [9]. Therefore, the sintering process had been progressively prohibited in fabricating artificial aggregates due to carbon emission restrictions and a more energy-intensive process. As a result, many researchers have investigated a new approach for producing low-carbon synthetic aggregates. In recent years, numerous initiatives have been devoted to advancing artificial aggregates through non-sintering techniques.

4.2 Autoclaving

Autoclaving is a process in which a chemical, such as cement, lime or gypsum, is introduced during agglomeration. Thus, the material acquires a bonding property. After curing in pressurized steam at 140ºC, the green pellets are extracted. This method reduces the bonding material used during particle formation and curing. However, the strength and durability characteristics remain primarily similar to conventional curing.

4.3 Cold Bonding

Using cold-bonding techniques on different industrial by-products is widely regarded as a sustainable approach for producing synthetic aggregates [10]. In cold-bonding, aggregates are cured at room temperature or in a confined space saturated with steam until the desired strength is attained. Conversely, the cold bonding technique requires cementitious materials and involves an extended duration of 28 days to produce aggregates possessing the required strength properties. Thus, long curing periods and inadequate mechanical development are the major drawbacks of synthetic aggregates produced through cold bonding techniques. Cold-bonded aggregate exhibits inferior properties compared to sintered aggregates. Similarly, synthetic aggregates produced through a cold bonding process comprising 80% fly ash and 20% OPC achieved only a compressive strength of 1.16 MPa following a curing period of 28 days, which fails to satisfy the engineering specifications [11]. Even though this process requires more time to accomplish the desired mechanical properties, this cold-bonding process is more economical than the sintering process due to its reduced energy demand. Among all the non-sintering methods, Ordinary Portland cement (OPC) was frequently used as a binder in the pelletizing process. The pelletization process utilizes cohesive and rotational forces to bind moistened particles to produce spherical artificial aggregate from pulverized material. Then, to impart the necessary strength to these unprocessed aggregates, cold-bonding or sintering techniques are implemented for their hardening. Conventional techniques such as geo polymerization, mechanical crushing, elevated curing temperature, calcination and additional reactive materials have been suggested to increase the compressive strength of produced synthetic aggregates. However, these techniques are restricted to materials exhibiting higher reactivity.

4.4 Pelletization

Pelletization is an efficient agglomeration technique that converts tiny wet particles into a solid matrix of a larger dimension. The strength of the developed pellet is enhanced by the collisions that occur due to the rotating motion within the pelletizer disc. The pelletizer demands only a tiny amount of operational space and enables better regulation of pellet size. Various factors, including particle size, shape, moisture content and specific surface area, impact the efficacy of pelletization. Furthermore, the period of the pelletization method and the velocity and angle of the pelletizer influence the efficacy. Centrifugal force governs the particle's motion at low velocities, whereas gravitational force exerts an influence at high speeds. Modifying the disc's speed and angle appropriately guarantees an appropriate collision between the moisturized finite particles, improving production efficiency [12].

4.5 Crushing Method

Crushing is the process of hammering the hardened paste substance to reduce it to the required particle size and then hardening the crushed particles through curing until the testing period. The crushing method facilitates the generation of artificial aggregates with an angular shape, ideally turning it to suit mass production compared to pelletization.

4.6 Mould Casting

The mixture is poured into moulds and cured under confined conditions until the testing age to obtain the desired size and shape aggregates. A frequently used method involves a plastic pipette to shape the reactive foam mixture into aggregates or a polystyrene mould to achieve a spherical shape. By using mould casting, Peyne et al., developed a novel type of aggregates using geopolymer formulations as a substitute for commercial lightweight aggregates (LWAs). The industrial sintering process of LWAs necessitates a substantial amount of energy. This study investigated the feasibility of producing geopolymer aggregates and revealed that this manufacturing process is more sustainable. Vignesh et al. [13] developed synthetic aggregates in different shapes and sizes by using silicon moulds.

4.7 Hand Shaping

The aggregates are manufactured manually, rendering them inappropriate for large-scale production.

5 Recent Innovations in Synthetic Aggregate Production

5.1 Material Innovations

- Development of geopolymer-based aggregates using fly ash and slag, providing high durability and chemical resistance.
- Use of plastic waste in aggregate production to address the global plastic crisis.

5.2 Technological Advancements

- 3D printing techniques for producing customized aggregate shapes.
- Advanced molding methods to enhance uniformity and performance.
- Nanotechnology for improving particle bonding and mechanical properties.

5.3 Energy Efficiency

- Adoption of low-temperature curing processes to reduce energy consumption.
- Use of renewable energy sources in production facilities.

5.4 *Carbon Capture and Utilization*

- Integration of CO_2 sequestration techniques into the production process, enabling carbon–neutral or carbon-negative aggregates.

5.5 *Advantages of Synthetic Aggregates*

Synthetic aggregates offer several benefits over natural aggregates, including:

- Lower density for lightweight construction.
- Reduced environmental impact by utilizing waste.
- Improved thermal insulation properties.

6 Performance and Applications

6.1 *Mechanical Properties*

Synthetic aggregates exhibit high compressive strength, low density and excellent thermal and acoustic insulation properties. A unique synthetic aggregate has been developed by using discarded plastic [14]. This study investigated the effect of fresh, hardened and durability properties of concrete by substituting synthetic aggregates for Lytag aggregate or natural pumice lightweight coarse aggregates. Additionally, it was observed that the usage of these produced aggregates in concrete resulted in reduced chloride penetration and water absorption compared to control mixtures. Similarly, shaped synthetic aggregates were developed by combining, cement, granulated blast furnace slag, metakaolin and a minor quantity of lime as 5% [15]. The density of produced aggregates was found to be higher and the water absorption value decreased as the percentage of the GGBFS content increased in the developed aggregates. Fibre reinforced synthetic aggregate concrete (FRSAC) exhibited good results for compressive strength, split tensile strength and durability properties, compared to conventional concrete. As a result, the properties of FRSAC were entirely satisfactory and could provide a sustainable solution to the construction sector, specifically for the production of green concrete.

6.2 *Durability Properties*

Makhloufi et al., [16] examined the findings of assessing the resistance of limestone mortars to magnesium sulphate solution ($MgSO_4$) when incorporating blast furnace slag, natural pozzolanas and limestone powder simultaneously. The findings indicated that mortars containing quaternary binders exhibited superior resistance to sulphate attack than mortars composed exclusively of ordinary Portland cement. On the other hand, [17] conducted an investigation to evaluate the acid resistance of quaternary blended concrete containing recycled aggregates in two grades of M40 and M60. Flyash, silica fume, nano silica along with cement were used to prepare quaternary blended concrete. This investigation utilized two distinct acids (HCl and H_2SO_4) at concentrations of 3% and 5%, respectively. In both the M40 and M60 grades of concrete, the test results indicated a slight reduction in weight and strength. However, ultrasonic pulse velocity (UPV) test results indicate that the overall quality of QBC remains satisfactory after acid exposure. The use of limestone filler as a binder (8–17%) facilitates the creation of quaternary blended binders with early-age reactivity and good durability properties [18].

6.3 Applications in Infrastructure

- Road construction: Lightweight aggregates reduce load on subgrades.
- Building materials: Used in lightweight concrete and insulation layers.
- Marine structures: Resistant to chemical attacks and corrosion.

7 Environmental and Economic Benefits

7.1 Life-Cycle Assessment (LCA)

Synthetic aggregates demonstrate lower carbon footprints compared to natural aggregates when evaluated over their lifecycle.

7.2 Reduction in Environmental Impact

- Reduction in landfill waste by utilizing industrial by-products.
- Lower energy requirements for production compared to quarrying and crushing natural aggregates.

7.3 Economic Viability

- Cost savings through the use of waste materials.
- Potential revenue generation from waste management and recycling industries.

8 Challenges and Limitations

8.1 Technical Barriers

- Scalability of production processes.
- Consistency in material properties due to variability in feedstock.

8.2 Regulatory and Standardization Issues

- Lack of universal standards for synthetic aggregates.
- Delayed adoption due to regulatory hurdles.

8.3 Public Perception and Market Adoption

- Resistance from stakeholders due to lack of awareness.
- Perceived risks associated with new materials.

9 Future Trends and Research Directions

- Exploration of bio-based aggregates using microbial processes.
- Development of smart aggregates with embedded sensors for real-time monitoring of infrastructure health.
- Enhanced automation and AI integration in production for improved efficiency.
- Greater adoption of synthetic aggregates in developing nations through knowledge dissemination and capacity building.

10 Conclusion

Synthetic aggregates represent a transformative approach to achieving sustainable infrastructure. With advancements in material science and technology, they have the potential to significantly reduce the environmental impact of construction while offering superior performance. Continued innovation, regulatory support and industry collaboration are critical for mainstream adoption and scaling of these materials. The production of synthetic aggregates is a promising solution for sustainable infrastructure, addressing environmental concerns and resource scarcity. Recent advancements in raw materials, manufacturing processes and performance improvements make synthetic aggregates a viable alternative to natural aggregates. However, further research, policy support and industry adoption are essential to maximize their potential in sustainable construction. This chapter provides a comprehensive overview of the production methods of synthetic aggregates, latest developments in synthetic aggregates, emphasizing their importance in the transition towards sustainable infrastructure.

References

1. Ren, P., Mehdizadeh, H., Ling, T. C. (2024). Performance investigation of the artificial aggregate by integrally recycling incineration bottom ash and fly ash. *Cement and Concrete Composites. 152*. https://doi.org/10.1016/j.cemconcomp.2024.105678.
2. Kisku, N., Joshi, H., Ansari, M., Panda, S. K., Nayak, S., Dutta, S. C. (2017) A critical review and assessment for usage of recycled aggregate as sustainable construction material, *Construction and Building Materials. 131* 721–740. https://doi.org/10.1016/j.conbuildmat.2016.11.029.
3. Chu, S. H., Jiang, Y., Kwan, A. K. H. (2019) Effect of rigid fibres on aggregate packing, *Construction and Building Materials. 224* 326–335. https://doi.org/10.1016/j.conbuildmat.2019.07.072.
4. Tuan, B. L. A., Hwang, C. L., Lin, K. L., Chen, Y. Y., Young, M. P. (2013). Development of lightweight aggregate from sewage sludge and waste glass powder for concrete, *Construction and Building Materials. 47* 334–339. https://doi.org/10.1016/j.conbuildmat.2013.05.039.
5. González-Corrochano, B., Alonso-Azcárate, J., Rodas, M., Barrenechea, J. F., Luque, F. J. (2011) Microstructure and mineralogy of lightweight aggregates manufactured from mining and industrial wastes, *Construction and Building Materials. 25* 3591–3602. https://doi.org/10.1016/j.conbuildmat.2011.03.053.
6. Muduli, R., Mukharjee, B. B. (2020). Performance assessment of concrete incorporating recycled coarse aggregates and metakaolin: A systematic approach, *Construction and Building Materials.* 233 117223. https://doi.org/10.1016/j.conbuildmat.2019.117223.
7. Bekkeri, G. B., Shetty, K. K., Nayak, G. (2023). Synthesis of artificial aggregates and their impact on performance of concrete: a review, *Journal of Material Cycles and Waste Management. 25* 1988–2011. https://doi.org/10.1007/s10163-023-01713-9.
8. Rashad, A. M. (2018). Lightweight expanded clay aggregate as a building material—An overview. *Construction and Building Materials. 170* 757–775. https://doi.org/10.1016/j.conbuildmat.2018.03.009.
9. Alqahtani, F. K., Rashid, K., Zafar, I., Iqbal Khan, M. (2021). Assessment of morphological characteristics and physico-mechanical properties of geopolymer green foam lightweight aggregate formulated by microwave irradiation. *Journal of Building Engineering 35* https://doi.org/10.1016/j.jobe.2020.102081.
10. Dong, B., Liu, J., Hong, S., Zhang, Y., Wang, Y., Xing, F. (2021). Core-Shell structured ceramsite made by excavated soil and expanded perlite through cold-bonded technology, *Construction and Building Materials. 306*. https://doi.org/10.1016/j.conbuildmat.2021.124941.
11. Zhang, C., Wang, G., Wu, C., Li, J., Wu, S., Jiang, W., Wang, X., Wang, W., Feng, M. (2021) Investigation of hierarchical porous cold bonded lightweight aggregates produced from red mud and solid-waste-based cementitious material, *Construction and Building Materials. 308*. https://doi.org/10.1016/j.conbuildmat.2021.124990.
12. Xu, L. Y., Qian, L. P., Huang, B. T., Dai, J. G. (2021). Development of artificial one-part geopolymer lightweight aggregates by crushing technique. *Journal of Cleaner Production. 315* 128200. https://doi.org/10.1016/j.jclepro.2021.128200.
13. Vignesh, R., Abdul-Rahim, A. (2023) New insights into the production of sustainable synthetic aggregates and their microstructural evaluation, Mater. *Construcción. 73* e324. https://doi.org/10.3989/mc.2023.328722.

14. Vignesh, R., Abdul Rahim, A. (2024). Emerging eco-friendly fiber-reinforced concrete with shaped synthetic aggregates using Taguchi grey relational analysis and utility concept. *Construction and Building Materials. 447*. https://doi.org/10.1016/j.conbuildmat.2024.138039.
15. Vignesh, R., Rahim, A. A. (2023) Multi-response Optimization of Fiber-Reinforced-Shaped synthetic aggregate concrete, *Arabian Journal for Science and Engineering*. https://doi.org/10.1007/s13369-023-08305-7.
16. Makhloufi, Z., Aggoun, S., Benabed, B., Kadri, E. H., Bederina, M. (2016) Effect of magnesium sulfate on the durability of limestone mortars based on quaternary blended cements. *Cement and Concrete Composites. 65* 186–199. https://doi.org/10.1016/j.cemconcomp.2015.10.020.
17. Rao, K. J., Keerthi, K., Vasam, S. (2018) Acid resistance of quaternary blended recycled aggregate concrete, *Case Studies in Construction Materials. 8* 423–433. https://doi.org/10.1016/j.cscm.2018.03.005.
18. Vignesh, R., Rahim, A. A. x. (2022) Mechanical and microstructural properties of quaternary binder system containing OPC-GGBS-Metakaolin-Lime. *Materials Today: Proceedings, 64*, 970–975.

A Critical Review on Mechanical and Microstructural Characteristics of Geopolymer Composites Reinforced with Industrial By-Products

Mathialagan Sumesh, U. Johnson Alengaram, Kim Hung Mo, Ravi Kiran, and C. Makendran

Abstract The developments in geopolymers made possible by the alkaline activation of industrial by-products for use in building construction are thoroughly reviewed in this paper. With an emphasis on their fresh and hardened properties, the microstructural characteristics, and their bonding qualities and mechanical performance, it examines the properties of pastes, mortars, and concrete based on geopolymers. This research presents an overview of the previous works carried out on different supplementary cementitious materials (SCMs) into the geopolymer materials. The workability, setting time, strength of geopolymers based on different by-products as well as comparison between grinding effect of the by-products and their effects in geopolymer products have been highlighted. This chapter summarizes the effects of incorporating various by-products on pozzolanic reaction, microstructure, mechanical and bond strength characteristics of paste/mortar and concrete. The performance of geopolymer composites has been reported to be comparable with the conventional repair materials. Furthermore, the research findings revealed that geopolymer

M. Sumesh (✉)
Department of Civil Engineering, Dayananda Sagar College of Engineering, Bengaluru 560111, Karnataka, India
e-mail: msumesh15@gmail.com; sumesh-cvl@dayanandasagar.edu

M. Sumesh · U. J. Alengaram
Department of Civil Engineering, Centre for Innovative Construction Technology (CICT), Faculty of Engineering, University of Malaya (UM), 50603 Kuala Lumpur, Malaysia

K. H. Mo
Department of Engineering, School of Engineering and Technology, Sunway University, Selangor Darul Ehsan, 47500 Petaling Jaya, Malaysia

R. Kiran
School of Sustainable Engineering and the Built Environment, Arizona State University, Phoenix, AZ 85281, USA

C. Makendran
Department of Civil Engineering, National Institute of Technology Tiruchirappalli, Tiruchirappalli 620015, Tamil Nadu, India

P. Danish et al. (eds.), *Innovative Technologies for Smart and Sustainable Infrastructure*,
https://doi.org/10.1007/978-3-031-92801-7_18

composites reinforced with various agro-based supplementary cementitious materials show the potential and can be used in concrete applications. Sustainability of the structures can be achieved through the geopolymer repair material which could be used for the defective components of the structures. Overall, development of high strength and high performance geopolymer product would reduce the maintenance cost and also gives a longer service life.

Keywords Geopolymer · Bond strength · Microstructure · Mechanical strength · Supplementary cementitious materials

1 Introduction

The second-largest source of greenhouse gas emissions is the cement sector. Every year, around one ton of cement is produced for each person on the planet. To safeguard the environment, a key focus on reducing CO_2 emissions could be achieved by lowering the amount of cement usage and its consumption in concrete production. As a result, the adoption of new technological materials such as geopolymers has been recommended [1, 2].

CO_2 reduction has emerged as a recent concern as it imparts on the tool for minimizing the Portland cement usage; In addition to the conventional cement replacement materials like fly ash (FA), ground granulated blast furnace slag (GGBS) and silica fume (SF), the agro-industrial by-products are also generated abundantly in the South-East Asian counties; these waste have been identified as potential cement replacement materials [3, 4]. In Malaysia, the large quantity of industrial/agro-based by-products, like fly ash (FA), bottom ash (BA), rice husk ash (RHA) and palm oil fuel ash (POFA) has motivated further research to utilize such waste into potential cement replacement materials. It is necessary to transform these wastes into substantial assets. Recently, extensive research work has been done on the use of alternative cementitious materials with high silicon and aluminum content [5–7].

Thus, there are abundant literatures available in this field, which focused mainly on cement mortar and concrete. There are few literatures which focused on geopolymer based on industrial by-products. Hence, this chapter provides a comprehensive summary of utilizing various industrial by-products on microstructural, mechanical properties of geopolymers composites.

2 Geopolymer Reinforced with Industrial By-Products

Finding the different key elements that have a major impact on the characteristics of geopolymers in relation to industrial by-products is crucial. The effect of various by-products such as FA, BA, GGBS, RHA, and POFA-based geopolymer composites has been studied and found significant findings by many researchers.

2.1 Microstructural Behavior—Geopolymer Paste/Mortar

The microstructural behavior of geopolymer samples along with scanning electron microscopy (SEM) were examined the specimens taken at surface (0–1 mm depth) for the exposure of chemical solutions (Fig. 1). In the microstructure, the additional reaction products/N-A-S-H gel appeared for all three specimens. It is clearly evident that the development of dense microstructure. Hence, a relatively dense microstructure was specifically observed in the 8FA specimen [8].

Fly ash and bottom ash-based specimens of geopolymer were analyzed for microstructure (Fig. 2). Solutions of NaOH and Na_2SiO_3 were employed as activators in this study. According to the XRD data, mullite and quartz make up the majority of the crystalline phases. It has been observed that fly ash produces a faster rate of geopolymerization and is more reactive than bottom ash [9]. The bottom ash is generally coarse in nature and the reactivity is low as compared to fly ash. To activate the bottom ash, it requires grinding to make larger specific surface area [9].

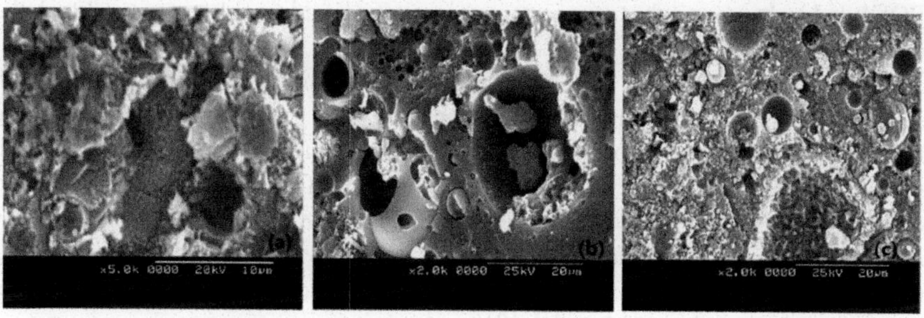

Fig. 1 Geopolymer Samples for SEM micrographs following two months to 5% Sodium Sulfate Chemical Exposure: **a** 8FA, **b** 8FASS, **c** 8FAK [8]

Fig. 2 XRD Patterns for fly ash and bottom Ash (Quartz-Q and Mullite-M) [9]

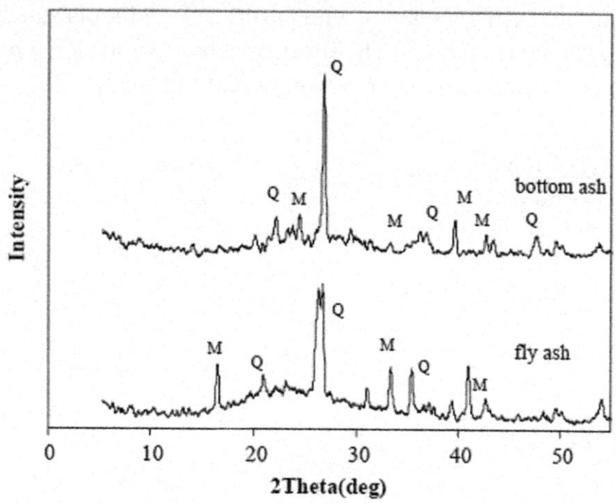

2.2 *Mechanical and Workability Behavior—Geopolymer Paste/Mortar*

POFA particles absorb a larger volume of water compared to other mixes. Their higher surface area and irregular shape increase water demand to achieve a required workability. In contrast, the spherical shape of FA particles facilitates a workable mix with a lower water requirement. A higher POFA component in the mixture results in a higher SiO_2/Al_2O_3 ratio, and the compressive failure behavior of POFA-based cube specimens was observed to be elastic behavior. This is due to higher surface area and reactivity. But, for FA-based geopolymer specimens brittle failure was observed [10].

The effect of GGBS, POFA, and FA on the compressive strength of geopolymer mortar was investigated, and it showed that increasing GGBS content enhanced the strength by up to 70%, while further increases did not yield significant improvements. POFA addition up to 30% with GGBS resulted in higher strength, making it suitable for achieving strengths beyond 60 MPa (Fig. 3). It was also observed that coarser POFA led to a more cohesive mortar, necessitating additional water to obtain the desired workability [11].

In terms of compressive strength, the geopolymer mortar containing bottom ash shows lower strength than those of specimens of fly ash geopolymer (Fig. 4). A large number porous exists with bottom ash particles and contributed to the lower strength to mortar. It was also observed that a higher amount of NaOH needs bottom ash to dissolve alumina and silica and facilitate geopolymerization. Since bottom ash has a chemical composition similar to fly ash, it can also be considered for geopolymer production. So, the quality of bottom ash can be improved by grinding to achieve the chemical reactivity and strength [9].

W/B ratio's impact on mortar-incorporated RHA and POFA was examined. Because of the hydration reaction, as the W/B ratio rises, mortar's compressive strength decreases. However, the strength of mortar based on RHA and POFA exhibits an increasing tendency with curing age and replacement level of cement due to the pozzolanic reaction. The strong reactivity of the RHA and POFA particles causes the enhanced pozzolanic reaction to build strength [12].

Fig. 3 GGBS's impact on compressive strength at 28-day [11]

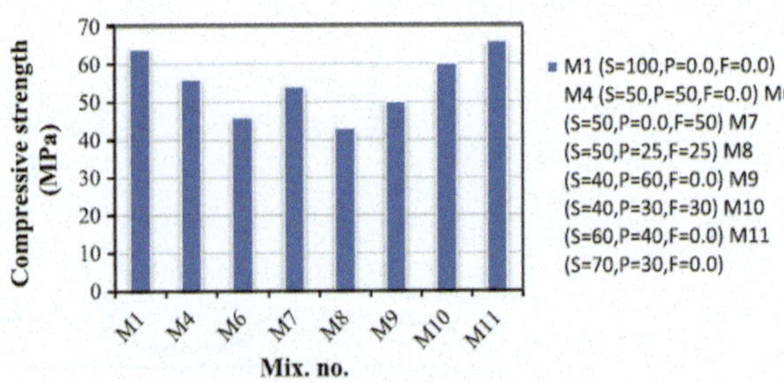

Legend: S – Slag (GGBS), P – POFA, F – Fly ash and mix compositions are shown in bracket in percentage (%)

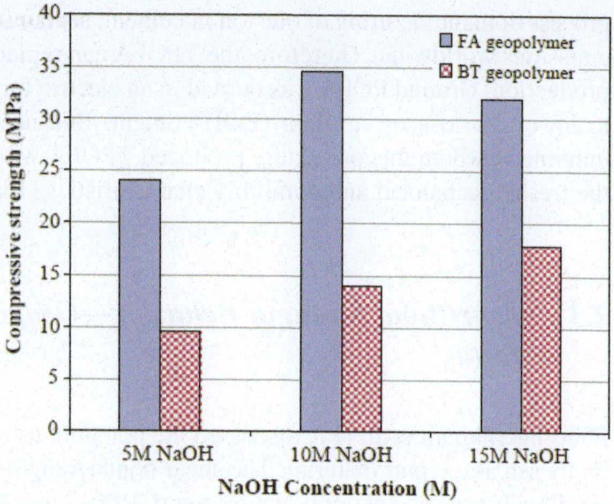

Fig. 4 Compressive strength of geopolymer mortars [9]

Various studies [13, 14] were reported that the addition of mineral additives, FA, and POFA to the geopolymer mortar enhanced the compressive strength. The modification of treated POFA (TPOFA) with mineral additives caused significant improvement in strength of 200.90% and 175.22% at 3 and 28 day, respectively. At the same time, the rise in concentration of NaOH affected the growth of calcium silicate hydrate (C–S–H), which is accountable for the strength development of geopolymer mortar due to the development of C–S–H component (Fig. 5).

Some of the previous research found that the TPOFA has better reactivity [15]. It is detrimental to the environment to use concrete with a high cement content. Besides, it is estimated that about 650–920 kg of CO_2 are released through the

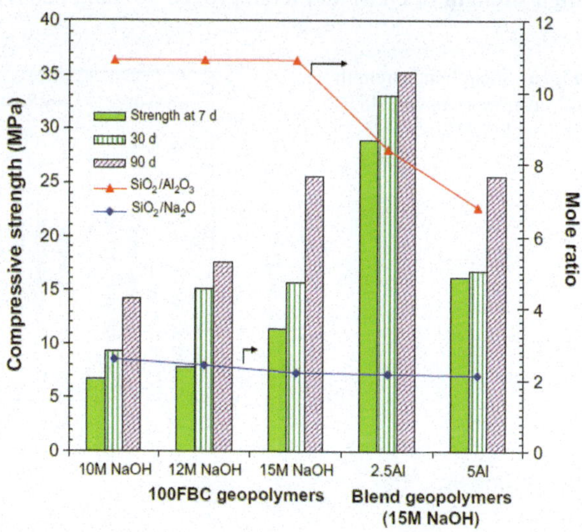

Fig. 5 Compressive strength of geopolymer specimens [14]

process of manufacturing of one ton of cement, accounting for roughly 7–8% of CO_2 emissions worldwide. Therefore, the TPOFA can replace the cement in geopolymer production. Ground POFA was burned in an electric furnace for two hours at 600 °C to lower the loss on ignition (LOI) content. Because of the reduced quantity of unburned carbon, this procedure produced TPOFA with little to no LOI, improving the fresh, mechanical and durability characteristics [15, 16].

2.3 Shear/Bond Strength Behavior—Geopolymer Paste/ Mortar

Phoo-ngernkham et al. [17] discussed the potential use of geopolymer mortar based on fly ash as a repair material. The shear bond strength of the plain cement concrete (PCC) substrate and geopolymer mortar (GPM) composite was examined in addition to a commercial repair material (RM). The findings demonstrated that shear bond strength rose in tandem with NaOH concentration and content of PC. The enhanced strength of fly ash-based specimens with a greater calcium concentration was due to the increased generation of reaction products. Additionally, the relationship between the GPM and the PCC substrate composite at the interface transition zone (ITZ) was strengthened by the development of C–S–H and Calcium Aluminosilicate Hydrate (C–A–S–H) gels formation in the geopolymer matrix (GPM) together with Sodium Aluminosilicate Hydrate (N–A–S–H) gel (Fig. 6).

Further, Phoo-ngernkham et al. [17] reported that the monolithic mode of failure in the diagonal shear bond prisms and reported that the cracks were developed through the GPM and PCC substrate. The shear bond strength at the ITZ showed comparatively very high. Monolithic type of failure was observed in GPM; this is due to the high strength of GPM and its improved bonding behaviors. The PCC substrate was

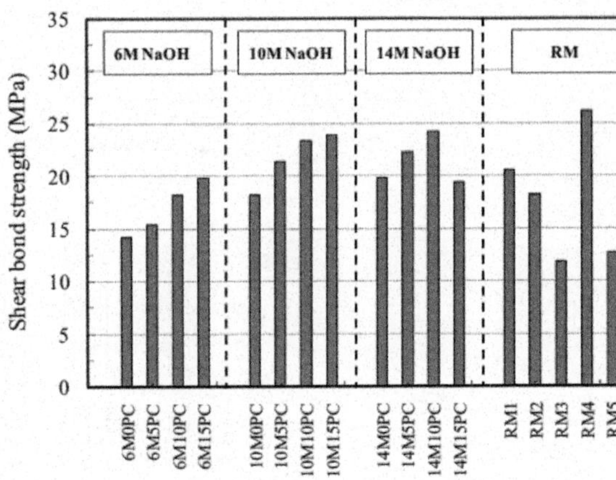

Fig. 6 Shear bond strength of repair mortars with 45° contact line [17]

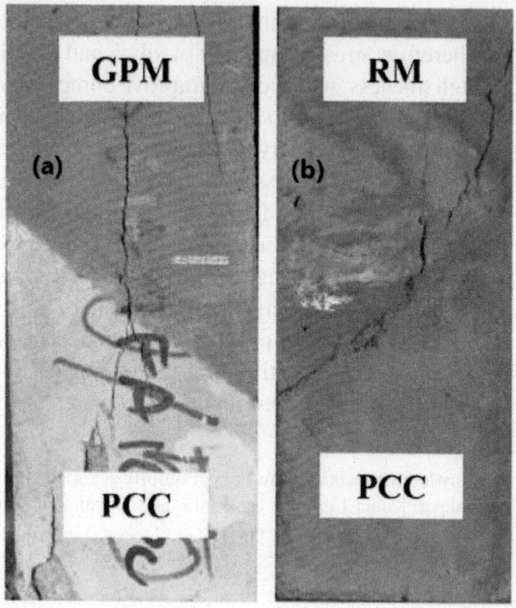

Fig. 7 Fracture mode between GPM and PCC substrate [17]

relatively intact in the case of RM specimen, but at the interface, RM showed signs of cracks and loss of bonding (Fig. 7).

3 Conclusions and Further Research

Based on the reviews, geopolymer composites hold significant potential for use as a construction material in various civil engineering applications. The various fundamental properties have been explored and comparable strengths behavior have been reported by many researchers. The production of geopolymer can be encouraged as it signifies one of the successful and novel product. A significant development is needed for the application of various supplementary cementitious materials to produce geopolymer products and it would be beneficial to gain more knowledge on microstructures, mechanical characteristics, bond strength between plain concrete and geopolymer product. Hence, the comprehensive overview of this chapter has highlighted the following significant key findings from this investigation.

- The addition of various supplementary cementitious materials in geopolymer products is not familiar and very few research works have been conducted.
- Bottom ash of higher fineness was found as more reactive than the coarser ashes and produced geopolymer mortars with higher strengths. Bottom ash can be grounded to a fine particle and become more effective and thus would enhance the binder reactivity. The improvement in strength could be greater for the different

ground materials as the fineness increases with enhanced chemical compositions. Therefore, greater amounts of silica and alumina can be dissolved by BA with a high fineness, which could improve compressive strength and geopolymerization.

- The supplementary source material or other additives materials are required to enhance the reactivity of binders derived from various by-product sources. Moreover, the effectiveness and reactivity of the binder depends on alkali activators in geopolymerization. The NaOH and Na_2SiO_3 solution could be used effectively as alkali activator for the production of geopolymer. The shear bond strengths improved with the higher NaOH concentration, and it was specifically due to the development of reaction products.
- Overall, this chapter proposes the appropriate potential research materials and mix details to produce structural grade geopolymer incorporating various industrial and agro-based by-products as binders.

Acknowledgements The authors gratefully acknowledge the support received from the University of Malaya, Kuala Lumpur, Malaysia, and Dayananda Sagar College of Engineering, Bangalore, Karnataka, India for this work.

References

1. Abdel-Gawwad, H., & Abo-El-Enein, S. (2014). A Novel Method to Produce Dry Geopolymer Cement Powder. *HBRC Journal., 12*, 13–24.
2. Sherif, Z., Sarfraz, S., Jolly, M., Salonitis, K. J. M. (2025). A critical review of the decarbonisation potential in the Uk cement industry. *18*, pp. 292.
3. Ranjbar, N., Mehrali, M., Alengaram, U. J., Metselaar, H. S. C., & Jumaat, M. Z. (2014). Compressive strength and microstructural analysis of fly Ash/Palm oil fuel ash based geopolymer mortar under elevated temperatures. *Construction and Building Materials, 65*, 114–121.
4. Sumesh, M., Alengaram, U. J., Nayaka, R. R. (2018) Effect of binder content and water-binder ratio in mortar developed using partial replacement of cement with palm oil clinker powder. *In IOP Conference Series: Materials Science and Engineering. 431*, pp 082007. IOP Publishing.
5. Foong, K. Y., Alengaram, U. J., Jumaat, M. Z., & Mo, K. H. (2015). Enhancement of the mechanical properties of lightweight Oil Palm shell concrete using rice Husk Ash and manufactured sand. *Journal of Zhejiang University Science A, 16*, 59–69.
6. Alengaram, U. J., Mohottige, N. H. W., Wu, C., Jumaat, M. Z., Poh, Y. S., & Wang, Z. (2016). Response of oil palm shell concrete slabs subjected to Quasi-Static and blast loads. *Construction and Building Materials, 116*, 391–402.
7. Liu, M. Y. J., Chua, C. P., Alengaram, U. J., Jumaat, M. Z. (2014) Utilization of palm oil fuel ash as binder in lightweight oil palm shell geopolymer concrete. *Advances in Materials Science and Engineering. 2014*, Article ID 610274, Wiley
8. Bakharev, T. (2005). Durability of geopolymer materials in sodium and magnesium sulfate solutions. *Cement and Concrete Research, 35*, 1233–1246.
9. Chindaprasirt, P., Jaturapitakkul, C., Chalee, W., & Rattanasak, U. (2009). Comparative study on the characteristics of fly ash and bottom ash geopolymers. *Waste Management, 29*, 539–543.
10. Ranjbar, N., Mehrali, M., Behnia, A., Alengaram, U. J., & Jumaat, M. Z. (2014). Compressive strength and microstructural analysis of Fly Ash/Palm oil fuel ash based geopolymer Mortar. *Materials & Design, 59*, 532–539.

11. Islam, A., Alengaram, U. J., Jumaat, M. Z., & Bashar, I. I. (2014). The development of compressive strength of ground granulated blast furnace slag-palm oil fuel Ash-Fly ash based geopolymer mortar. *Materials & Design, 56*, 833–841.
12. Sata, V., Tangpagasit, J., Jaturapitakkul, C., & Chindaprasirt, P. (2012). Effect of W/B ratios on pozzolanic reaction of biomass ashes in portland cement matrix. *Cement and Concrete Composites, 34*, 94–100.
13. Mijarsh, M., Johari, M. M., & Ahmad, Z. A. (2015). Compressive strength of treated palm oil fuel ash based geopolymer mortar containing calcium hydroxide, aluminum hydroxide and silica fume as mineral additives. *Cement and Concrete Composites, 60*, 65–81.
14. Chindaprasirt, P., Thaiwitcharoen, S., Kaewpirom, S., & Rattanasak, U. (2013). Controlling ettringite formation in FBC Fly Ash geopolymer concrete. *Cement and Concrete Composites, 41*, 24–28.
15. Alsubari, B., Shafigh, P., & Jumaat, M. Z. (2015). Development of Self-Consolidating high strength concrete incorporating treated Palm oil fuel Ash. *Materials, 8*, 2154–2173.
16. Hasan, K., Roy, S., Yahaya, F. M. (2024) An experimental and numerical investigation of using palm oil fuel ash as a partial cement replacement in steel fiber-reinforced concrete. *Innovative Infrastructure Solutions, 9*, pp 361. Springer
17. Phoo-ngernkham, T., Sata, V., Hanjitsuwan, S., Ridtirud, C., Hatanaka, S., & Chindaprasirt, P. (2015). High calcium fly ash geopolymer mortar containing portland cement for use as repair material. *Construction and Building Materials, 98*, 482–488.

Evaluation of Geopolymer Concrete Exposed to Elevated Temperatures: A Review of Microstructure and Physical–Mechanical Characteristics

Shimol Philip and M. Nidhi

Abstract The service life of concrete structures has been severely impacted by the significant rise in fire risk and the growing frequency of fire accidents. Despite the widespread perception that ordinary Portland cement (OPC) concrete is naturally fire-resistant, its strength and durability are affected by changes in its physical and chemical characteristics when exposed to elevated temperatures. There is a chance of cracking and spalling, which can result in financial losses and extreme conditions, even resulting in casualties. Additionally, manufacturing concrete adversely affects the environment by depleting natural resources and increasing the amount of CO_2 released into the atmosphere. Researchers worldwide acknowledge the need for an eco-friendly concrete substitute. Geopolymer concrete (GPC) is gaining attention as a sustainable alternative to OPC concrete. However, it is necessary to assess its performance at elevated temperatures to develop heat-resistant GPC structural elements. Therefore, this chapter reviews the physical, mechanical and microstructural properties of GPC when exposed to elevated temperatures. The key parameters analysed in this study include visual inspection, weight loss, cracking behaviour, volume shrinkage, spalling resistance, strength properties, bond strength, dynamic modulus of elasticity and microstructural properties. The literature review revealed that GPC possesses superior thermal resistance compared to OPC concrete under high-temperature exposure. However, precise mix design is necessary for GPC to obtain significant chemical stability, strength endurance, low volume shrinkage and spalling resistance. This study demonstrates that the high-temperature behaviour of GPC is greatly influenced by factors such as the selection of precursor materials, aggregate properties, fibre reinforcement, the characteristics of the alkali activator, curing conditions, etc. Therefore, these factors must be selected carefully to achieve optimal performance of GPC in elevated heat conditions. The chapter concludes with recommendations for future investigations on the fire resilience characteristics

S. Philip · M. Nidhi (✉)
Department of Civil Engineering, National Institute of Technology Puducherry, Karaikal, India
e-mail: nidhi.m@nitpy.ac.in

S. Philip
e-mail: shimolphilip16@gmail.com

P. Danish et al. (eds.), *Innovative Technologies for Smart and Sustainable Infrastructure*,
https://doi.org/10.1007/978-3-031-92801-7_19

of GPC, which are essential for the sustainable development of civil engineering structures.

Keywords Elevated temperature · Geopolymer concrete · Weight loss · Strength reduction · Spalling resistance

1 Introduction

Fire safety is an essential consideration in building planning and design, as buildings may be exposed to elevated heat levels in regular operation or the event of a catastrophe [1]. Fire is widely recognized as one of the most destructive factors causing damage to civil infrastructures. Although the frequency of fire accidents in a building may be relatively low, they can lead to substantial negative impacts on concrete infrastructures. The effect of fire exposure can severely degrade the physicomechanical properties of concrete [2]. Due to the risk of fires, the building industry has been driven to search for innovative materials with enhanced resistance to fire suitable for use in the construction sector to safeguard the stability of structural elements during a severe fire.

OPC is the most conventional and widely utilized binding material in concrete manufacturing. OPC-based concrete is widely regarded as an exceptionally fire-resistant construction material. However, numerous recent investigations on concrete made with OPC have revealed significant deterioration at high temperatures [3]. OPC-based concrete demonstrates relatively good fire resistance up to a certain temperature threshold. However, as the heat, flame or temperature progressively increases, it results in damages such as a decline in mass and strength, pore expansion, fracture propagation and spalling. The strength loss of conventional concrete is usually insignificant, up to 200 °C, with only slight changes to its **microstructure.** However, there is a higher chance of strength degradation, microstructural deterioration and spalling in conventional concrete after 300 °C. When concrete is exposed to intense heat, it causes the mixture to decompose, altering its morphology and initiating thermal cracks, which lead to surface deformation. The dehydration and destruction of calcium silicate hydrate (C-S–H) gel and related crystalline hydrates are the principal contributors to conventional OPC concrete's failure under extreme temperatures. Moreover, the phase composition of OPC consists of compounds that lack mineral polymers, which can lead to durability issues, particularly upon exposure to high heat [3]. Furthermore, a substantial quantity of energy is used in the manufacture of conventional concrete, which has a major impact on global warming and indirectly contributes to environmental pollution. The production of one tonne of OPC is estimated to emit the same amount of CO_2 into the environment. Therefore, adopting innovative alternative cementitious materials is essential for mitigating global warming issues.

Recent studies found that GPC could be a potential solution for enhancing the heat-resistant characteristics of concrete under extreme heat conditions [4]. Geopolymer has been identified as the most suitable eco-friendly alternative to cement, as it incorporates materials from industrial by-products as source materials, which are activated by alkaline solutions. GPC production utilizes industrial by-products or agricultural residues as precursors [5] and eliminates the use of cement. Another critical factor affecting GPC production is the alkaline activator solution, typically a combination of sodium or potassium-based silicates and hydroxides. The geopolymer binder significantly lowers greenhouse gas (CO_2) emissions and energy consumption during manufacture by utilizing industrial by-products effectively. In addition to the superior mechanical and durability characteristics, another essential characteristic of GPC is its resistance to temperature exposure. The spalling resistance and stability of GPC are affected by several parameters, including the mix proportioning, the precursor material used, the molarity of the alkaline solution and the alkaline solution to binder proportion [6].

Therefore, one of the most essential prerequisites for building design is the implementation of suitable fire safety measures to guarantee the security of its occupants. An in-depth review of the existing literature is necessary to assess the performance of GPCs subjected to extreme temperatures and fire scenarios. This chapter examines the characteristics of GPC exposed to elevated thermal environments, focusing on its physical, mechanical and microstructure characteristics. Several promising research directions are also proposed for future research to encourage broader implementation of GPC in advanced construction practices. This chapter aims to enhance understanding and optimize the performance of geopolymers in high-temperature conditions by analysing their properties under extreme conditions.

2 Overview of Geopolymer Concrete

Geopolymer is formed by reacting silica- and alumina-rich source material with an alkaline solution, resulting in a three-dimensional Si–O–Al network structure. Due to its low greenhouse gas emissions and minimal energy requirements for manufacturing, geopolymers have attracted considerable attention from researchers [7]. The formation of aluminosilicate oligomers occurs initially during the geopolymerization process. Afterwards, the final three-dimensional structure, Na_2O–Al_2O_3–SiO_2–H_2O (N-A-S–H), is produced by polycondensation between these oligomers [8]. Geopolymers address the challenges associated with OPC production and manage industrial waste by recycling it. Furthermore, their development and efficient utilization could result in an 80% decline in greenhouse gas emissions. The selection of an alkaline activator, the aluminosilicate source material and the curing conditions used all affect GPC performance [9]. Fly ash (FA), sugarcane bagasse ash (SBA), alccofine, metakaolin (MK), ground granulated blast furnace slag (GGBS), palm oil fuel ash (POFA), silica fume, and rice husk ash (RHA) are examples of source materials from industrial processes or agricultural residues containing abundant silica and alumina, suitable for efficient utilization to produce geopolymer. Also, the most frequently used alkali activator is sodium or potassium hydroxide mixture and the

Fig. 1 Materials and process flow for geopolymer concrete production

corresponding silicate [10]. The higher early strength, quicker setting time, enhanced durability, increased resistance to fire and acid, reduced thermal conductivity and less shrinkage are just a few of the advantages of geopolymer over OPC [11]. Figure 1 illustrates the schematic representation of the composition and formulation process of GPC.

3 Impact of Elevated Temperatures on GPC: A Comprehensive Review

A thorough understanding of the inherent properties and performance of GPC under extreme heat exposure is essential, as GPC has the potential to enhance fire resistance in building applications. The present study summarizes the physical, mechanical and microstructure characteristics of GPC subjected to extreme heat conditions. Moreover, the effect of precursors and other factors influencing the performance of GPC exposed to extreme heat environments is presented. The high-temperature behaviour of GPC can be optimized by carefully selecting and optimizing its constituents, thereby promoting the formation of a microstructure with the necessary thermal and mechanical properties. The GPC produced from different precursors and the tests performed on GPCs to determine their elevated temperature performance at varying temperatures, as reported in numerous studies, are compiled and presented in Table 1.

Table 1 Summary of studies on GPC performance under high-temperature conditions

Ref	Precursors used	Fibre used	Temperature range	Tests performed
Matalkah et al. [12]	Wheat straw ash, MK and coal FA	–	100 °C–800 °C	$f_{c,r}$
Nuaklong et al. [13]	High calcium FA, nano-silica and RHA	–	821 °C, 925 °C, 986 °C and 1029 °C	Visual inspection, spalling, mass loss, $f_{c,r}$
Rahmati & Toufigh [14]	GGBS and fly ash	–	100 °C–1200 °C	f_c
Kanagaraj et al. [15]	GGBS and FA	–	821 °C, 925 °C, 986 °C and 1029 °C	Physical appearance, mass loss, crack pattern, spalling, image analysis, analytical modelling
Prasad et al. [16]	GGBS and FA	SF, PF, BF, SF + PF, SF + BF	821 °C, 925 °C, 986 °C and 1029 °C	$f_{c,r}$, bond strength, crack pattern, SEM, EDAX, image analysis
Tanu and Unnikrishnan [17]	GGBS and SBA	–	100 °C–800 °C	Mass loss, physical appearance, $f_{c,r}$, SEM, EDAX, XRD, FTIR
Liu et al. [18]	MK, glass powder and RHA	–	300 °C–900 °C	Physical appearance, crack pattern, volume shrinkage, f_c, SEM
Kanagaraj et al. [4]	GGBS and FA	–	821 °C, 925 °C, 986 °C and 1029 °C	Crack pattern, mass loss, $f_{c,r}$, SEM
Raj et al. [19]	GGBS and dolomite	–	100 °C–700 °C	f_c, mass loss, UPV, dynamic modulus of elasticity, SEM, XRD, TGA, DTA, MIP, TEM, DSC, isothermal scanning calorimetry analysis
Hussin et al. [3]	Pulverized fuel ash, FA and POFA	–	200 °C–800 °C	Mass loss, crack pattern, f_c, XRD, FTIR, TGA
Guler and Akbulut [20]		–	200 °C–800 °C	Mass loss, physical appearance, crack pattern, f_c, flexural strength, SEM
Aygormez et al. [21]	MK, silica fume and colemanite	PP	300 °C–900 °C	Mass loss, physical appearance, crack pattern, f_c, flexural strength, SEM, XRD

(continued)

Table 1 (continued)

Ref	Precursors used	Fibre used	Temperature range	Tests performed
Kong & Sanjayan [22]	FA	–	ambient temperature to 800 °C	Mass loss, crack pattern, f_c, spalling
Yasaswini and Rao [23]	FA	–	200 °C	f_c and flexural strength
Saavedra and Gutierrez [24]	FA and GGBS	–	25 °C–1100 °C	f_c, crack pattern, physical appearance, DTG analysis, pore volume, SEM, XRD
Shaikh and Vimonsatit [25]	FA	–	200 °C–800 °C	f_c, physical appearance
Albidah et al. [26]	MK	–	200 °C– 600 °C	f_c, physical appearance, weight loss, water absorption, Residual modulus of elasticity, SEM, EDS
Zhang et al. [27]	FA, MK, nano-silica	–	ambient temperature to 700 °C	f_c, split tensile strength, sorptivity test, spalling, XRD
Chidhambar & Manjunath [28]	GGBS and SBA	–	200 °C–800 °C	f_c, split tensile and flexural strength, weight loss
Zhang et al. [29]	FA	–	20 °C–1000 °C	Mass loss, crushing index of coarse aggregate, image J analysis, f_c, crack pattern, SEM, TGA

3.1 Physical Characteristics

Visual Inspection. Visual inspection of GPC post-exposure to high temperatures is one method for analyzing the effects of thermal exposure on its properties. When GPC samples are exposed to elevated heat, the moisture content present in the samples evaporates, leading to colour changes and the formation of cracks. Numerous factors, including the aggregate types, fibres, alkaline activator, heat transfer rate, cooling technique and heating duration, significantly impact the concrete's appearance and surface colour variations [16].

SF (steel fibre); PP (polypropylene fibre); BF (basalt fibre); DTA (differential thermal analysis); SEM (scanning electron microscopy); XRD (X-ray diffraction); DSC (differential scanning calorimetry); FTIR (Fourier transform infrared spectroscopy); MIP (mercury intrusion porosimetry); TEM (transmission electron microscopy); UPV (ultrasonic pulse velocity); DTG (differential thermogravimetry);

EDS (energy dispersive spectroscopy); EDAX (energy dispersive X-ray analysis); TGA (thermogravimetric analysis); fc,r (residual compressive strength); fc (compressive strength).

Nuaklong et al. [13] found that the surface colour of GPC made from high-calcium fly ash, nano-silica and RHA changed from dark brown to red when upon exposure to extreme temperatures of approximately 800–1000 °C after 30 min. Iron oxide content in the fly ash is responsible for the colour variation under thermal exposure. A comparable colour variation is reported by Zhang et al. [29] in their investigation on the behaviour f low-calcium fly ash-based GPC (FA-GPC) at elevated heat levels. In another study, Saavedra and Gutierrez [24] also reported a colour change from black to red in fly ash-incorporated alkali-activated concrete (AAC) when exposed to 1100 °C. Similarly, Guler and Akbulut [20] found that fibre-reinforced fly ash-based geopolymer mortar changes its surface colour from dark grey at room temperature to lighter shades of brown and red at high temperatures, especially after 800 °C. Contrastingly, Kanagaraj et al. [15] found that GPC made from fly ash and GGBS showed a significant colour change from grey at ambient temperature to light brown when exposed to temperatures above 1000 °C, especially when the GGBS content was lower. However, when an equal proportion of GGBS and fly ash was used, no notable colour variation was detected on the exterior of the GPC specimens.

Liu et al. [18] observed that the exterior colour of metakaolin-based GPC changed to a black texture when heated above 500 °C. This colour variation is due to the calcination of the surface of the GPC samples due to limited oxygen availability. Aygormez et al. [21] reported that geopolymer mortar made from metakaolin and a binary combination of metakaolin with silica fume and colemanite did not show a notable colour change when exposed to 300 °C. However, a colour change was visible at 600 °C, and after 900 °C, the colour change became more significant. According to Albidah et al. [26], the metakaolin-based GPC experienced a little decolourization when exposed to 200 °C, and the colour changed from orange to white-orange at this temperature. However, upon heating to 400 °C and 600 °C, the GPC specimen became grey.

Weight Loss. GPC often experiences a mass loss when concrete is exposed to intense heat. The mass loss in GPC could result from the moisture loss caused by evaporation from the pores and the subsequent breakdown of the binder media. GPC specimens' weight loss is calculated based on the difference in mass before and after thermal exposure. This process may cause crack formation, which could even result in explosive spalling [30]. According to past studies, the weight reduction in GPC samples up to 200 °C is attributed to free water loss from the pores and, to some extent, the release of chemically bound water resulting from geopolymer gel dehydration. Dehydration of the geopolymer gel and de-hydroxylation of the hydroxyl groups were responsible for the weight loss under heating conditions exceeding 200 °C. GPC specimens with high strength generally experience higher mass loss compared to GPC specimens with low strength when exposed to elevated temperatures [15].

The precursor combinations used for producing GPC significantly impact the loss of mass in GPC under elevated heat conditions. For instance, binary GPC developed from GGBS and dolomite exhibits higher mass loss than ternary GPC developed from fly ash, GGBS and dolomite at high-temperature exposure. The weight loss in GPC under elevated heat, especially above 300 °C, is due to the decomposition of its constituents. Notably, the reaction products from slag are more susceptible to decomposition at elevated temperatures (>300 °C) than those formed from fly ash [19]. The experimental outcomes reported by Hussin et al. [3] show that the mass loss of GPC made from POFA and pulverized fuel ash (PFA) at 800 °C is much lower than that of concrete produced with OPC at the same temperature. This higher mass loss of concrete made with OPC is due to the reaction of CH present in it at elevated temperatures. This reaction induces stresses in the OPC samples, forming cracks and surface scaling in the concrete.

The water-to-binder ratio (W/B) is an essential factor affecting the weight loss of GPC samples under elevated thermal conditions. Kanagaraj et al. [4] reported that GPC with a lower W/B exhibit smaller weight loss at elevated temperatures than GPC samples with a higher W/B. An increased W/B indicates the presence of more water relative to the binder. When GPC samples with a high W/B value are exposed to elevated heat conditions, the water in the concrete turns into steam. As steam generation increases, the internal pressure within the GPC samples rises, exerting greater stress on the interfacial transition zone (ITZ). This results in the expansion of the ITZ, making it less effective at transferring loads. Consequently, the strength of the GPC samples decreases, leading to weight loss. The sodium silicate-to-sodium hydroxide (SS/SH) ratio also affects the weight loss percentage of GPC exposed to extreme heat conditions. This study by Albidah et al. [26] suggested that a decrease in the SS/SH ratio of metakaolin-based GPC increased the weight loss percentage. The curing conditions influence the mass deterioration of GPC samples exposed to high-temperature conditions. The test results of Zhang et al. [29] on low-calcium FA-GPC revealed that samples cured at room temperature exhibited more significant weight loss than those cured under heating conditions.

The incorporation of fibres significantly impacts the percentage decrease in mass of GPC specimens after being subjected to intense heat. The weight loss of fibre-reinforced GPC is generally lower than that of GPC without fibre reinforcement when exposed to extreme thermal conditions. According to Prasad et al. [16], basalt fibre-reinforced GPC (GPC-BF) made from GGBS and fly ash exhibits lower weight loss when exposed to elevated temperatures than polypropylene and steel fibre-reinforced GPC. Moreover, steel fibre-reinforced GPC (GPC-SF) and hybrid steel and basalt fibre-reinforced GPC also exhibited lower weight loss than GPC without fibre reinforcement. Rapid mass reduction in polypropylene fibre-reinforced GPC (GPC-PF) may be caused by the fibre's strong decomposition-induced reinforcing effect, driven by the fibre-matrix interfacial reaction, which results in significant fibre degradation. This implies that the interfacial fibre and the geopolymer matrix have a relatively weak connection. As a result, as the temperature rises, the geopolymer composites dehydrate, causing weight loss, brittle cracks and microstructural damage. Similar conclusions were drawn by Güler and Akbulut in their study on fly ash-derived geopolymer mortar reinforced with glass and basalt fibres [20]. The mass loss of GPC reinforced with glass and basalt fibres showed a lower mass loss compared to

GPC without fibres. Moreover, glass fibre-reinforced GPC exhibited a slightly lower mass loss than GPC-BF. This occurs because glass fibres melt at high temperatures, opening up more passageways than basalt fibres to remove bound and free water from the matrix. The free and chemically bound water present in the geopolymer matrix can exit the matrix through these passage channels, significantly decreasing any possible internal pore pressures. Aygörmez et al. [21] found that the loss in mass percentage of metakaolin-based geopolymer mortar incorporating polypropylene fibres was higher than that of geopolymer mortar without reinforcement. A dehydration reaction arises when the temperature rises, causing moisture to migrate towards the sample surface and exit. This damages the microstructure internally and, as a result, causes the geopolymer composite to lose weight. The free and structured water causes the geopolymer sample to lose weight rapidly during the early heating stages. The variation in weight loss of GPC from different source materials when exposed to extreme temperatures, as reported in various studies, is illustrated in Fig. 2.

Cracking Behaviour. The damage to GPC caused by being subjected to high temperatures can be assessed by examining its surface, with the analysis of heat-induced cracks serving as an approach to measure the extent of the damage. Numerous factors, such as the curing conditions, properties of aggregates and the W/B value, influence the crack width formation of GPC under extreme heat exposure. For instance, GPC samples developed from GGBS and SBA, upon exposure to 400 °C, showed no signs of cracking on their surface [17]. However, when the temperature reached 600 °C, hairline cracks developed on the samples' surface, and the GPC samples' colour became slightly darker. This indicates that the decomposition and debonding of components in the concrete are significant during this phase of heating. GPC specimens characterized by a W/B value of 0.4 exhibit relatively wider cracks compared

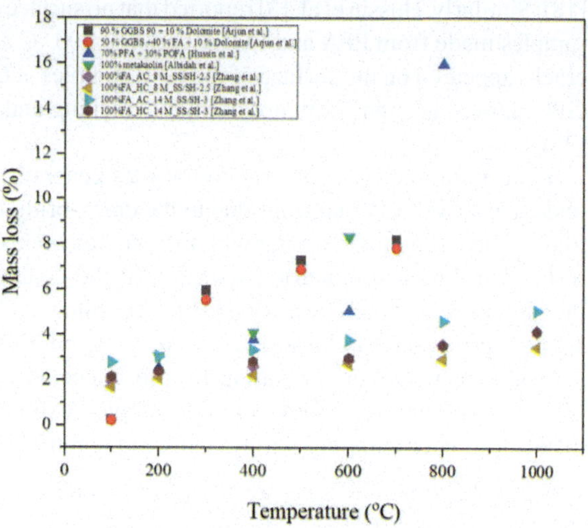

Fig. 2 Variation in GPC weight loss at high temperatures across different source materials [3, 19, 26, 29]

Fig. 3 Crack development on GPC surface after high-temperature exposure [15]

to GPC specimens possessing a W/B value of 0.6 [4]. Kanagaraj et al. [15] found that after exposure to 821 °C, small hairline cracks appeared on the surface of GPC developed from fly ash and GGBS, as depicted in Fig. 3. The cracks widened with increasing temperature and exposure time. The cracks became more pronounced at 925 °C, with their spread clearly visible at 1029 °C. The failure was attributed to internal pore water pressure and cracks that developed at the hot surface due to the specimens' internal temperature gradient. Longer heating times were shown to increase crack width, highlighting the adverse effects of high temperatures on GPC.

Saavedra and Gutierrez [24] compared the crack patterns in fly ash-based AAC and OPC concrete. According to this study, the concrete samples exhibited visible cracking at 500 °C. The differing thermal expansion properties of the aggregate materials and the matrix caused these surface cracks. Moreover, OPC concrete did not exhibit any cracks on its surface; instead, spalling and fissures were observed on the edges of the concrete at 1100°. GPC developed from metakaolin showed slight shrinkage and cracking between 300 °C and 700 °C temperatures. The crack pattern enlarged significantly when the temperature reached 900 °C. Moreover, when glass powder or RHA were incorporated into the metakaolin-based GPC, up to 20% replacement, T-shaped shrinkage and cracking were observed on the surface at 900 °C [18]. Similarly, Hussin et al. [3] reported that no surface cracks were found on the GPC samples made from PFA and POFA between 200 °C and 400 °C. However, hairline cracks appeared on the surface of the GPC samples at 600 °C. The metakaolin-based GPC showed no signs of major or minor cracking under heat exposure up to 600 °C [26].

Fibre-reinforced geopolymer composites generally exhibit high thermal stability under elevated heat conditions due to the crack-bridging capability of heat-resistant fibres. The strong interlocking with the concrete matrix and the heat-resistant fibres resulted in lower volumetric expansion of the concrete and reduced dehydration, thereby reducing crack formation [16]. According to Guler and Akbulut, large-small spongy spaces and capillary cracks were found on the surface of the fibre-reinforced geopolymer mortar incorporating fly ash, especially after 800 °C. However, none of the fibre-reinforced GPC mortar samples exhibited fragmentation, breakage, disintegration or chipping during high-temperature exposure [20].

Volume Shrinkage. GPC exposed to elevated temperatures may experience volume shrinkage due to bound water dehydration and the geopolymer matrix's densification. Liu et al. found that the volume shrinkage of metakaolin-based GPC exhibited approximately the same volume shrinkage at 500 °C and 700 °C, which was lower than that at 300 °C. This occurs as a result of pore collapse induced by the evaporation of bound water between 250 °C and 800 °C, with the most significant pore collapse occurring predominantly at 500 °C. However, the volume shrinkage rate reached its maximum at 900 °C. Additionally, metakaolin-based GPC with 10% glass powder and metakaolin-based GPC with 10% RHA exhibited lower volume shrinkage, indicating that adding RHA and glass powder up to 10% improved the volume stability of the GPC samples. This could be due to the dissolved SiO_2 in the RHA and glass powder, which exhibit high reactivity and contribute to the filling effect.

Spalling Resistance. Concrete may experience spalling due to fire action in addition to internal cracking. Concrete spalling is caused by a sudden increase in vapour pressure inside the concrete driven by fire heat; however, this tendency will be reduced if the pressure can be quickly released into the atmosphere. Another aspect that contributes to concrete spalling is the aggregate's restraining impact. Since the matrix and aggregate are bonded, any shrinkage caused by the surrounding aggregates will result in internal stress in the aggregate. The pop-outs of concrete elements were caused by the combined influence of rising pore pressure and thermal stress. The size of the remaining heat-damaged samples is generally used to evaluate the concrete's resistance to spalling. Concrete with more permeability has less spalling since the vapour pressure in the composite is lowered. Hence, moisture can escape more quickly through the porous structure than the denser one [13]. The exposure of GPC specimens to higher temperatures results in the development of thermal compressive stresses on the surface of the concrete specimen, along with tensile stresses. If the combined stresses surpass the concrete's tensile strength, fragments of the concrete may detach from the primary specimen [15]. Figure 4 illustrates the schematic representation of the stresses that develop within and around the concrete specimen during temperature exposure.

Zhang et al. [27] reported that GPC samples from fly ash and metakaolin exhibited better spalling resistance than OPC concrete. In this study, only one high-strength GPC mix experienced spalling around 500 °C. The permeability and pore volume of GPC is increased by the sintering reaction that takes place at about 700 ^{0}C. In GPC, the well-connected pore network and its exceptional strength retention during exposure to high temperatures help lower the spalling chance. Kanagaraj et al. [15] observed no spalling in GPC specimens developed from fly ash and GGBS exposed to thermal conditions ranging from 821 °C to 1029 °C. This tendency indicates that the developed GPC showed superior crack and spalling resistance. Aggregate size substantially influences the spalling characteristics of GPC under extreme temperature conditions. According to Kong and Sanjayan [22], FA-GPC that employs small-sized aggregates, especially those below 10 mm, increases spalling and extensive cracking under high-heat conditions. Moreover, GPC incorporating large-sized aggregates, specifically those above 10 mm, was more stable under extreme heat conditions. The

Fig. 4 Stress profile in concrete specimens subjected to elevated temperatures

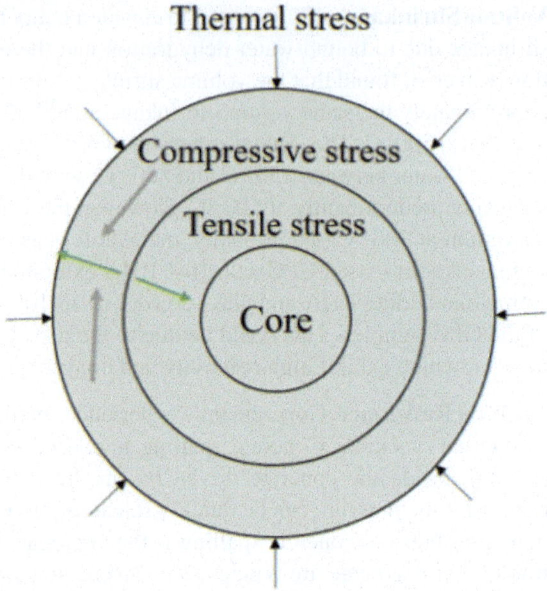

aggregate controls the matrix shrinkage, which is more rigid than the geopolymer paste. Around the aggregate particles, this restraining influence results in tangential and radial stress. Due to this restraining effect, the aggregate particles are subject to tangential and radial stresses. These stresses are enhanced by the shrinkage imposed by moisture loss, which causes large cracks to form. Concretes with smaller aggregates exhibit this shrinkage trend. The absence of irregularly shaped smaller aggregates restricts the creation of efficient aggregate bridging. Contrastingly, according to Nuaklong et al. [13], the aggregate type has a lesser impact on the spalling behaviour of GPC when exposed to elevated thermal conditions. Albidah et al. [26] investigated the effect of varying Na_2O/Al_2O_3 and SiO_2/Al_2O_3 molar ratios on metakaolin-based GPC tested under thermal conditions ranging from 200 °C to 600 °C. The test results revealed that none of the GPC samples exhibited corner breakage or spalling.

3.2 Mechanical Characteristics

Compressive Strength (f_c). Residual compressive strength ($f_{c,r}$) is generally used to measure the performance of GPC subjected to extreme heat. It indicates the strength measured after being exposed to extreme heat. Elevated temperature exposure alters the physical and chemical characteristics of GPC samples, leading to a loss in strength. Numerous factors, such as temperature exposure, binder composition, the molarity of the alkali activator, curing conditions, alkali activator-to-binder ratio, fibre reinforcement, concrete age, etc., affect the $f_{c,r}$. The molarity of GPC samples significantly influenced the $f_{c,r}$. Yasaswini and Rao [23] investigated the effect of three molarities (12, 16 and 20 M) on the strength behaviour of GPC developed from a blend of GGBS and fly ash at 200 °C. The results revealed that concrete exhibited higher f_c at 16 molar than other molarities. Moreover, GPC with an SS/SH ratio of 2 exhibited higher f_c than GPC with a 2.5 ratio. The f_c of FA-GPC with high molarity of sodium hydroxide solutions (NaOH) was higher than that of low molar NaOH solutions at all elevated temperatures. Interestingly, it has been shown that GPC with a high molar NaOH solution also exhibits a high rate of f_c loss at high temperatures. This may be explained by the high sodium oxide content of the high molar NaOH solution, which enhances the alumina-silicate networks during the polymerization reaction and raises the f_c [25]. According to Tanu and Unnikrishnan [17], GPC samples from GGBS and SBA retained strength at higher molarities, especially at 12 M, compared to 8 M and 10 M.

Similarly, the binder combination significantly affected the $f_{c,r}$. As the SBA addition increased, the f_c reduction increased at elevated temperatures [25]. The drop in strength of GPC under elevated heat exposure is due to the debonding of materials. Similarly, Raj et al. [19] reported that adding fly ash to GGBS- and dolomite-based GPC improved the f_c. The production of sulfoaluminate gel is the main factor responsible for the greater f_c observed in the GGBS-fly ash-dolomite (GFD) combinations exposed to elevated temperatures. The production of this gel significantly improved the mechanical characteristics in conjunction with the C-S–H and C-A-S–H gels already present in the GFD mix. The performance of binary blended GPC and OPC concrete was compared at elevated temperatures in a study by Hussin et al. [3]. The binary GPC exhibited a rise in f_c up to 400 °C. However, the OPC concrete exhibited maximum f_c at 200 °C. This study indicates that GPC has a more stable and durable matrix than conventional cement concrete at elevated temperatures. The low diffusion coefficient of Na + at high temperatures, which raises the melting temperature of the geopolymer, is the reason for an increase in the f_c of blended ash GPC exposed to temperatures rising up to 400 °C. Saavedra and Gutierrez [24] compared the f_c performance after high-temperature exposure of FA-GGBS and FA-OPC-based AAC with OPC concrete. All mixtures lose their f_c, and the drop was significant after 700 °C. The breakdown of the C-S–H gel and the decarbonation of the calcite, particularly the combustion of the residual carbon in FA, can be linked to the loss of f_c in alkaline-activated concrete. This leads to the formation of pores in the material, reducing its mechanical performance. Unlike OPC concrete, which drastically loses strength at elevated temperatures, AAC, particularly formulated with GGBS and fly ash, demonstrated significantly better f_c retention. The development of new crystalline phases and decreased porosity in the alkaline-activated concrete, which support their structural integrity under heat stress, are responsible for this improved performance.

According to Albidah et al. [26], the f_c reduction in metakaolin-based GPC was higher when heated from ambient conditions to 200 °C. However, when the GPC specimens were heated from 200 °C to 400 °C and from 400 °C to 600 °C, the f_c loss was lower compared to the loss observed up to 200 °C. The strength loss is primarily attributed to the loss of internal and chemically bound water from the GPC samples due to evaporation, which occurs mainly in the 100–300 °C temperature range. Beyond this range, evaporation diminishes, and consequently, the strength reduction caused by evaporation decreases. This study also highlighted the impact of the Na_2O/Al_2O_3 proportion on the $f_{c,r}$. According to this study, a higher Na_2O/Al_2O_3 ratio decreased $f_{c,r}$ when heated to 400 °C. However, this ratio did not significantly influence the $f_{c,r}$ when heated to 600 °C. Liu et al. [18] tested the elevated temperature performance of metakaolin-based GPC and metakaolin-based GPC incorporated with glass powder and RHA. The study revealed that glass powder-incorporated metakaolin GPC exhibited higher f_c than other GPC mixes up to 700 °C. The rise of SiO_2/Al_2O_3 and SiO_2/Na_2O due to the addition of glass powder may be responsible for increased f_c. GPC developed from coal fly ash, wheat straw ash and metakaolin exhibited a rise in f_c up to 500 °C, and the strength reduced after 600 °C [12]. This reduction in strength after 600 °C occurs because of the varying thermal expansion between limestone aggregates and the geopolymer binder. The higher levels of crystalline compounds at elevated temperatures also lead to a reduction in strength. For the temperature range 700 °C and 800 °C, unreacted silicates present in the geopolymer undergo swelling and sintering, which results in a harmful expansion that damages the geopolymer paste. The variation in f_c of GPC under high-temperature exposure, developed from various source materials, is illustrated in Fig. 5. From the graph, it is clear that the f_c of GPC decreases as the temperature rises for most of the precursors, while some exhibit an increase in f_c until reaching 600 °C.

The W/B ratio substantially influences the f_c of GPC subjected to high-temperature environments. For instance, GPC developed from fly ash and GGBS with a W/B value of 0.4 demonstrated a lower fc reduction than GPC samples with a W/B value 0.6[4]. Curing conditions are another factor influencing the f_c of GPC under elevated thermal exposure. Low-calcium FA-GPC is thermally stable after exposure to 600 °C for 2 h. For low-calcium FA-GPC, the f_c increases up to 600 °C. Heat-cured GPC samples exhibit higher strength than ambient-cured GPC samples after exposure to extreme heat conditions. Moreover, normal-strength FA-GPC has shown higher

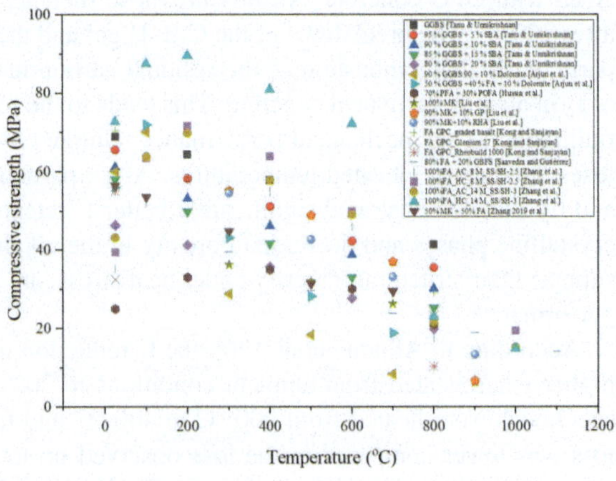

Fig. 5 Variation of $\mathbf{f_c}$ of GPC at elevated temperatures from different source materials [3, 17–19, 22, 24, 27, 29]

f_c growth and lower strength reduction than high-strength GPC [29]. Yasaswini and Rao [23] observed that GPC specimens cured at 60 °C exhibited higher f_c than those cured at 70 °C when exposed to extreme heat conditions. Similarly, Shaikh and Vimonsatit [25] found that FA-GPC with longer heat-curing durations exhibited higher f_c when exposed to high temperature conditions. This increase in f_c was attributed to the additional geopolymerization process that occurred in the FA-GPC during the extended heat-curing periods. This study also highlighted the influence of extra water on f_c under high heat conditions. Adding extra water resulted in a decline in f_c due to the formation of weak polymerization products in the GPC caused by excess water. Dash et al. [31] performed an experimental evaluation of the performance of GGBS-based GPC in high-heat environments. This study also developed a machine-learning prediction model for f_c at elevated temperatures and performed a sensitivity analysis. The results highlighted that exposure temperature and curing days significantly influenced the f_c. According to the study by Rahmati and Toufigh [14], the $f_{c,r}$ of GPC made from GGBS and fly ash is significantly affected by the time and temperature of exposure to heat. Furthermore, among the properties of concrete, sodium silicate (Na_2SiO_3) and curing time significantly impacted the $f_{c,r}$ of GPC following exposure to high-heat conditions. The heating duration is another factor that substantially affects the $f_{c,r}$. The damage to GPC increases with prolonged heating exposure. According to a study by Nuaklong et al. [13], fly ash and RHA-incorporated GPC specimens exhibited a reduction in $f_{c,r}$ by 73–74% of their ambient temperature strength after 30 min of exposure to elevated temperatures. Moreover, the GPC samples completely decomposed after 90 min of heat exposure at 1000 °C. This phenomenon can be attributed to the rising shrinkage of the matrix, which elevates the thermal stresses surrounding the aggregates and initiates the formation of internal cracks.

The aggregate size also correlates with the f_c of GPC exposed to extreme heat conditions. The crushing index serves as a parameter to evaluate the strength of coarse aggregates. Elevated temperature exposure affects the crushing index as coarse aggregates expand under elevated temperature conditions. Zhang et al. [29] reported that the threshold value for the crushing index of coarse aggregates in low-calcium-based fly ash GPC is 7.7%, indicating that the aggregates can retain their strength up to this percentage. However, this study shows that beyond 600 °C, the temperature gradient induces cracks in the coarse aggregates, leading to strength loss in concrete. The strength of the FA-GPC with 20-mm coarse aggregates was significantly lower than that of the concrete with 10-mm coarse aggregates at elevated temperatures. The delayed development of microcracks in the ITZ of the 10-mm coarse aggregate GPC may also be the reason for its marginally higher f_c than that of the 20-mm aggregate GPC [25]. Kong and Sanjayan [22] investigated the effect of coarse aggregate size on the f_c of FA-GPC under extreme high-temperature conditions. According to this study, GPC specimens with higher incompatibility exhibited more significant strength loss. The test results showed that GPC incorporated with slag-based coarse aggregate exhibited a more significant reduction in f_c than basalt-based coarse aggregate GPC under elevated heat conditions. The reduction in f_c of GPC specimens at high temperatures is most likely caused by the thermal incompatibility of the aggregate components and the geopolymer matrix. The aggregates' thermal expansion rate governs the high-temperature behaviour of GPC. This study also investigated the elevated temperature performance of FA-GPC using two different superplasticizers, namely Glenium 2 and Rheobuild 1000. The test results demonstrated that the reduction in f_c for GPC with superplasticizers was higher than that for GPC without superplasticizers when exposed to extreme temperatures.

Fibre addition significantly influences the $f_{c,r}$ under extreme temperature conditions. GPC-BF exhibits higher $f_{c,r}$ than GPC-SF and GPC-PF [16]. This increase in the f_c of GPC-BF can result from silica and lime in basalt fibre. The proportion of calcium compared to silicon in GPC is the influencing factor behind the improved C-(A)-S–H, C-S–H and N-(A)-S–H gels, which results in the formation of dense structure and raises the f_c of GPC-BF. The graph in Fig. 6 illustrates the impact of fibre inclusion on the $f_{c,r}$ of GPC under elevated thermal exposure. Guler and Akbulut [20] also reported the impact of fibres on the $f_{c,r}$ in their investigation of fly ash-derived geopolymer mortar reinforced with glass and basalt fibres. According to this study, glass fibre-reinforced geopolymer mortar exhibited slightly higher $f_{c,r}$ than basalt fibre-reinforced geopolymer mortar. This is primarily caused by the fact that glass fibres are more agglomerated in the matrix than basalt fibres, which are unable to distribute uniformly in three dimensions and effectively control cracks. Additionally, geopolymer mortar developed with a hybrid of 12 mm and 24 mm glass fibres, or 12 mm and 24 mm basalt fibres, exhibited higher $f_{c,r}$ than geopolymer mortar reinforced with individual glass or basalt fibres having a length of 24 mm. However, Aygormez et al. [21] reported that the incorporation of 0.8% by volume of polypropylene fibres into geopolymer mortar made from metakaolin, as well as binary blends of metakaolin with silica fume and metakaolin with colemanite, resulted in a decrease in f_c under extreme heat. The low density of polypropylene fibres and their flocculation, which result in reduced unit weight, lower f_c, and higher water absorption, are the two leading causes of the loss in f_c. This led to the observation that adding polypropylene fibre reduced the f_c values in all mixes.

Split Tensile Strength. The split tensile strength of GPC also decreased when exposed to extreme temperatures, similar to the reduction in f_c. However, the reduction of tensile strength is more pronounced than that of f_c. This is associated with the crack formation. According to Zhang et al. [27], the change in the splitting tensile strength of GPC developed from fly ash and metakaolin follows a trend similar to that of f_c when exposed to extreme heat conditions. However, the degradation rate of split tensile strength between 500 °C and 700 °C is higher than the reduction in f_c. This study also highlighted that beyond 300 °C, the degradation rate of split tensile strength in GPC is lower than that of OPC concrete. Chidhambar and Manjunath [28] found that GPC made from GGBS and SBA exhibited an increase in splitting tensile strength as the NaOH molarity increased.

Flexural Strength. The concrete structure exposed to extreme thermal conditions not only loses its f_c and split tensile strength but also lowers its flexural resistance. Generally, greater f_c in GPC structures corresponds to increased flexural resistance.

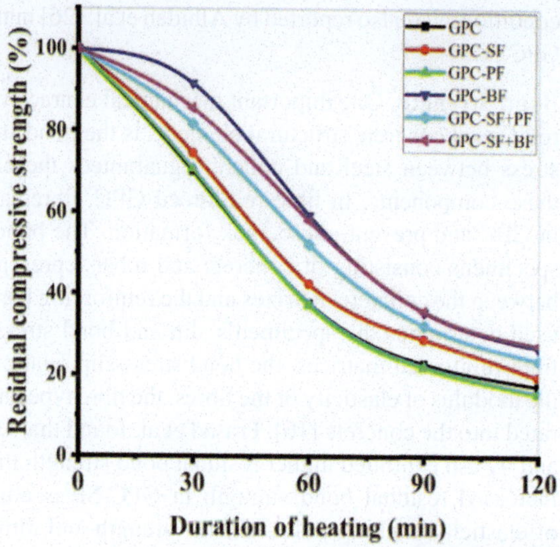

Fig. 6 Effect of different fibres on the $f_{c,r}$ [16]

Moreover, a reduction in flexural resistance is associated with an early deterioration in the structure's durability [32]. The experimental results by Guler and Akbulut [20] demonstrated that geopolymer mortar with fibre reinforcement exhibited higher residual flexural strength than mortar without fibre reinforcement. Moreover, glass fibre is more effective than basalt fibre in resisting flexural loads at elevated temperatures. The findings presented by Aygormez et al. [21] indicated that the decline in the flexural strength of geopolymer mortar made from metakaolin, as well as binary combinations of metakaolin with silica fume and colemanite, was more significant than the reduction in f_c when exposed to extreme heat. This resulted from the flexural strength being more susceptible to internal microstructure flaws, such as fracture propagation and the formation of porous structures. Yasaswini and Rao [23] found that GPC with a higher SS/SH ratio exhibited higher flexural strength when tested at 200 °C.

Dynamic Modulus of Elasticity (DME). DME of GPC is a key mechanical property that defines its stiffness when subjected to dynamic or oscillatory loads. The type and proportion of binders, alkali activators, aggregate properties, curing conditions, concrete age, presence of fibres, microstructure, porosity, etc., influence the DME of GPC. An increase in temperature adversely impacts the DME of GPC specimens [19]. The investigation on GPC made from GGBS, fly ash and dolomite indicated that, compared to the f_c of GPC, the reduction in DME was more prominent at elevated temperatures. The rapid degradation of the DME at 100 °C suggests that high temperatures caused significant damage to the geopolymer matrix. The leading cause of the sudden drop in the DME after 100 °C was the lower resistance of the C–S–H and C-A-S–H gels made from the slag under high heat conditions. The higher rate of dehydration of the modulus of elasticity compared to f_c under high heat conditions was also reported by Albidah et al. [26] in their study on metakaolin-based GPC.

Bond Strength. One important mechanical characteristic considered while building reinforced concrete structural elements is the bond stress of the concrete. This bond stress between steel and concrete guarantees the effective load transfer between these components. In fibre-reinforced GPC, fibres aid in strengthening the matrix in GPC and prevent microcrack formation. The bond stress-slip behaviour in bond specimens consisting of concrete and rebar represents the extent of stress transfer between the concrete matrixes and the reinforcing bar. The type of steel and concrete used determines the specimen's slip and bond stress relationship. Furthermore, in fibre-reinforced matrices, the bond stress-slip relationship is primarily governed by the modulus of elasticity of the fibres, the fibre type and the number of fibres incorporated into the concrete [16]. Prasad et al. found that GPC-SF developed from GGBS and fly ash exhibited higher residual bond strength than GPC-BF and GPC-PP. This increased residual bond strength in GPC-SF is attributed to the higher modulus of elasticity, melting point, tensile strength and ultimate elongation of steel fibres compared to those of basalt and polypropylene fibres.

3.3 *Microstructural Properties*

The variation in the mechanical characteristics of GPC is attributed to the variations in its microstructure. Therefore, it is essential to analyse the the microstructural changes in GPC exposed to elevated heat to provide insight into its thermal stability and other thermo-mechanical properties. Numerous studies have employed SEM analysis to understand the changes in the microstructural properties occurring during high-temperature exposure [3, 15, 18, 21, 33–35]. Microstructural studies on GGBS-SBA-based GPC revealed significant degradation at elevated thermal conditions, including the formation of interface gaps and cracks due to the debonding of materials, particularly at 800 °C. Additionally, the decomposition of the C-A-S–H gel was observed, which is critical for the material's strength, alongside changes in elemental composition confirmed by EDS and XRD analyses [17]. GPC produced with metakaolin showed no damage to its microstructure after exposure to 300 °C. After 500 °C, the GPC specimen became more porous due to mass loss, decomposition of the geopolymer matrix and phase transformation occurring beyond 500 °C. The porosity of the GPC specimen declined at 600 °C compared to the corresponding porosity level at 300 °C. Silica-rich gels are produced at high temperatures, resulting in reduced pore volume. This increases the densification of the concrete and causes it to collapse, especially between 600 °C and 750 °C [21]. Hussin et al. [3] pointed out that calcium-rich gels, like C-S–H, have adverse effects on the high-temperature behaviour of OPC concrete. On the other hand, N-A-S–H gel at an equivalent concentration had a relatively lower impact on geopolymer material structure under high heat exposure.

4 Comparison of OPC and GPC Concrete at High Temperatures

Despite being commonly used, conventional concrete frequently lacks sufficient fire protection, which could result in property and human casualties. OPC concrete has comparatively good heat resistance to a certain degree, but as the temperature, flame, or heat progressively increases, damages such as strength reduction, mass reduction, pore expansion, fracture propagation and spalling happen. Numerous studies have shown that the high-temperature behaviour of concrete made with OPC is inferior to that of GPC. The $f_{c,r}$ test results obtained by Matalkah et al. [12] for OPC concrete exposed to extreme thermal conditions are illustrated in Fig. 7. The internal structural damage and the strength loss are usually insignificant up to 200 ^{0C}. However, the risk of strength degradation, microstructural damage and spalling increases after 300 ^{0}C. Concrete deterioration at high temperatures is caused by three main factors: the increase of internal vapour pressure, the breakdown of C-S–H and the thermal incompatibility of the concrete mix components. It appears that the ITZ, the weakest link in the chain of ingredients in concrete, is more susceptible to heat cracks. At very high temperatures, the binder may shrink while the fillers often expand. The physical, chemical and hardening characteristics of concrete are all harmed by this incompatibility, which affects the ITZ. Furthermore, the building sector is susceptible to many environmental sustainability-related issues. The huge amount of Portland cement consumption in traditional concrete production contributes significantly to CO_2 emissions.

Geopolymers are inorganic binders formed by the chemical reaction between pozzolanic or aluminosilicate substances and alkaline solutions. Due to their ceramic-like characteristics, geopolymers exhibit exceptional fire resilience [22]. Davidovits has demonstrated that an alkaline activator consisting of potassium and sodium can

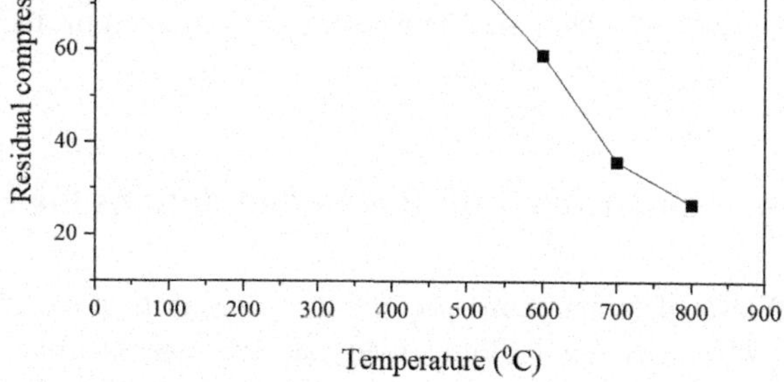

Fig. 7 Variation of residual compressive strength percentage with temperature for OPC

achieve thermal resistance up to 1200 °C [36]. The binding matrix of GPC is rich in alumina and silica, which allows it to perform exceptionally well at high temperatures. At temperatures between 600 to 800 °C, the geopolymer matrix retains thermal stability while maintaining a significant portion of its initial strength. Because of its compact microstructure and decreased porosity and thermal expansion, GPC is less likely to crack and flake on the surface. It is also highly resistant to heat-induced damage because of its excellent thermal stability and lower free water content, which makes it the perfect material for high-temperature and fire-resistant applications. Thus, GPC is a sustainable alternative to OPC concrete because of its remarkable heat-resistant and environmentally beneficial features.

5 Challenges in Evaluating GPC Performance

The following challenges exist for comprehensively investigating GPC performance at elevated temperatures.

- Currently, no proven solutions exist to counteract the detrimental effects of high temperatures on GPC, and no reliable design prediction models are available.
- The spalling behaviour of GPC is still not completely understood due to its complex nature. There is currently no standardized testing methodology or universal consensus on the spalling mechanism.
- The microstructure of GPC changes significantly at high temperatures, including phase transitions, cracking and dehydration. Because of the intricate relationships between different components of GPC, it is difficult to accurately analyze these changes and their effects on mechanical properties.
- Measuring thermal conductivity, specific heat and thermal diffusivity during heating and cooling cycles is challenging. These properties depend on temperature and need precise experimental setups and advanced equipment.
- The wide range of binder compositions seen in GPC formulations (such as slag, fly ash and metakaolin) makes it challenging to generalize study results. Because each type of binder has a different thermal behaviour, every formulation requires extensive testing.
- There are no standardized testing procedures for assessing GPC performance at high temperatures. Differences in heating rates, maximum temperatures and cooling techniques between investigations limit the comparability and repeatability of research findings.

6 Conclusion and Recommendations for Future Research

Elevated temperature performance is a critical aspect to be taken into account when selecting building materials for construction. This study investigated the elevated temperature behaviour of GPC, and the results indicate that GPC has shown promising results. GPC shows enhanced behaviour under extreme temperature conditions, and the reduction in f_c is modest compared to OPC-based concrete. However, its effectiveness may differ depending on the material types employed and the selected mix design. The difference in characteristics of GPC under elevated thermal conditions arises from the intrinsic nature of the geopolymer binder and the components included in its formulation. The literature review on GPC provides the following conclusions about its performance under elevated temperatures.

- GPC demonstrates exceptional durability and resistance when subjected to harsh environmental conditions and can endure temperatures from 800 °C to 1000 °C. This heat resistance makes the GPC a more reliable alternative to traditional concrete in high-temperature applications.
- The decrease in GPC weight under high-temperature exposure directly correlates with the exposure temperature. GPC with high strength experiences more significant weight loss than GPC with lower strength. Moreover, GPC cured under ambient conditions experiences more substantial weight loss than those cured under heat curing.
- Visual inspection of GPC following exposure to high temperatures indicates that surface cracks and discolouration are essential indicators of thermal degradation. The type of aggregate, fibre inclusion, alkaline activator, heating period, cooling method and raw material composition affect the colour change of GPC specimens under high-temperature conditions.
- The loss in mass of GPC under high-temperature conditions is mainly caused by the breakdown of reaction by-products, drying out of the geopolymer matrix and evaporation of pore moisture. The degree of weight reduction depends on several factors, including the composition of the raw materials, the binder-to-water ratio, the curing techniques and the fibre reinforcement. Fibres, especially steel and basalt fibres, improve GPC's heat resistance by decreasing mass loss and microstructural degradation, making fibre-reinforced GPC a superior choice for high-temperature applications.
- Surface cracks, impacted by aggregate characteristics, curing processes and the W/B ratio, can be used to assess the degree of damage in GPC subjected to high temperatures. Fibre-reinforced GPC is a more stable choice for high-temperature applications because of its exceptional heat resistance, which reduces crack formation and structural deterioration.
- The bound water dehydration and matrix densification affect GPC's volume shrinkage at high temperatures. The additions like glass powder and RHA increase volume stability by improving reactivity and the filling effect.
- Compared to OPC concrete, GPC exhibits superior resistance to spalling. This is because of its interconnected pore structure and superior mechanical property retention at elevated temperatures. Larger aggregates increase structural stability, while smaller aggregates increase the risk of cracking and material spalling during excessive heat exposure. Aggregate size is a critical factor in spalling characteristics.

- The factors, including the concentration of the NaOH solution, binder composition and exposure conditions, significantly impact the $f_{c,r}$ of GPC; high molarity of NaOH solutions increases initial strength but also causes strength loss at higher temperatures.
- The split tensile and flexural strength of GPC exposed to extreme temperatures also decreases, similar to the f_c loss. However, the degradation rate for these strengths is higher than that of the corresponding f_c.
- The bond strength of GPC with fibre reinforcement plays a crucial role in ensuring effective load transfer between concrete and reinforcement. The fibre selection highly contributes to the high-temperature performance of GPC. Specifically, GPC reinforced with steel fibre demonstrates superior residual bond strength compared to other fibre types.
- The microstructural changes during high-temperature exposure are responsible for the variation in GPC's mechanical properties. The thermal stability and strength of GPC are directly impacted by significant degradation, including the formation of cracks and the breakdown of critical gel phases.

Further research is needed to analyze the effects of novel materials, such as nanomaterials, fibres and waste-based additives, on improving the performance of GPC at elevated temperatures. Limited studies have examined the prediction modelling of GPC exposed to extreme thermal conditions. This study recommends advanced machine learning models incorporating SHAP (Shapley Additive Explanations) analysis to predict the properties of GPC under extreme thermal conditions and to identify the factors influencing the predicted properties. The long-term performance of GPC during repeated heat cycles should be investigated in future studies to assess the material's behaviour over prolonged exposure times. The impact of temperature cycling on GPC stability in practical applications could also be examined. Studies could also evaluate the role of various parameters, such as source material type, the molarity of alkaline activator, SS/SH ratio, alkaline activator to binder ratio and heating duration, on the behaviour of GPC exposed to extreme thermal conditions.

References

1. Kanagaraj, B., Anand, N., Jerry, R., et al. (2023). Axial compressive behaviour and physical characteristics of high strength self-compacting geopolymer concrete (HSGC) columns exposed to elevated temperature. *Construction and Building Materials, 401*, 1–22. https://doi.org/10.1016/j.conbuildmat.2023.132866
2. Anand, N., Thanaraj, D. P., Andrushia, D., et al. (2023). Microstructure investigation, strength assessment, and thermal modelling of concrete exposed to different heating cooling regimes. *Journal of Thermal Analysis and Calorimetry, 148*, 3221–3247. https://doi.org/10.1007/s10973-023-11998-5
3. Hussin, M. W., Bhutta, M. A. R., Azreen, M., et al. (2015). Performance of blended ash geopolymer concrete at elevated temperatures. *Materials and Structures/Materiaux et Constructions, 48*, 709–720. https://doi.org/10.1617/s11527-014-0251-5

4. Kanagaraj, B., Anand, N., Andrushia, D., & Kodur, V. (2023). Residual properties of geopolymer concrete for post-fire evaluation of structures. *Materials, 16*, 1–22. https://doi.org/10.3390/ma16176065
5. Philip, S., Nidhi, M. (2023). A review on the material performance of geopolymer concrete as green building materials. *Mater Today Proc* 1–9. https://doi.org/10.1016/j.matpr.2023.04.110.
6. Kanagaraj, B., Anand, N., Diana Andrushia, A., et al. (2023). Pull-Out behavior and microstructure characteristics of binary blended self-compacting geopolymer concrete subjected to elevated temperature. *Alexandria Engineering Journal, 76*, 469–490. https://doi.org/10.1016/j.aej.2023.06.055
7. Saludung, A., Azeyanagi, T., Ogawa, Y., & Kawai, K. (2023). Mechanical and microstructural evolutions of fly ash/slag-based geopolymer at high temperatures: Effect of curing conditions. *Ceramics International, 49*, 2091–2101. https://doi.org/10.1016/j.ceramint.2022.09.175
8. Godani, P., Priya, T. S., & Alengaram, U. J. (2024). Evaluation of chemo-physio-mechanical characteristics of GGBS-MBS-based ready-mix geopolymer under ambient curing condition. *Construction and Building Materials, 449*, 1–14. https://doi.org/10.1016/j.conbuildmat.2024.138202
9. Nath, S. K., Maitra, S., Mukherjee, S., & Kumar, S. (2016). Microstructural and morphological evolution of fly ash based geopolymers. *Construction and Building Materials, 111*, 758–765. https://doi.org/10.1016/j.conbuildmat.2016.02.106
10. Li, J., Tao, Y., Zhuang, E., et al. (2022). Optimal amorphous oxide ratios and multifactor models for binary geopolymers from metakaolin blended with substantial sugarcane bagasse ash. *Journal of Cleaner Production, 377*, 1–15. https://doi.org/10.1016/j.jclepro.2022.134215
11. Reddy, S. R. K., & Murugan, S. B. (2020). Experimental and microstructural assessment of ternary blended geopolymer concrete with different Na2SiO3-to-NaOH volume ratios. *Innovative Infrastructure Solutions, 5*, 1–14. https://doi.org/10.1007/s41062-020-0279-z
12. Matalkah, F., Soroushian, P., Asce, F., et al. (2016). Characterization of Alkali-Activated Nonwood Biomass Ash-Based Geopolymer Concrete. *Journal of Materials in Civil Engineering, 29*, 1–9. https://doi.org/10.1061/(ASCE)MT.1943-5533
13. Nuaklong, P., Jongvivatsakul, P., Pothisiri, T., et al. (2020). Influence of rice husk ash on mechanical properties and fire resistance of recycled aggregate high-calcium fly ash geopolymer concrete. *Journal of Cleaner Production, 252*,. https://doi.org/10.1016/j.jclepro.2019.119797
14. Rahmati, M., & Toufigh, V. (2022). Evaluation of geopolymer concrete at high temperatures: An experimental study using machine learning. *Journal of Cleaner Production, 372*,. https://doi.org/10.1016/j.jclepro.2022.133608
15. Kanagaraj, B., Anand, N., Andrushia, A. D., & Lubloy, E. (2022). Investigation on engineering properties and micro-structure characteristics of low strength and high strength geopolymer composites subjected to standard temperature exposure. *Case Studies in Construction Materials, 17*, 1–20. https://doi.org/10.1016/j.cscm.2022.e01608
16. Vijaya Prasad, B., Anand, N., Kanagaraj, B., et al. (2023). Investigation on residual bond strength and microstructure characteristics of fiber-reinforced geopolymer concrete at elevated temperature. *Case Studies in Construction Materials, 19*, 1–24. https://doi.org/10.1016/j.cscm.2023.e02526
17. H.M T, Unnikrishnan S,. (2023). Durability and elevated temperature behaviour of geopolymer concrete developed with ground granulated blast furnace slag and sugarcane bagasse ash. *Journal of Building Pathology and Rehabilitation, 8*, 1–18. https://doi.org/10.1007/s41024-023-00354-7
18. Liu, H., Jing, W., Qin, L., et al. (2022). Thermal stability of geopolymer modified by different silicon source materials prepared from solid wastes. *Construction and Building Materials, 315*, 1–12. https://doi.org/10.1016/j.conbuildmat.2021.125709
19. Raj, P. K. A., Joshi, R., D S, et al. (2025). Enhancing elevated temperature performance of alkali activated slag-dolomite concrete with Fly ash. *Physics and Chemistry of the Earth, 138*, 1–19. https://doi.org/10.1016/j.pce.2025.103857

20. Guler, S., & Akbulut, Z. F. (2022). Effect of high-temperature on the behavior of single and hybrid glass and basalt fiber added geopolymer cement mortars. *Journal of Building Engineering, 57*, 1–12. https://doi.org/10.1016/j.jobe.2022.104809
21. Aygörmez, Y., Canpolat, O., Al-mashhadani, M. M., & Uysal, M. (2020). Elevated temperature, freezing-thawing and wetting-drying effects on polypropylene fiber reinforced metakaolin based geopolymer composites. *Construction and Building Materials, 235*, 1–32. https://doi.org/10.1016/j.conbuildmat.2019.117502
22. Kong, D. L. Y., & Sanjayan, J. G. (2010). Effect of elevated temperatures on geopolymer paste, mortar and concrete. *Cement and Concrete Research, 40*, 334–339. https://doi.org/10.1016/j.cemconres.2009.10.017
23. Yasaswini, K., & Venkateshwara Rao, A. (2020). Behaviour of geopolymer concrete at elevated temperature. *Mater Today Proc, 33*, 239–244. https://doi.org/10.1016/j.matpr.2020.03.833
24. Valencia Saavedra, W. G., & Mejía de Gutiérrez, R. (2017). Performance of geopolymer concrete composed of fly ash after exposure to elevated temperatures. *Construction and Building Materials, 154*, 229–235. https://doi.org/10.1016/j.conbuildmat.2017.07.208
25. Shaikh, F. U. A., & Vimonsatit, V. (2015). Compressive strength of fly-ash-based geopolymer concrete at elevated temperatures. *Fire and Materials, 39*, 174–188. https://doi.org/10.1002/fam.2240
26. Albidah, A., Alqarni, A. S., Abbas, H., et al. (2022). Behavior of Metakaolin-Based geopolymer concrete at ambient and elevated temperatures. *Construction and Building Materials, 317*, 1–12. https://doi.org/10.1016/j.conbuildmat.2021.125910
27. Zhang, H. Y., Qiu, G. H., Kodur, V., & Yuan, Z. S. (2020). Spalling behavior of metakaolin-fly ash based geopolymer concrete under elevated temperature exposure. *Cement and Concrete Composites, 106*,. https://doi.org/10.1016/j.cemconcomp.2019.103483
28. Manjunath DrYM, E. C. (2019). Experimental investigation on geopolymer concrete subjected to elevated temperature. *International Journal of Advance Research, Ideas and Innovations in Technology, 5*, 290–296.
29. Zhang, H., Li, L., Yuan, C., et al. (2020). Deterioration of ambient-cured and heat-cured fly ash geopolymer concrete by high temperature exposure and prediction of its residual compressive strength. *Construction and Building Materials, 262*, Article 120924. https://doi.org/10.1016/j.conbuildmat.2020.120924
30. Kanagaraj, B., Anand, N., Andrushia, D, et al. (2023). mechanical properties and microstructure characteristics of self-compacting concrete with different admixtures exposed to elevated temperatures. *Jordan Journal of Civil Engineering 17*:1–9. https://doi.org/10.14525/jjce.v17i1.01.
31. Kumar Dash, P., Kumar Parhi, S., Kumar Patro, S., & Panigrahi, R. (2023). Efficient machine learning algorithm with enhanced cat swarm optimization for prediction of compressive strength of GGBS-based geopolymer concrete at elevated temperature. *Construction and Building Materials, 400*,. https://doi.org/10.1016/j.conbuildmat.2023.132814
32. Rihan, M. A. M., Onchiri, R. O., Gathimba, N., & Sabuni, B. (2024). Effect of elevated temperature on the mechanical properties of geopolymer concrete: A critical review. *Discover Civil Engineering, 1*, 1–12. https://doi.org/10.1007/s44290-024-00028-4
33. Paul Thanaraj, D., Kiran, T., Kanagaraj, B., et al. (2023). Influence of heating-cooling regime on the engineering properties of structural concrete subjected to elevated temperature. *Buildings, 13*, 1–20. https://doi.org/10.3390/buildings13020295
34. Hakeem, I. Y., Amin, M., Agwa, I. S., et al. (2023). Using a combination of industrial and agricultural wastes to manufacture sustainable ultra-high-performance concrete. *Case Studies in Construction Materials, 19*, 1–14. https://doi.org/10.1016/j.cscm.2023.e02323
35. Mostafa, S. A., Tayeh, B. A., & Almeshal, I. (2022). Investigation the properties of sustainable ultra-high-performance basalt fibre self-compacting concrete incorporating nano agricultural waste under normal and elevated temperatures. *Case Studies in Construction Materials, 17*, Article e01453. https://doi.org/10.1016/j.cscm.2022.e01453
36. Davidovits Joseph. (2020). Geopolymer : chemistry and applications. Institut Géopolymère

Challenges Associated with Alkali-Activated Materials in Infrastructure Development

MD Ikramullah Khan, V. Vinayaka Ram, and Vipulkumar Ishvarbhai Patel

Abstract Alkali-activated binders (AABs) have gained significant research interest as promising alternative to ordinary Portland cement (OPC). Several technical challenges encountered during the manufacturing stage and processing of alkali activated materials (AAMs) such as variability in raw material sources, mix design complexities and the high alkalinity of activators continue to hinder their large-scale production as a viable replacement for OPC. The urgent need for alternatives is underscored by OPC's substantial environmental impact, contributing over 8% of global carbon dioxide emissions, a figure expected to increase with ongoing infrastructure development. Green binder alternatives, such as AAMs, are essential to conserving natural resources and reducing environmental emissions. The principles of AAMs are introduced in this chapter, along with a comparison and discussion of their characteristics. The present issues impeding the widespread adoption and extensive use of AAMs were examined in this analysis. It was found that AAMs could compete as competitive alternatives to OPC with the development of user-friendly activators with minimal embodied energy and carbon footprint.

Keywords Alkali activated materials · Ordinary portland cement alternative · CO_2 emissions · Setting time · Drying shrinkage · Sustainable binders · Green construction materials · Alkali activation

M. I. Khan · V. Vinayaka Ram (✉)
Department of Civil Engineering, Birla Institute of Technology and Science, Pilani, Hyderabad Campus, Hyderabad, Telangana, India
e-mail: vinayak@hyderabad.bits-pilani.ac.in

M. I. Khan
e-mail: p20200431@hyderabad.bits-pilani.ac.in

V. I. Patel
School of Computing, Engineering and Mathematical Sciences, La Trobe University, Bendigo, VIC, Australia
e-mail: v.patel@latrobe.edu.au

P. Danish et al. (eds.), *Innovative Technologies for Smart and Sustainable Infrastructure*,
https://doi.org/10.1007/978-3-031-92801-7_20

1 Introduction

Concrete, primarily composed of ordinary Portland cement (OPC), is one of the most widely used construction materials globally. In 2023 alone, approximately 4.1 billion tons of OPC was produced, with an expected annual growth rate of 3%. This vast production volume, however, has serious environmental repercussions, as the OPC manufacturing process is responsible for significant CO_2 emissions [1], contributing to as much as 5–6% of the global total each year [2–4]. The high emissions associated with OPC stem from several energy-intensive steps, including the de-acidification of limestone, the thermal activation of clinker raw materials through calcination at temperatures exceeding 1400 °C, fuel consumption in kilns, and power generation. Together, these processes emit an average of 0.8 tons of CO_2 per ton of OPC produced. In response, the government and the industry stakeholders have been actively exploring sustainable alternatives to OPC to mitigate these substantial pollution levels.

The degradation issues evident in existing OPC structures further highlight the need for alternative materials [5–7]. Structures constructed with OPC binders over past decades exhibit a range of durability concerns due to environmental exposures and material limitations. The primary causes of OPC based concrete degradation include chloride exposure, carbonation, freeze–thaw deterioration, chemical attacks from acids, salts, alkalis, or sulfates, alkali-aggregate reactions, surface abrasion, high-temperature effects, and structural overloads.

Alkali-activated materials (AAMs) have emerged as promising alternatives to the traditional cementitious materials [8–10]. The alkali activation process involves reacting aluminosilicate-rich solids in an alkaline environment, resulting in a hardened binder based on hydrous alkali aluminosilicate phases [11–13]. Often termed "geopolymers," AAMs that use low-calcium aluminosilicate sources, are characterized by an alkali aluminosilicate gel with a three-dimensional tetrahedral structure of silicon and aluminum, which offers resistance to water dissolution [14]. Notably, AAMs possess mechanical properties comparable to those of OPC, with superior strength and durability characteristics under certain conditions, such as resistance to chloride, acid, and sulfate attacks. Most importantly, AAMs have demonstrated the potential to reduce carbon dioxide emissions by 50–70% compared to traditional OPC concrete [15–17].

Another sustainability advantage of AAMs is their reliance on industrial by-products such as fly ash (FA) and slag as aluminosilicate precursors, which helps conserve natural resources like limestone. However, despite these environmental and durability benefits, several challenges hinder the widespread adoption of AAMs in commercial settings. Key issues include high drying shrinkage, rapid setting times, and handling difficulties associated with alkaline activators, which can complicate mixing, placement, and finishing processes.

AAMs can be categorized based on the type of aluminosilicate precursors used. One variant, alkali-activated slag (AAS), utilizes ground-granulated blast furnace slag (GGBFS) as its aluminosilicate source. The fundamental differences in chemistry between AAMs and OPC resulting from the unique mineral compositions and formation processes in AAMs yield distinct material properties, influencing their behavior in concrete applications [18, 19].

2 Research Significance

The primary aim of this chapter is to investigate the key limitations associated with AAMs and to explore possible approaches to address these issues. As interest in AAMs grows due to their potential for carbon footprint reduction, understanding their challenges in both fresh and hardened states becomes crucial for advancing their adoption. This review is designed to provide researchers, construction managers, and engineers with an in-depth technical assessment of the fundamental challenges that AAMs currently face. AAMs exhibit several limitations in their fresh properties, which impact their workability and applicability in construction. Specifically, these materials are known for their rapid setting times, which significantly limits the timeframe for placement and finishing. Unlike OPC, which typically offers a longer initial setting period, AAMs can begin to harden within minutes, depending on the type and concentration of alkaline activator used, as well as the prevailing temperature and humidity conditions. This rapid set behavior complicates the logistics of handling, transporting, and placing the AAMs, particularly for large-scale applications where delays in placement can result in poor surface finishes or compromised structural integrity.

In addition, AAMs tend to have limited flowability, as the high concentration of alkaline solutions required for activation can increase viscosity, reducing the ease with which the material can flow and fill the formwork. This restricted workability may lead to voids within the structure, resulting in weakness that could diminish durability over time. In the hardened state, AAMs present additional challenges. A key issue is limited volume stability, primarily attributed to high drying shrinkage rates. Drying shrinkage in AAMs often exceeds that observed in OPC-based systems due to the inherent composition of alkali-activated binders, which tend to have a higher water demand for activation. The internal drying shrinkage results in volumetric changes that can lead to cracking, reducing both durability and service life. This shrinkage issue is exacerbated under variable environmental conditions, making AAMs potentially less suitable for certain climates or applications without further modification. Furthermore, the pore structure of hardened AAMs, often is highly variable and influenced by the factors like alkali type and concentration, affects their susceptibility to environmental stressors such as freeze–thaw cycles and chemical attacks.

Addressing these issues is essential for developing AAMs into a more practical and viable alternative to OPC in construction. The significance of this research lies in its potential to inform industry stakeholders about both the benefits and the current technical limitations of AAMs. By examining factors like activator selection, mixing techniques, and the design of curing processes, this review aims to shed light on pathways for mitigating these drawbacks. Identifying solutions to improve the fresh and hardened properties of AAMs will be crucial for their successful implementation as a sustainable construction material.

3 Application of AAMs

The application of AAMs in civil engineering is in a nascent stage, yet it shows promising potential for sustainable infrastructure development. Researchers have conducted limited but insightful studies on the performance of precursors such as FA, metakaolin, and GGBFS when activated by concentrated alkaline hydroxide or silicate solutions. These precursors, which are rich in aluminosilicates, react with alkaline activators like sodium hydroxide (NaOH) or sodium silicate (Na_2SiO_3) to form complex, high-strength binding gels within the matrix. Studies, including that of Mehta et al. [20], have demonstrated the efficacy of alkali-activated low-calcium and high-calcium precursors as sources of amorphous silica and alumina. The reaction in AAM systems often produces N–A–S–H (sodium aluminosilicate hydrate) or C–A–S–H (calcium aluminosilicate hydrate) gels as shown in Fig. 1, both of which contribute to enhanced compressive strength and durability. In particular, N–A–S–H gel is common in low-calcium systems, providing a robust, amorphous gel network that resists environmental degradation. In high-calcium systems, C–A–S–H gel formation occurs, often resulting in denser and more compact microstructures that enhance mechanical performance and durability.

High-calcium FA, in particular, has been shown to accelerate early-age strength development in AAMs. This advantage arises due to the reactive calcium content in high-calcium FA, which promotes the rapid formation of C–A–S–H gel, facilitating an early increase in compressive strength. Previous studies have also evaluated the viability of incorporating industrial by-products such as GGBFS and FA as sustainable, cost-effective alternatives to traditional OPC [21–23]. Researchers, including Khan et al. [24], have shown that highly alkaline solutes, such as NaOH or Na_2SiO_3, significantly enhance the durability and mechanical properties of AAMs, making them suitable for a wide range of civil applications, including pavements, structural components, and precast elements.

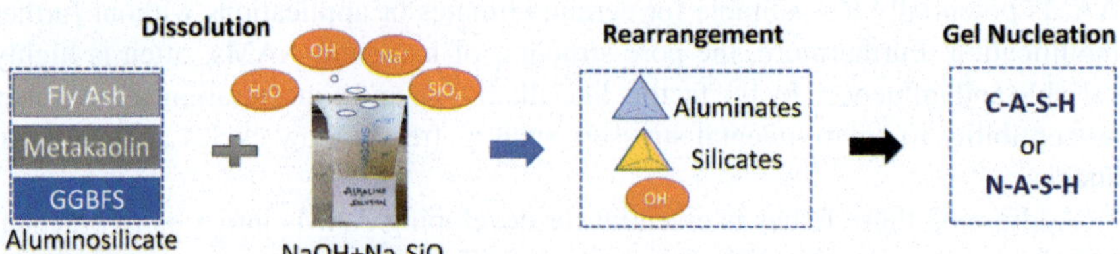

Fig. 1 AAMs reaction mechanisms

4 Challenges Associated with AAMs

4.1 Quick Setting Time

The setting time of AAMs is primarily influenced by two factors: the type of alkaline activator used and its concentration. Sodium silicate (Na_2SiO_3) is commonly employed as an activator but is known to induce a rapid, often challenging, setting time in AAMs [25]. Research indicates that Na_2SiO_3 activation can lead to flash setting [26], a phenomenon where the material hardens almost immediately upon mixings activated solely with Na_2SiO_3 show a significantly shorter setting time than those activated with sodium hydroxide (NaOH), which exhibits a more gradual reaction profile. When Na_2SiO_3 and NaOH are used in combination, the setting time is generally extended compared to Na_2SiO_3 alone, providing a more workable mix. This variation is attributed to the distinct reaction mechanisms of each activator. Na_2SiO_3 activation promotes the rapid formation of calcium silicate hydrate (C–S–H) phases, which accelerates the setting process [27, 28]. In contrast, NaOH tends to yield calcium carbonate and polysilicate hydrate phases, which result in a more controlled setting profile with an extended initial set time [29]. Elevated temperatures expedite the setting of AAMs, especially those activated with Na_2SiO_3, by enhancing the reaction kinetics and accelerating gel formation. Interestingly, however, the combined use of NaOH and Na_2SiO_3 as activators appears to be less sensitive to temperature variations, suggesting that different activators may have unique interactions with environmental conditions due to their distinct chemical compositions. These findings emphasize the importance of controlling activator types, proportions, and ambient conditions to optimize the setting characteristics of AAMs in various construction applications [30, 31].

4.2 High Curing Temperature

AAMs formulated with low-calcium aluminosilicate precursors, such as class F FA, often require elevated curing temperatures to accelerate the dissolution of silicate and aluminate monomers. This high-temperature requirement, generally above 70 °C, is crucial for achieving compressive strengths comparable to those of OPC [32, 33]. Elevated temperatures enhance the polymerization reaction rate, promoting the formation of aluminosilicate gels (such as N–A–S–H) that develop a dense and durable matrix necessary for strength development. However, the energy-intensive curing process adds significantly to the total embodied energy and carbon emissions associated with AAMs, limiting their environmental advantage over OPC and potentially offsetting the intended carbon reduction benefits using high-calcium precursors like GGBFS often do not require high-temperature curing to achieve comparable strength. The calcium content in slag accelerates the reaction even at ambient temperatures, facilitating the formation of calcium silicate hydrate (C–S–H) phases, which contribute to early-age strength gain. While this can be beneficial for reducing curing energy demand, slag-based AAMs are prone to significant drying shrinkage, which can lead to cracking and durability issues in large-scale applications [34]. Research indicates that elevated curing temperatures can mitigate shrinkage in slag-based AAMs by reducing free water and promoting early densification of the matrix. The high-temperature curing remains a practical constraint, especially in large-scale applications where controlled curing conditions are challenging. However, steam curing presents a feasible solution for precast applications, where production occurs in controlled factory environments. Despite the efficiency gains in precast settings, high-temperature curing still increases the embodied energy of AAMs. This factor has driven ongoing research to identify alternative activator formulations or curing regimes that can minimize temperature reliance, thus optimizing both energy consumption and emissions in AAM production [35].

4.3 High Drying Shrinkage

High drying shrinkage is a significant limitation of AAMs, especially those activated with Na_2SiO_3 and utilizing slag as the primary aluminosilicate precursor. Studies have shown that the Na_2SiO_3-activated slag-based AAMs shrink more than traditional OPC concrete. This pronounced shrinkage in slag-based AAMs is primarily due to the formation of a silica-rich gel, which results from the high concentration of silicate in the activating solution. This gel structure retains large amounts of moisture that, upon drying, leads to notable moisture loss, shrinkage, and ultimately microcracking within the matrix [34, 36]. This shrinkage creates microstructural weaknesses, potentially reducing the material's durability and leading to crack propagation under environmental stresses. One of the primary factors contributing to this shrinkage is the higher mesoporosity of AAMs relative to OPC concrete [37]. The mesoporous structure—containing pore sizes in the range of 2–50 nm—enhances moisture retention and evaporative drying, leading to a larger volumetric change as water exits the system. Consequently, AAMs tend to develop shrinkage-induced micro-cracks more readily than OPC concrete, which compromises their structural integrity over time, especially in environments where exposure to cyclic moisture variations is frequent. To address these shrinkage issues, air-entraining agents (AEAs) have shown promise in stabilizing AAMs [24]. AEAs, such as MASTERAIR 721, contain surfactant molecules with dual functionalities: a hydrophilic end and a hydrophobic end. During the mixing process, AEA molecules disperse throughout the material, and the hydrophobic ends adsorb at the interface of microscopic air bubbles. This adsorption stabilizes the bubbles by lowering the surface tension around them, creating a consistent distribution of air voids within the hardened AAM matrix. A controlled air content of 4–7% is considered optimal, as it helps to offset shrinkage by providing micro-scale "relief spaces" within the matrix that can accommodate internal stresses from drying.

4.4 Environmentally Unfriendly Activators

The environmental impact and logistical limitations of commonly used alkaline activators, specifically Na_2SiO_3 and NaOH, pose substantial challenges to the sustainable application of AAMs [15, 38–41]. While these activators facilitate the geopolymerization process, they are also associated with high embodied energy, substantial carbon emissions, and elevated costs. The embodied energy in Na_2SiO_3 and NaOH is notably high due to their intensive production processes: Na_2SiO_3, for example, is produced by melting sand and soda ash at high temperatures, while NaOH is created through energy-demanding electrochemical methods, often utilizing fossil fuel-based power sources. Consequently, the carbon footprint of AAMs activated with these compounds approaches or even exceeds that of traditional OPC-based materials, undermining their environmental advantages [16, 42]. The cost factor is another substantial limitation. Na_2SiO_3 and NaOH, though effective, are considerably more expensive than the binders used in OPC production, making AAMs less economically viable for widespread use. The higher material costs, coupled with specific requirements for safe handling and storage due to the caustic and corrosive nature of these chemicals, further deter construction industries from adopting AAMs in construction. Sodium hydroxide, for instance, is highly corrosive and requires specialized storage and transport facilities to ensure worker safety and compliance with environmental regulations, significantly increasing operational expenses.

4.5 Long-Term Durability and Performance

The durability of concrete is closely linked to its permeability and, by extension, its water absorption rate. For AAMs, water absorption levels are particularly relevant because they influence susceptibility to moisture-driven deterioration. Some studies report that AAMs exhibit higher water absorption rates than OPC concrete, potentially compromising durability under prolonged exposure to aggressive environments like freeze–thaw cycles and chemical attacks. This is largely attributed to the distinct microstructure of AAMs, which includes a network of interconnected pores that allows moisture ingress [43, 44]. On the contrary, certain studies have found that AAMs can resist harmful environmental factors, such as sulfate and chloride attacks, due to their denser gel matrix and reduced calcium content, which mitigates expansive reactions within the material [45, 46]. One contentious aspect affecting AAM durability is the alkali-aggregate reaction (AAR), which traditionally occurs in high-alkali OPC concrete due to reactive aggregates. In OPC concrete, AAR is mitigated by limiting the equivalent sodium oxide content to 0.8%. However, AAMs typically contain higher levels of alkalis due to their use of sodium-based activators, which raises concerns about potential AAR risks in AAM systems. Paradoxically, some studies suggest that the alkalis in AAMs become chemically bound within the hydration products, particularly in the form of sodium aluminosilicate and calcium aluminosilicate hydrates. This chemical binding reduces the free alkali concentration in the pore solution, theoretically mitigating the potential for AAR. Yet, experimental evidence on AAR mitigation in AAMs remains inconclusive, as factors such as precursor composition, type of activator, and curing conditions vary widely across studies [47, 48].

4.6 Local Market for Raw Ingredients

The local market for raw ingredients in alkali-activated materials (AAMs) significantly influences their viability and scalability across regions. The variability in aluminosilicate precursors, which are essential for AAM production, depends on local industrial by-products and geological resources. Unlike OPC, which predominantly relies on abundant limestone, AAMs require aluminosilicate sources like fly ash, blast furnace slag, metakaolin, and rice husk ash. Fly ash, a by-product of coal combustion, is readily available in regions with coal-fired power plants, such as the United States, China, and India. Conversely, in countries where coal-fired power generation is scarce, like many in Europe and Canada, the limited availability of fly ash poses a significant hurdle to AAM production. In addition, blast furnace slag, another key aluminosilicate precursor, is primarily available in regions with large steel industries. For example, countries with well-established steel production, like South Korea, Japan, and Germany, have an advantage in slag availability, whereas regions without a robust steel industry may face difficulties in sourcing this material locally. Similarly, rice husk ash, abundant in agricultural regions like India and Southeast Asia, has limited availability in colder climates where rice cultivation is less common. These regional differences in raw material availability result in supply-chain challenges and cost implications, as transporting aluminosilicate materials over long distances to production sites can reduce the environmental and economic benefits of AAMs. The situation is further complicated by the varying availability and cost of activators like NaOH and Na_2SiO_3, which are essential for the alkali-activation process. These activators are often industrially produced and require specialized handling due to their corrosive properties, making them relatively costly and regionally inconsistent. For instance, regions with well-established chemical industries produce and distribute NaOH and Na_2SiO_3 more economically, while areas with limited chemical manufacturing infrastructure face higher costs and logistical challenges in sourcing activators.

4.7 No Standard Codes for AAMs

The absence of standardized codes for AAMs presents a significant challenge, especially when relying on established concrete codes that were developed for traditional OPC systems. One example of the discrepancy is in carbonation testing, where accelerated carbonation tests commonly applied to OPC tend to overestimate the carbonation depth of AAMs. The accelerated method typically involves higher CO_2 concentrations, which AAMs, especially those based on AAS, respond to differently compared to OPC. Research indicates that AAS concretes have reduced carbonation rates in real conditions, likely due to the denser microstructure and lower porosity in some AAMs. However, under accelerated conditions, the increased CO_2 exposure induces atypical carbonation behavior, resulting in an inaccurately high predicted carbonation rate. This discrepancy arises because the hydration products in AAMs, such as N–A–S–H or C–A–S–H gels, do not carbonate in the same manner as the C–S–H gel in OPC. Consequently, these accelerated carbonation tests, while reliable for OPC, may yield misleading results when applied to AAMs, necessitating the development of AAM-specific carbonation testing protocols [49].

4.8 High Efflorescence

Efflorescence, a common phenomenon in AAMs, manifests as white crystalline deposits on the surface due to the reaction between leached alkalis and carbon dioxide in the atmosphere as shown in Fig. 2. This process is typically exacerbated in AAMs activated with high concentrations of sodium-based activators, such as NaOH or Na_2SiO_3, leading to a higher propensity for alkali leaching. When sodium ions migrate to the surface and react with atmospheric CO_2, bicarbonate crystals, often sodium bicarbonate ($NaHCO_3$), form. Although minor efflorescence is generally considered superficial and unlikely to affect the structural integrity of AAMs, a high accumulation can potentially degrade both the material's appearance and some mechanical properties over time [50]. In addition to bicarbonate formation, AAMs may also exhibit alkali sulfate deposits, particularly when sodium sulfate (Na_2SO_4) is used as an activator or additive. Sodium sulfate is less soluble than other sodium compounds, causing it to crystallize on the surface under certain environmental conditions, particularly when exposed to high humidity or wetting–drying cycles. The presence of alkali sulfate on the surface not only influences appearance but can also increase the rate of surface degradation, especially in freeze–thaw environments where salt crystallization cycles induce mechanical stress. This sulfate efflorescence can accelerate microcrack formation on the surface, leading to reduced resistance to freeze–thaw damage and potentially impacting the durability of AAMs [51]. To address efflorescence, research has explored methods to reduce alkali leaching, such as optimizing the molar ratios of the activating solutions and adding supplementary materials that bind free alkalis within the binder matrix. For instance, calcium-based additives like calcium carbonate have been found to reduce efflorescence by encouraging the formation of stable calcium silicate and calcium aluminosilicate phases, which can sequester sodium ions and reduce their migration to the surface. Lowering the water-to-binder ratio, using denser binder systems, or applying surface sealers are additional strategies that can mitigate efflorescence by limiting the pathways for moisture and alkalis to reach the surface.

Fig. 2 Efflorescence of AAMs

5 Possible Remedies to AAM Difficulties

Addressing the challenges associated with AAMs requires targeted adjustments to material compositions, curing processes, and the introduction of various admixtures. For instance, to manage rapid setting times in AAS systems, which can hinder workability, retarding admixtures similar to those in OPC are effective. Both sulfonated naphthalene formaldehyde and polycarboxylic ether admixtures have shown promise in extending the setting time for AAS, enabling easier handling and placement. An alternative approach to slow down the setting time is to replace sodium silicate with sodium carbonate as the primary activator. Sodium silicate is known for its rapid-setting characteristics, whereas sodium carbonate activates AAMs at a rate comparable to OPC, providing additional time for placement and reducing the likelihood of premature stiffening [52, 53]. A primary issue with AAMs is drying shrinkage, particularly in slag-based systems. Using sodium carbonate as an activator instead of sodium silicate has demonstrated potential in reducing AAS shrinkage, aligning its behavior more closely with that of OPC concrete. Other strategies to combat shrinkage include the incorporation of shrinkage-reducing admixtures (SRAs) and wrapped curing techniques, which retain moisture within the concrete and help prevent rapid drying [9]. Industrial by-products rich in calcium and silica, such as blast furnace slag, can also contribute to reduced shrinkage by promoting the formation of stable binder phases that are less susceptible to volumetric changes [54].

AEAs used in OPC concrete are another effective remedy. These agents introduce and stabilize microscopic air bubbles in the concrete, which act as micro-expansion spaces to counteract shrinkage [24]. In addition, the partial replacement of slag with FA has been found to reduce AAM shrinkage. FA has lower reactivity and can alter the binder matrix to provide a more stable microstructure, reducing the shrinkage potential. A small percentage (around 10%) of OPC has also been introduced to AAM mixes to improve shrinkage resistance, blending the performance characteristics of OPC with those of AAMs for a balanced response to drying conditions [54]. Furthermore, high aggregate content, the inclusion of lightweight aggregates, and fiber reinforcement can further improve the performance of AAMs by enhancing dimensional stability and reducing microcracking tendencies that lead to shrinkage. These aggregates and fibers distribute stresses and help to prevent crack propagation, especially in conditions prone to drying shrinkage [55]. Efflorescence, the formation of unsightly bicarbonate crystals on the surface, is another concern in AAMs, particularly when using high alkali content activators. This issue can be minimized by selecting activators with lower alkali concentrations or through wrapped curing, which limits the exposure of the concrete surface to CO_2 during curing [50]. Reducing alkali leaching to the surface not only diminishes efflorescence but also helps retain the aesthetic integrity and durability of AAMs.

6 Conclusions

The comprehensive analysis in this review highlights that AAMs can be an effective and sustainable alternative to traditional cement-based concrete. However, the full realization of their potential relies on mitigating certain drawbacks inherent to AAM systems. The following conclusions underscore the critical areas and advancements required to harness AAMs' durability and environmental benefits in practical applications:

- Drying shrinkage in AAMs, particularly in slag-based systems, presents a significant challenge to long-term stability and structural integrity. Proven techniques such as wrapped curing, the use of SRAs, and optimizing mix designs with a high coarse aggregate content show promise in effectively minimizing shrinkage. These measures help to create a stable matrix, controlling volumetric changes and microcracking that may arise from drying, thus aligning AAM performance closer to that of OPC concrete under similar conditions.
- The selection and formulation of activators play a vital role in addressing AAMs' fresh and hardened properties. Alternative activators, such as sodium carbonate, have shown potential in reducing rapid setting issues, shrinkage, and high efflorescence observed with sodium silicate and NaOH activators. These alternative activators not only moderate setting times but also contribute to lowering AAMs' embodied energy and carbon footprint, enhancing their sustainability. Continued development in activator chemistry is essential to fine-tune these properties, making AAMs more practical for diverse construction environments.
- Widespread adoption of AAMs as a sustainable construction material hinges on the collaborative efforts of stakeholders across the concrete industry, including researchers, manufacturers, regulatory bodies, and end-users. Developing industry standards specific to AAMs, increasing market accessibility to raw materials, and providing education on best practices for AAM handling and curing are necessary steps to ensure confidence in AAM performance. Standardized guidelines address the specific chemical and mechanical behavior of AAMs, driving acceptance and providing contractors with a structured approach to incorporating AAMs in construction projects.

In conclusion, while AAMs show considerable promise as a low-carbon, durable alternative to traditional cementitious materials, their practical viability depends on addressing these technical challenges. Continued research into mix design, activator formulation, and curing methods, along with industry-standardization efforts, establish AAMs as a reliable and environmentally responsible choice for future construction needs.

Acknowledgements The authors want to acknowledge the Department of Civil Engineering, BITS-Pilani, Hyderabad Campus, for providing the necessary facilities and support throughout the research.

References

1. Fapohunda, C., Akinbile, B., & Shittu, A. (2017). Structure and properties of mortar and concrete with rice husk ash as partial replacement of ordinary Portland cement—A review. *International Journal of Sustainable Built Environment, 6*, 675–692. https://doi.org/10.1016/j.ijsbe.2017.07.004
2. Wang, H., Xu, J., Yang, X., Miao, Z., & Yu, C. (2015). Organic Rankine cycle saves energy and reduces gas emissions for cement production. *Energy, 86*, 59–73. https://doi.org/10.1016/j.energy.2015.03.112
3. Thwe, E., Khatiwada, D., & Gasparatos, A. (2021). Life cycle assessment of a cement plant in Naypyitaw, Myanmar. *Cleaner Environmental Systems., 2*, Article 100007. https://doi.org/10.1016/j.cesys.2020.100007
4. Huang, H., Wang, T., Kolosz, B., Andresen, J., Garcia, S., Fang, M., & Maroto-Valer, M. M. (2019). Life-cycle assessment of emerging CO_2 mineral carbonation-cured concrete blocks: Comparative analysis of CO2 reduction potential and optimization of environmental impacts. *Journal of Cleaner Production, 241*, Article 118359. https://doi.org/10.1016/j.jclepro.2019.118359
5. Khan, M. D. I., Abdy Sayyed, M. A., Yadav, G. S., Varma, S. H. (2021) The impact of fly ash and structural fiber on the mechanical properties of concrete, Materials Today: Proceedings. *39* 508–512. https://doi.org/10.1016/j.matpr.2020.08.242.
6. Khan, I., Challa, B., Haripriya Varma, S., Sayyed, M. A. A. (2019) Sorptivity and durability assessment of dolomite impregnated ternary concrete. *International Journal of Recent Technology and Engineering. 8* 5676–5681. https://doi.org/10.35940/ijrte.B2896.078219.
7. Khan, M. I., Abdy Sayyed, M. A., Ali, M. M. A. (2021) Examination of cement concrete containing micro silica and sugarcane bagasse ash subjected to sulphate and chloride attack. *Materials Today: Proceedings. 39* 558–562. https://doi.org/10.1016/j.matpr.2020.08.468.
8. Xie, J., Wang, J., Rao, R., Wang, C., & Fang, C. (2019). Effects of combined usage of GGBS and fly ash on workability and mechanical properties of alkali activated geopolymer concrete with recycled aggregate. *Composites. Part B, Engineering, 164*, 179–190. https://doi.org/10.1016/j.compositesb.2018.11.067
9. Khan, M. D. I., Ram, V. V., Patel, V. I. (2023). Influence of pretreatment methods for coarse recycled aggregates on the performance of alkali-activated concrete. *Arabian Journal for Science and Engineering*. 1–19. https://doi.org/10.1007/s13369-023-08436-x.
10. Khan, M. I., Ram, V. V., & Patel, V. I. (2025). Unveiling the potential of pretreated recycled aggregates and alkali-activated binders in the development of pavement-quality concrete. *Journal of Materials in Civil Engineering, 37*, 1–16. https://doi.org/10.1061/JMCEE7.MTENG-18660
11. Provis, J. L., Van Deventer, J. S. J. (2013). Alkali activated materials: state-of-the-art report, RILEM TC 224-AAM, *Springer Science & Business Media.*

12. Tang, Z., Li, W., Hu, Y., Zhou, J. L., & Tam, V. W. Y. (2019). Review on designs and properties of multifunctional alkali-activated materials (AAMs). *Construction and Building Materials, 200*, 474–489. https://doi.org/10.1016/j.conbuildmat.2018.12.157
13. Yusslee, E., & Beskhyroun, S. (2022). The potential of one-part alkali-activated materials (AAMs) as a concrete patch mortar. *Science and Reports, 12*, 1–10. https://doi.org/10.1038/s41598-022-19830-0
14. Kumar Das, S., Mishra, J., Mohammed Mustakim, S. (2018). Rice husk ash as a potential source material for geopolymer concrete: a review. *International Journal of Applied Engineering Research. 13* 81–84. http://www.ripublication.com.
15. Gopalakrishna, B., & Dinakar, P. (2024). Life cycle assessment (LCA) and the influence of alkaline activator content on mechanical and microstructural properties of geopolymer mortar. *Journal of Engineering Research.* https://doi.org/10.1016/j.jer.2024.01.010
16. Gopalakrishna, B., & Dinakar, P. (2024). The evaluation of the life cycle and corrosion properties of recycled aggregate geopolymer concrete incorporating fly ash and GGBS. *Journal of Building Engineering., 94*, Article 109977. https://doi.org/10.1016/j.jobe.2024.109977
17. Alsalman, A., Assi, L. N., Kareem, R. S., Carter, K., & Ziehl, P. (2021). Energy and CO2 emission assessments of alkali-activated concrete and Ordinary Portland Cement concrete: A comparative analysis of different grades of concrete. *Cleaner Environmental Systems., 3*, Article 100047. https://doi.org/10.1016/j.cesys.2021.100047
18. Jun, Y., Kim, J. H., Han, S. H., & Kim, T. (2021). Influence of seawater on alkali-activated slag concrete. *Materials, Structures, and Construction., 54*, 1–16. https://doi.org/10.1617/s11527-021-01719-5
19. Morsy, A. M., Abd-Elmoaty, A. E. M., Hassan Ali, A., Shalan, A. H. (2023) Study on the physical and mechanical properties of Alkali-Activated concrete bricks. Construction and Building Materials. *408* 133487. https://doi.org/10.1016/j.conbuildmat.2023.133487.
20. Mehta, A., Siddique, R., Ozbakkaloglu, T., Uddin Ahmed Shaikh, F., Belarbi, R. (2020) Fly ash and ground granulated blast furnace slag-based alkali-activated concrete: Mechanical, transport and microstructural properties, *Construction and Building Materials. 257* 119548. https://doi.org/10.1016/j.conbuildmat.2020.119548.
21. Ding, Y., Dai, J. G., & Shi, C. J. (2016). Mechanical properties of alkali-activated concrete: A state-of-the-art review. *Construction and Building Materials, 127*, 68–79. https://doi.org/10.1016/j.conbuildmat.2016.09.121
22. Dindi, A., Quang, D. V., Vega, L. F., Nashef, E., Abu-Zahra, M. R. M. (2019) Applications of fly ash for CO_2 capture, utilization, and storage, J. CO_2 Util. *29* 82–102. https://doi.org/10.1016/j.jcou.2018.11.011.
23. Jin, Y., Feng, W., Zheng, D., Dong, Z., & Cui, H. (2020). Structure refinement of fly ash in connection with its reactivity in geopolymerization. *Waste Management, 118*, 350–359. https://doi.org/10.1016/j.wasman.2020.08.049
24. Khan, M. I., Ram, V. V., Patel, V. I. (2024) Durability performance of alkali-activated concrete with pre-treated coarse recycled aggregates for pavements, Scientific Reports. *14*. https://doi.org/10.1038/s41598-024-64506-6.
25. Taghvayi, H., Behfarnia, K., & Khalili, M. (2018). The effect of alkali concentration and sodium silicate modulus on the properties of alkali-activated slag concrete. *Journal of Advanced Concrete Technology, 16*, 293–305. https://doi.org/10.3151/jact.16.293
26. Pol Segura, I., Ranjbar, N., Juul Damø, A., Skaarup Jensen, L., Canut, M., Arendt Jensen, P. (2023) A review: Alkali-activated cement and concrete production technologies available in the industry. *Heliyon. 9*. https://doi.org/10.1016/j.heliyon.2023.e15718.
27. Elibol, C., & Sengul, O. (2016). Effects of activator properties and ferrochrome slag aggregates on the properties of alkali-activated blast furnace slag mortars. *Arabian Journal for Science and Engineering, 41*, 1561–1571. https://doi.org/10.1007/s13369-015-1910-8
28. Karahan, O., & Yakupoğlu, A. (2011). Resistance of alkali-activated slag mortar to abrasion and fire. *Advances in Cement Research, 23*, 289–297. https://doi.org/10.1680/adcr.2011.23.6.289

29. Dai, X., Ren, Q., Aydin, S., Yardimci, M. Y., & De Schutter, G. (2023). Accelerating the reaction process of sodium carbonate-activated slag mixtures with the incorporation of a small addition of sodium hydroxide/sodium silicate. *Cement and Concrete Composites, 141*, Article 105118. https://doi.org/10.1016/j.cemconcomp.2023.105118
30. Liu, C., Yao, X., & Zhang, W. (2020). Controlling the setting times of one-part alkali-activated slag by using honeycomb ceramics as carrier of sodium silicate activator. *Construction and Building Materials, 235*, Article 117091. https://doi.org/10.1016/j.conbuildmat.2019.117091
31. Ahmadi, S. (2021). Alkali-Activated slag cement and concrete: a review. *Advanced Ceramics Progress., 7*, 1–9.
32. Masi, G., Filipponi, A., & Bignozzi, M. C. (2021). Fly ash-based one-part alkali activated mortars cured at room temperature: Effect of precursor pre-treatments. *Open Ceram., 8*, Article 100178. https://doi.org/10.1016/j.oceram.2021.100178
33. Shi, S., Li, H., Zhou, Q., Zhang, H., Basheer, P. A. M., & Bai, Y. (2023). Alkali-activated fly ash cured with pulsed microwave and thermal oven: A comparison of reaction products, microstructure and compressive strength. *Cement and Concrete Research, 166*, Article 107104. https://doi.org/10.1016/j.cemconres.2023.107104
34. Sadeghian, G., Behfarnia, K., & Teymouri, M. (2022). Drying shrinkage of one-part alkali-activated slag concrete. *Journal of Building Engineering., 51*, Article 104263. https://doi.org/10.1016/j.jobe.2022.104263
35. Ramagiri, K. K., & Kar, A. (2019). Effect of high-temperature on the microstructure of alkali-activated binder. *Materials Today: Proceedings., 28*, 1123–1129. https://doi.org/10.1016/j.matpr.2020.01.093
36. Ye, H., Cartwright, C., Rajabipour, F., & Radlińska, A. (2017). Understanding the drying shrinkage performance of alkali-activated slag mortars. *Cement and Concrete Composites, 76*, 13–24. https://doi.org/10.1016/j.cemconcomp.2016.11.010
37. Li, Z., Chen, Y., Provis, J. L., Cizer, Ö., Ye, G. (2023) Autogenous shrinkage of alkali-activated slag: A critical review. Cement and Concrete Research. *172*. https://doi.org/10.1016/j.cemconres.2023.107244.
38. Sasui, S., Kim, G., Nam, J., Koyama, T., Chansomsak, S. (2020) Strength and microstructure of class-C fly ash and GGBS blend geopolymer activated in NaOH & NaOH + Na_2SiO_3, *Materials (Basel). 13*. https://doi.org/10.3390/ma13010059.
39. Çevik, A., Alzeebaree, R., Humur, G., Niş, A., & Gülşan, M. E. (2018). Effect of nano-silica on the chemical durability and mechanical performance of fly ash based geopolymer concrete. *Ceramics International, 44*, 12253–12264. https://doi.org/10.1016/j.ceramint.2018.04.009
40. Qureshi, M. N., & Ghosh, S. (2014). Effect of Silicate Content on the Properties of Alkali-Activated Blast Furnace Slag Paste. *Arabian Journal for Science and Engineering, 39*, 5905–5916. https://doi.org/10.1007/s13369-014-1172-x
41. Gugulothu, V., Gunneswara Rao, T. D. (2020) Effect of binder content and solution/binder ratio on alkali-activated slag concrete activated with neutral grade water glass. *Arabian Journal for Science and Engineering. 45* 8187–8197. https://doi.org/10.1007/s13369-020-04666-5.
42. Tempest, B., Sanusi, O., Gergely, J., Ogunro, V., Weggel, D. (2009) Compressive strength and embodied energy optimization of fly ash based geopolymer concrete. *3rd World Coal Ash, WOCA Conf. - Proc.* 1–17.
43. Parthiban, K., Saravana Raja Mohan, K. (2017) Influence of recycled concrete aggregates on the engineering and durability properties of alkali activated slag concrete. *Construction and Building Materials. 133* 65–72. https://doi.org/10.1016/j.conbuildmat.2016.12.050.
44. Wardhono, A., Law, D. W., & Molyneaux, T. C. K. (2015). Long term performance of alkali activated slag concrete. *Journal of Advanced Concrete Technology, 13*, 187–192. https://doi.org/10.3151/jact.13.187
45. Mengasini, L., Mavroulidou, M., & Gunn, M. J. (2021). Alkali-activated concrete mixes with ground granulated blast furnace slag and paper sludge ash in seawater environments. *Sustain. Chem. Pharm., 20*, Article 100380. https://doi.org/10.1016/j.scp.2021.100380
46. Baščarević, Z., Komljenović, M., Miladinović, Z., Nikolić, V., Marjanović, N., & Petrović, R. (2014). Impact of sodium sulfate solution on mechanical properties and structure of fly ash

based geopolymers. *Materials, Structures, and Construction., 48*, 683–697. https://doi.org/10.1617/s11527-014-0325-4
47. Li, N., Zhang, Z., Shi, C., Zhang, J. (2018) Some progresses in the challenges for geopolymer. *IOP Conference Series: Materials Science and Engineering. 431*. https://doi.org/10.1088/1757-899X/431/2/022003.
48. Cyr, M., Pouhet, R. (2015) Resistance to alkali-aggregate reaction (AAR) of alkali-activated cement-based binders, Handb. *Alkali-Activated Cem. Mortars Concr*. 397–422. https://doi.org/10.1533/9781782422884.3.397.
49. Zhang, X., Long, K., Liu, W., Li, L., Long, W. J. (2020) Carbonation and chloride ions' penetration of alkali-activated materials: a review. *Molecules. 25*. https://doi.org/10.3390/molecules25215074.
50. Liang, K., Cui, K., Sabri, M. M. S., Huang, J. (2022) Influence factors in the wide application of alkali-activated materials: a critical review about efflorescence. *Materials (Basel). 15*. https://doi.org/10.3390/ma15186436.
51. Criado, M., Jiménez, A. F., & Palomo, A. (2010). Effect of sodium sulfate on the alkali activation of fly ash. *Cement and Concrete Composites, 32*, 589–594. https://doi.org/10.1016/j.cemconcomp.2010.05.002
52. Awoyera, P., & Adesina, A. (2019). A critical review on application of alkali activated slag as a sustainable composite binder. *Case Studies in Construction Materials., 11*, Article e00268. https://doi.org/10.1016/j.cscm.2019.e00268
53. Abdalqader, A. F., Jin, F., & Al-Tabbaa, A. (2016). Development of greener alkali-activated cement: Utilisation of sodium carbonate for activating slag and fly ash mixtures. *Journal of Cleaner Production, 113*, 66–75. https://doi.org/10.1016/j.jclepro.2015.12.010
54. Humad, A. M., Kothari, A., Provis, J. L., & Cwirzen, A. (2019). The effect of blast furnace slag/fly ash ratio on setting, strength, and shrinkage of alkali-activated pastes and concretes. *Frontiers in Materials., 6*, 1–10. https://doi.org/10.3389/fmats.2019.00009
55. Dheyaaldin, M. H., Mosaberpanah, M. A., & Alzeebaree, R. (2022). Shrinkage behavior and mechanical properties of alkali activated mortar incorporating nanomaterials and polypropylene fiber. *Ceramics International, 48*, 23159–23171. https://doi.org/10.1016/j.ceramint.2022.04.297

The manufacturer's authorised representative in the EU is Springer Nature Customer Service Centre GmbH, Europaplatz 3, 69115 Heidelberg, Germany. If you have any concerns regarding our products, please contact ProductSafety@springernature.com

Printed and bound by CPI Group (UK) Ltd, Croydon, CR0 4YY

30/09/2026

02225571-0011